IEE CIRCUITS AND SYSTEMS SERIES 7

Series Editors: Dr D. G. Haigh
               Dr R. S. Soin

MMIC DESIGN

Other volumes in this series:

Volume 1 **GaAs technology and its impact on circuits and systems**
D. G. Haigh and J. Everard (Editors)
Volume 2 **Analogue IC design: the current-mode approach**
C. Toumazou, F. J. Lidgey and D. G. Haigh (Editors)
Volume 3 **Analogue-digital ASICs** R. S. Soin, F. Maloberti and
J. Franca (Editors)
Volume 4 **Algorithmic and knowledge-based CAD for VLSI**
G. E. Taylor and G. Russell (Editors)
Volume 5 **Switched-currents: an analogue technique for digital technology**
C. Toumazou, J. B. Hughes and N. C. Battersby (Editors)
Volume 6 **High frequency circuits** F. Nibler and co-authors

MMIC DESIGN

Edited by I. D. Robertson

The Institution of Electrical Engineers

Published by: The Institution of Electrical Engineers, London,
United Kingdom

© 1995: The Institution of Electrical Engineers

This publication is copyright under the Berne Convention and the Universal Copyright Convention. All rights reserved. Apart from any fair dealing for the purposes of research or private study, or criticism or review, as permitted under the Copyright, Designs and Patents Act, 1988, this publication may be reproduced, stored or transmitted, in any forms or by any means, only with the prior permission in writing of the publishers, or in the case of reprographic reproduction in accordance with the terms of licences issued by the Copyright Licensing Agency. Inquiries concerning reproduction outside those terms should be sent to the publishers at the undermentioned address:

The Institution of Electrical Engineers,
Michael Faraday House,
Six Hills Way, Stevenage,
Herts. SG1 2AY, United Kingdom

While the editor and the publishers believe that the information and guidance given in this work is correct, all parties must rely upon their own skill and judgment when making use of it. Neither the editor nor the publishers assume any liability to anyone for any loss or damage caused by any error or omission in the work, whether such error or omission is the result of negligence or any other cause. Any and all such liability is disclaimed.

The moral right of the authors to be identified as authors of this work has been asserted by them in accordance with the Copyright, Designs and Patents Act 1988.

British Library Cataloguing in Publication Data

A CIP catalogue record for this book
is available from the British Library

ISBN 0 85296 816 7

Printed in England by Short Run Press Ltd., Exeter

Contents

Preface	xv
Foreword	xvii
Contributors	xix
Acknowledgements	xxi

1 Introduction *I. D. Robertson*   1
- 1.1 Introduction   1
- 1.2 A brief history of MMIC technology   2
- 1.2 Advantages and disadvantages of MMICs   5
  - 1.2.1 Cost   6
  - 1.2.2 Performance   7
  - 1.2.3 Investment required   7
  - 1.2.4 Reproducibility   8
  - 1.2.5 Reliability   9
  - 1.2.6 Size and mass   9
- 1.3 MMIC technologies   9
- 1.4 Applications   11
- 1.5 Types of MMIC   16
- 1.6 A guided tour through the book   21
- 1.7 References   23

2 Active devices and fabrication technology *A. A. Rezazadeh and C. Sansom*   27
- 2.1 Introduction   27
- 2.2 The GaAs MESFET   28
  - 2.2.1 Principles of operation   30
  - 2.2.2 High frequency capability   31
  - 2.2.3 Noise analysis   32
- 2.3 The high electron mobility transistor (HEMT)   33

2.4	The heterojunction bipolar transistor (HBT)	36
	2.4.1 Principles of operation	38
	2.4.2 High frequency capability	41
	2.4.3 HBT material systems	45
2.5	Passive components	50
	2.5.1 Capacitors	51
	2.5.2 Inductors	52
	2.5.3 Resistors	54
2.6	MMIC fabrication	55
2.7	Yield and process control	60
2.8	References	65

3 Passive components *M. Gillick and I. D. Robertson* **67**

- 3.1 Introduction — 67
- 3.2 Inductors — 68
  - 3.2.1 Loop inductors — 69
  - 3.2.2 Spiral inductors — 69
  - 3.2.3 Stacked spirals — 70
  - 3.2.4 Spiral inductor models — 72
  - 3.2.5 Planar spiral transformers — 73
- 3.3 Capacitors — 74
  - 3.3.1 Overlay capacitors — 74
  - 3.3.2 Interdigital capacitors — 76
- 3.4 Resistors — 77
  - 3.4.1 GaAs resistors — 77
  - 3.4.2 Thin film resistors — 78
  - 3.4.3 Resistor models — 79
- 3.5 Via-holes and grounding — 79
- 3.6 Microstrip components — 82
  - 3.6.1 Microstrip couplers and power splitters — 82
  - 3.6.2 Special MMIC realisations — 84
- 3.7 Other transmission-line media — 87
  - 3.7.1 Transitions and baluns — 87
- 3.8 References — 88

4 Advanced CAD techniques *R. G. Arnold and D. M. Brookbanks* **91**

- 4.1 Introduction — 91
- 4.2 Design integration — 92
- 4.3 Simulator customisation — 93
- 4.4 Schematic capture — 94
  - 4.4.1 Hierarchical design — 95
  - 4.4.2 Control parameters — 96
  - 4.4.3 Back annotation — 98
- 4.5 Layout-based CAD — 98
  - 4.5.1 Hierarchical design — 105

4.6	Completion of the design	106
4.7	Limitations in foundry-based MMIC design	107
4.8	Electromagnetic field simulators	109
4.9	The spectral domain approach	109
	4.9.1 Transmission-line solution	111
	4.9.2 Three-dimensional field solution	112
	4.9.3 Spectral operator expansion approximation	115
4.10	Transmission-line simulators	117
4.11	Three-dimensional electromagnetic field simulators	122
4.12	Electromagnetic field analysis and its use in MMIC design	124
4.13	Latest commercial software packages	129
	4.13.1 User (custom) model capability	130
	4.13.2 Time-domain analysis	130
	4.13.3 System simulation	131
	4.13.4 Electromagnetic simulation and field theoretical methods	131
4.14	Summary	131
4.15	References	132

5 Amplifiers *I. D. Robertson and M. W. Geen*   133

5.1	Introduction	133
5.2	Classical stability and gain analysis	134
	5.2.1 Constant gain circles	138
	5.2.2 The practical implications of the theory	141
	5.2.3 Amplifier design under conditional stability conditions	141
	5.2.3.1 Resistive loading	141
	5.2.3.2 Parallel feedback	142
	5.2.3.3 Series feedback	142
	5.2.3.4 The balanced amplifier	142
5.3	Matching techniques	144
	5.3.1 Lumped element matching	144
	5.3.1.1 L-networks	144
	5.3.1.2 Matching two complex impedances with L-networks	144
	5.3.1.3 T- and π-networks	146
	5.3.2 Distributed matching networks	147
	5.3.2.1 Series line and parallel stub	147
	5.3.2.2 The quarter-wave transformer	150
	5.3.2.3 The short transformer	150
5.4	DC bias injection	150
	5.4.1 Stacked bias	154
	5.4.2 Off-chip components	155
	5.4.3 RFOW testing considerations	158
5.5	Reactively matched amplifier design	158
	5.5.1 Multi-stage design	158
5.6	Lossy matching	161
5.7	FET feedback amplifier	162

5.8	Distributed amplifiers	163
	5.8.1	Gate and drain-line losses	166
	5.8.2	Equalisation of gate and drain-line phase velocities	170
	5.8.3	Distributed amplifier with constant-R networks	171
	5.8.4	Cascode distributed amplifier	173
	5.8.5	Single-section distributed amplifier	173
	5.8.6	Matrix distributed amplifier	174
	5.8.7	Practical design guidelines	175
5.9	Active matching	177
	5.9.1	Common-gate/common-source/common-drain amplifier	177
	5.9.2	Darlington pair	178
	5.9.3	DC-coupled amplifiers	178
5.10	Power amplifiers	180
	5.10.1	Device characterisation	181
	5.10.2	Power combining and cluster matching	182
	5.10.3	Class B operation	186
	5.10.4	High power distributed amplifiers	187
		5.10.4.1 The capacitively-coupled distributed amplifier	187
		5.10.4.2 The split gate-line distributed amplifier	188
		5.10.4.3 The tapered drain-line technique	188
5.11	Low noise amplifiers	189
	5.11.1	Noise matching	190
	5.11.2	Simultaneous match LNA	191
5.12	Summary	194
5.13	References	194

6 Mixers *S. J. Nightingale*   201

6.1	Introduction	201
	6.1.1	The three basic types of mixer	202
		6.1.1.1 Single-ended mixers (SEMs)	202
		6.1.1.2 Single-balanced mixers (SBMs)	202
		6.1.1.3 Double-balanced mixers (DBMs)	202
6.2	Mixer terminology	203
	6.2.1	Mixer bandwidth	203
	6.2.2	Narrowband mixer	203
	6.2.3	Wideband mixer	203
	6.2.4	Broadband mixer	203
	6.2.5	Harmonic mixer	203
	6.2.6	Image recovery mixer (IRM)	203
	6.2.7	Image rejection mixer	203
6.3	Mixer analysis	204
	6.3.1	General analysis	204
	6.3.2	Restricted analysis	204
6.4	Background reading	204
6.5	Analysis of mixer circuits	206

		6.5.1 Analysis of a simple single loop circuit	206
		6.5.2 General non-linear analysis	206
		6.5.2.1 Conversion loss	207
		6.5.2.2 Input and output impedance	207
		6.5.2.3 Isolation	208
		6.5.3 Conversion loss matrix method	208
		6.5.3.1 Conversion loss	209

6.6 Diode mixers — 209
    6.6.1 Introduction — 209
    6.6.2 Equivalences — 213
    6.6.3 Noise figure — 215
    6.6.4 A simple single-ended mixer — 216
    6.6.5 Design of a 94 GHz mixer — 218
        6.6.5.1 Introduction — 218
        6.6.5.2 General background to a 94 GHz mixer design — 222
        6.6.5.3 Details of a 94 GHz single-balanced mixer design — 224

6.7 FET mixers — 231
    6.7.1 Design considerations for FET mixers — 231
    6.7.2 Pump-drain, RF-gate — 232
    6.7.3 Pump-gate, RF-drain — 232
    6.7.4 Pump-gate, RF-drain — 233
    6.7.5 Pump-source, RF-gate — 233
    6.7.6 General comments on FET and HEMT mixer design — 233

6.8 Coupling structures — 236
    6.8.1 Passive coupling structures — 237
    6.8.2 180° 3 dB coupler — 237
    6.8.3 90° 3dB coupler — 240
    6.8.4 Wilkinson with high-pass and low-pass sections — 242
    6.8.5 Planar transformers — 244
    6.8.6 Active coupling structures — 244

6.9 Some final comments and observations — 244

6.10 A general theory and representation of balanced mixers — 246
    6.10.1 Diode lattice mixer — 247

6.11 References — 250
    6.11.1 General — 250
    6.11.2 Diode mixers — 251
    6.11.3 Subharmonic mixers — 254
    6.11.4 FET mixers — 254
    6.11.5 Baluns — 256
    6.11.6 Miscellaneous MMIC and MIC references — 257

7 Phase shifters *S. Lucyszyn and J. S. Joshi* — **259**

7.1 Introduction — 259

7.2 Analogue implementations — 261
    7.2.1 Introduction to analogue phase shifters — 261

	7.2.2	Single-stage reflection-type phase shifters	262
		7.2.2.1 General design approaches	263
		7.2.2.2 Optimum design approaches	263
	7.2.3	Single-stage reflection-type delay lines	265
	7.2.4	Cascaded-match reflection-type phase shifter	267
		7.2.4.1 One octave bandwidth (6.5-13 GHz) design example	269
		7.2.4.2 Two octave bandwidth (4.4-16.1 GHz) design example	274
	7.2.5	Ultra-wide bandwidth CMRTPS implementations	275
		7.2.5.1 Decade bandwidth (1-19 GHz) hybrid-MIC design example	275
		7.2.5.2 Decade bandwidth (0.5-5.5 GHz) MMIC design example	278
	7.2.6	Millimetre-wave implementations	280
	7.2.7	Dual-gate MESFET	282
	7.2.8	Phase splitter-power combiner	283
		7.2.8.1 Passive design	283
		7.2.8.2 Active design	284
7.3	Digital implementations		284
	7.3.1	Switched-line	285
	7.3.2	Reflection-type	286
		7.3.2.1 Delay lines	286
		7.3.2.2 Phase shifters	288
	7.3.3	Loaded-line	289
	7.3.4	Switched-filter	291
	7.3.5	Intrinsic phase shifters	295
		7.3.5.1 Boire phase shifter	295
		7.3.5.2 Switched active balun	297
		7.3.5.3 Switched-coupler	297
7.4	Summary		298
7.5	References		299

8 Switches and attenuators *J. S. Joshi* 305

8.1	Introduction		305
8.2	GaAs FET MMIC switches		306
	8.2.1	GaAs FET switching mechanism	306
	8.2.2	Switch FET equivalent circuit	309
	8.2.3	Implementation of MMIC switches	310
8.3	Digital attenuators		316
	8.3.1	Design approaches	316
	8.3.2	Segmented dual-gate FET technique	317
	8.3.3	Switched attenuators	318
	8.3.4	Switched scaled FETs	321
	8.3.5	Switched bridged-T attenuator	321
	8.3.6	Switched T- or π-attenuators	323
8.4	Attenuator design example		324

	8.4.1 Higher attenuation bits	324
	8.4.2 Lower attenuation bits	324
	8.4.3 Overall chip design and layout	327
	8.4.4 Measured results	329
8.5	Analogue attenuators	332
	8.5.1 Analogue reflection-type attenuator	332
	8.5.2 Other analogue attenuators	334
8.6	Summary	335
8.7	References	335

9 Oscillators *K. K. M. Cheng and U. Karacaoglu* — 337

- 9.1 Introduction — 337
- 9.2 Design principles — 339
  - 9.2.1 Feedback oscillators — 339
  - 9.2.2 Negative resistance oscillators — 340
- 9.3 Active device — 341
  - 9.3.1 GaAs MESFET versus silicon bipolar transistors — 341
  - 9.3.2 Heterojunction bipolar transistors — 342
  - 9.3.3 Biasing and spurious response suppression — 342
- 9.4 Resonators — 343
  - 9.4.1 Dielectric resonator — 344
  - 9.4.2 Transmission-line resonators — 345
  - 9.4.3 Superconductive resonators — 349
- 9.5 CAD techniques for non-linear oscillators — 350
  - 9.5.1 Time-domain methods — 350
  - 9.5.2 Harmonic balance method — 350
  - 9.5.3 Volterra series analysis — 351
- 9.6 Oscillator noise analysis — 355
  - 9.6.1 Noise analysis of negative resistance oscillator — 356
  - 9.6.2 Linear noise analysis of feedback oscillators — 358
  - 9.6.3 Optimum Q_L for minimum phase noise — 359
  - 9.6.4 Effects of gain compression — 359
  - 9.6.5 Effects of non-zero open loop phase error — 360
- 9.7 MMIC VCO design example — 362
  - 9.7.1 Varactor diode modelling — 363
  - 9.7.2 VCO design — 363
- 9.8 References — 367

10 Silicon millimetre-wave ICs *J. -F. Luy* — 371

- 10.1 Millimetre-wave circuits on high resistivity silicon — 371
- 10.2 The Si/SiGe material system — 373
- 10.3 Transit-time devices — 376
  - 10.3.1 A unified transit-time diode model — 376
  - 10.3.2 The IMPATT diode — 379
  - 10.3.3 Heterojunction transit-time diodes — 384

xii Contents

	10.3.4 A SiGe Gunn device?	385
10.4	PIN diodes	387
10.5	Schottky diodes	389
10.6	SiGe heterojunction bipolar transistor	390
	10.6.1 Design considerations	392
	10.6.2 SiGe oscillator circuit on silicon substrate	393
10.7	Remarks and conclusions	393
10.8	References	395

11 Measurement techniques *S. Lucyszyn* — **399**
- 11.1 Introduction — 399
- 11.2 Test fixture measurements — 400
  - 11.2.1 Two tier calibration — 402
    - 11.2.1.1 Time-domain gating — 402
    - 11.2.1.2 In-fixture calibration — 405
    - 11.2.1.3 Equivalent circuit modelling — 406
  - 11.2.2 One tier calibration — 408
  - 11.2.3 Test fixture design considerations — 409
- 11.3 Probe station measurements — 410
  - 11.3.1 Passive microwave probe design — 411
  - 11.3.2 Prober calibration — 413
  - 11.3.3 Measurement errors — 415
  - 11.3.4 DC biasing — 415
  - 11.3.5 MMIC layout considerations — 415
  - 11.3.6 Low-cost multiple DC biasing technique — 418
- 11.4 Thermal and cryogenic measurements — 419
  - 11.4.1 Thermal measurements — 419
  - 11.4.2 Cryogenic measurements — 421
- 11.5 Experimental field probing techniques — 422
  - 11.5.1 Electromagnetic-field probing — 422
  - 11.5.2 Magnetic-field probing — 423
  - 11.5.3 Electric-field probing — 424
    - 11.5.3.1 Electron-beam probing — 424
    - 11.5.3.2 Photo-emissive sampling — 424
    - 11.5.3.3 Opto-electronic sampling — 425
    - 11.5.3.4 Electro-optic sampling — 425
    - 11.5.3.4 Electrical sampling scanning-force microscopy — 427
- 11.6 Summary — 427
- 11.7 References — 428

12 Advanced techniques *S. Lucyszyn and I. D. Robertson* — **435**
- 12.1 Introduction — 435
- 12.2 Uniplanar techniques — 437
  - 12.2.1 Transmission lines for uniplanar MMICs — 437
    - 12.2.1.1 Coplanar waveguide — 437

	12.2.1.2 Thin-film microstrip	438
	12.2.1.3 Slotline	439
	12.2.1.4 Coplanar strips	439
12.2.2	CAD for CPW and multilayer MMICs	440
	12.2.2.1 Xgeom™ and em™ software	440
12.2.3	CPW circuits	441
	12.2.3.1 Branch-line couplers	442
	12.2.3.2 Two-stage balanced Ku-band CPW MMIC amplifier	443
12.2.4	Multilayer circuits	443
	12.2.4.1 Multilayer directional couplers	444
	12.2.4.2 Amplifiers using thin-film microstrip	444
12.3	Electronically tunable elements	446
12.3.1	Varactor diodes	446
	12.3.1.1 Heterojunction-based varactor diodes	449
12.3.2	Tunable active inductors	449
12.3.3	Variable resistance elements	451
12.3.4	Tunable negative resistance elements	453
12.4	Active filters	453
12.4.1	Active inductor based filters	454
12.4.2	Actively-coupled passive resonators	454
12.4.3	Techniques using negative resistance elements and active loops	454
12.4.4	Transversal and recursive filters	456
12.4.5	Practical applications	457
12.5	Active splitters, combiners, baluns, isolators and circulators	457
12.5.1	Active power dividers and combiners	457
12.5.2	Active baluns	458
12.5.3	Active isolators	460
12.5.4	Active circulators	461
12.6	Modulators, demodulators and frequency translators	467
12.7.1	Frequency translators	468
12.7.2	SSB modulators	470
12.7.3	Serrodyne modulators	471
12.7.4	Vector modulators	473
12.7.5	Modulators and demodulators for digital communications	474
12.8	Linearisers and high efficiency power amplifiers	476
12.8.1	Linearisation techniques	477
12.9	Active antennas and quasi-optical techniques	478
12.9.1	Optimally-shaped on-chip radiators	481
12.9.2	Using the rear face of the chip	481
12.9.3	Multi-level techniques	482
12.9.4	Antennas with novel switching and tuning functions	482
12.9.5	Using a patch as both a radiating and feedback element	483
12.10	Optoelectronics and microwaves	483
12.10.1	Applications	483

12.10.2 Optically controlled circuits   484
    12.10.2.1 Phase shifters   484
    12.10.2.2 Switches   485
    12.10.2.3 Amplifiers and attenuators   485
    12.10.2.4 Oscillators   485
    12.10.2.5 Mixers   486
12.10.3 Circuits for optoelectronic applications   486
12.10.4 MMIC measurement anomalies   487
12.11 References   488

Index   **499**

Preface

I have been fortunate in that my involvement with MMIC technology has spanned a period covering the 1980s and 1990s in which many particularly exciting developments were made in a number of areas such as device technology, microwave measurements, computer-aided design, circuit techniques, and personal computing. I have witnessed many major breakthroughs, including the first microwave heterojunction transistors, the first truly automatic network analyser (the HP 8510™), the first RF-on-wafer probing systems (developed at both Plessey Research Caswell and Tektronix), the first PC-based microwave CAD package (Touchstone™), and even the first RISC workstations. My first MMIC layout at Caswell (in 1984) was drawn with a pencil and ruler on a piece of A0 graph paper, and yet only 10 years later the same task can be accomplished almost automatically using SMART™ library elements in a CAD package, and substantial parts of the layout can be directly simulated electromagnetically. My first measurements at Caswell were on spiral transformers, and were carried out on an HP 8410™. This took several minutes to measure 20 frequency points and could only measure up to 18 GHz. Today, one can make on-wafer measurements up to 110 GHz, with 801 points measured almost in real time.

The reason I mention these early efforts of mine is to illustrate that, in the last 10 years, MMIC technology has progressed very rapidly and has now reached a state of maturity. The rationale behind this book is that since so many different circuit techniques have evolved, and since the technology is now entering commercial applications more and more, it is very appropriate to publish a single work which brings together the many circuit techniques which have been demonstrated up to now. The emphasis of the book is on how to design MMICs, but essential information on the applications, technology, CAD, and measurement techniques is also included. Sufficient background information is given to make the book accessible to those unfamiliar with MMICs, but a basic knowledge of microwave principles (the Smith chart, S-parameters, transmission line theory, and microstrip circuits) is a prerequisite. It was not deemed necessary to cover these fundamentals since there are already a number of excellent texts that do so very well. As a result, in the academic world this book would be ideal background reading for undergraduates on general third-year microwaves courses, but is only intended to be used as the main course text in more specialised third-year, fourth-year, and MSc courses. In research laboratories and in industry the book should be highly suitable since the material represents the state-of-the-art at the time of writing and will be of considerable use to experienced microwave designers who want reference material and a guide to the design of the different circuit functions.

The book is divided into twelve chapters. Chapter 1 gives an introduction to the technology and applications of MMICs. Chapter 2 describes the active devices and gives an introduction to the fabrication technology. Chapter 3 describes the passive components used in MMIC design. Chapter 4 describes the modelling and computer aided design techniques used for MMIC design. Chapters 5 to 9 describe the design of the different circuit functions, these being amplifiers, phase shifters, switches and gain control circuits, mixers, and oscillators. Chapter 10 describes the technology and design of silicon millimetre-wave monolithic circuits based on high resistivity substrates. Chapter 11 describes the measurement techniques used for MMICs. Finally, Chapter 12 describes a wide range of advanced techniques including coplanar waveguide and multi-layer circuits, active filters, active circulators, integrated millimetre-wave antennas, direct microwave implementations of modulators and demodulators, and the application of MMICs in opto-electronic systems.

Ian Robertson

Foreword

In the following pages of this book you will be taken step by step from the fabrication technology of microwave devices through their applications and finally to a vision of the future. Who would have believed that this technology area would have progressed, albeit painstakingly slowly over the last 30 years or so, from very primitive beginnings to the sophisticated multi-hundred million dollar market segment it enjoys today.

I was there right at the very beginning and it never ceases to amaze me how things have progressed. Clever people now understand how the MESFET really works and can do incredible things with it in a variety of circuit functions occupying an ever decreasing area of GaAs and at ever increasing frequencies!

The early devices produced in 1964 were, by today's standards, very crude with gate lengths of 24 microns or more. This was however quite an achievement bearing in mind the very 'Heath-Robinson' lithography equipment available at the time. A tin can with a UV lamp screwed into one end and a non-ideal hand operated slide shutter at the other was the best available at the time. Nevertheless, this technology was soon producing four micron gate length devices allowing microwave (circa 1 GHz) performance to be achieved.

At this point, however, theory and practice began to diverge quite alarmingly. Classical theory predicted 10 GHz operation from a two micron device, but this was not to be, as hot electron effects, not predicted by the early theories, produced a less optimistic relationship between gate length and operating frequency. This meant that you had to work even harder to reduce the gate length of the transistor necessitating a move to electron beam lithography, where one micron and shorter gate lengths could be resolved.

First 'home built' and then purpose built electron beam machines were able to reproducibly produce gate stripes of one micron or less, and X-Band (10 GHz) operation was achieved. It was not until 1976, however, that this technology, coupled with a very primitive IC process, produced the first microwave monolithic circuit – an amplifier with gain at 10 GHz. At this point the world sat up and took notice and soon, from every part of the globe, came papers on just about every microwave function you could imagine.

The MESFET story is by no means over, and it is only now that many companies across the world are beginning to seriously exploit the technology in telecommunications, automobiles, television etc. in a rapidly expanding market which by the end of this century is predicted to reach in excess of $2Bn/annum.

This book is devoted to further advancing our understanding of this exciting technology. It even dares to mention **that other** technology – this is entirely reasonable as there is no doubt that both silicon and gallium arsenide will cohabit in many of the future microwave and millimetre-wave systems, to the benefit of both communities.

Please read this book carefully, it is a source of all the wonderful things possible in the world of microwaves. When you have completed it, I hope it will inspire you to join the rapidly growing band of people who are devoted to ensuring that in every future high frequency system there will be at least one small piece of that exquisite material – gallium arsenide!

Jim Turner

Contributors

I. D. Robertson
A. A. Rezazadeh
S. Lucyszyn
K. K. M. Cheng
U. Karacaoglu
M. Gillick

> Department of Electronic & Electrical Engineering
> King's College London
> Strand
> London, WC2R 2LS

C. Samson
D. M. Brookbanks
R. G. Arnold
M. W. Geen

> GEC Marconi Materials Technology Ltd.
> Caswell
> Towcester
> Northants, NN12 8EQ

S. J. Nightingale

> Racal Radar Defence Systems Ltd.
> Racal Thorn Crawley
> Manor Royal
> Crawley
> West Sussex, RH10 2PZ

J. S. Joshi

> British Aerospace (Space Systems) Ltd.
> Gunnels Wood Road
> Stevenage
> Hertfordshire, SG1 2AS

J.-F. Luy

> Daimler-Benz AG
> Research Centre Ulm
> Wilhelm-Runge Str. 11
> D-89081 Ulm
> Germany

Acknowledgements

I would most of all like to thank all the authors for their excellent contributions. In every case I was pleasantly surprised by the breadth of coverage achieved in each chapter. I would also like to thank all the past and present lecturers on the annual MMIC design course at King's College, all of whom have contributed indirectly to the contents of this book through the knowledge they have imparted to me over the years. I would like to acknowledge the financial support provided by the Engineering and Physical Sciences Research Council.

On a personal note I would like to thank my wife Tracy for her support and encouragement throughout this project and would like to thank Prof. A. H. Aghvami for his contribution to the success of the MMIC research at King's and for his constant help and advice.

C. Sansom, R. G. Arnold, D. M. Brookbanks, and M. W. Geen would like to thank GEC Marconi Materials Technology Ltd. for granting permission to publish the material in Chapters 2, 4 and 5.

J.-F. Luy would like to express thanks to his colleagues J. Buechler, U. Güttich, H. Jorke, A. Klaaßen, F. Schäffler, E. Sasse, A. Schüppen, K. M. Strohm, M. Wollitzer and to E. Biebl and P. Russer from the Technical University of Munich. He is grateful to H. Dämbkes for supporting this work.

J. S. Joshi would like to acknowledge Philips Microwave for the use of their switch FET model and data for the FET switch performance.

S. Lucyszyn would like to acknowledge the financial support of the Engineering and Physical Sciences Research Council and would like to thank Professor C. W. Turner at King's College London, Professor H. C. Reader at the University of Stellenbosch, and C. Stewart at Cascade Microtech Europe for their assistance with the material in Chapter 11.

Chapter 1

Introduction

I. D. Robertson

1.1 Introduction

A monolithic microwave integrated circuit (MMIC) is a microwave circuit in which the active and passive components are fabricated on the same semiconductor substrate. The frequency of operation can range from 1 GHz to well over 100 GHz, and a number of different technologies and circuit approaches can be used. The additional term 'monolithic' is necessary to distinguish them from the established microwave integrated circuit (MIC) which is a hybrid microwave circuit in which a number of discrete active devices and passive components are integrated onto a common substrate using solder or conductive epoxy. These became known as integrated circuits because the alternative is to employ hollow metal waveguides, where the inclusion of active devices is a matter requiring considerable mechanical design and workshop machining. Strictly speaking, monolithic circuits operating at over 30 GHz should perhaps be called monolithic millimetre-wave integrated circuits, but a number of alternative abbreviations have been used (such as MMWIC or M^3IC) and there doesn't seem to be general agreement as yet. In this book the abbreviation MMIC is used almost exclusively, and it should be understood to include millimetre-wave circuits as well.

Fig. 1.1 shows a three-stage MMIC low-noise amplifier developed at Caswell for a satellite application (the chip size is 3.5×1.5 mm). The input and output are at the left and right of the chip, respectively, and the pads are in a ground-signal-ground pattern to enable RF-on-wafer (RFOW) measurements to be made. This circuit contains 3 transistors, 9 capacitors, 10 spiral inductors, 7 resistors, and a number of microstrip interconnections. This heavy usage of passive components is necessary because matching networks are essential in order to achieve usable gain from the transistors at microwave frequencies. A further feature of design at microwave frequencies is that the interconnecting tracks have a major effect on the amplifier, and both the interconnections and any discontinuities (bends and tee-junctions for example) must be modelled as microstrip elements using microwave CAD. Whilst MMICs may have rather lower apparent complexity than other integrated circuits, they do offer the highest frequencies of operation. This means that MMICs have a special range of applications and that MMIC design is very different to conventional VLSI design, in which CAD offers a high degree of layout automation. The important transmission-line nature of interconnections on an MMIC requires far more involvement from the designer in the layout process.

The purpose of this chapter is to provide a general background in the history, applications, and technology of MMICs. The advantages and disadvantages of MMICs compared with hybrid microwave integrated circuits are discussed in detail. The large range of available device technologies is introduced, and the major applications which have driven the development of MMIC technology are described. Finally, there is a brief guide to the material presented in each chapter of the book.

1.2 A brief history of MMIC technology

Gallium arsenide (GaAs) has been used extensively in the development of MMICs because of its suitability for both high frequency transistors and low loss passive components. The technique that made the fabrication of stable high resistivity GaAs material possible was the development of the liquid-encapsulated Czochralski (LEC) method of growing GaAs ingots [1]. Much of the early research into the growth of GaAs using boron trioxide encapsulant was carried out at the Royal Radar Establishment at Malvern [2], and this method is still in use. The horizontal Bridgman method, whilst producing low dislocation densities, did not directly yield the high resistivity required to make the GaAs a suitable substrate for microwave components. The research into the fabrication of transistors using GaAs was first carried out by Jim Turner at Plessey Research (Caswell), where the earlier disappointing experiences with the GaAs bipolar transistor led, in 1962, to interest in the GaAs field-effect transistor [3]. The first devices had a 24 µm gate length and produced power gain in the VHF band. In 1967 a 4 µm gate length device was fabricated, producing 10 dB gain at 1 GHz, and this became the first commercial device, called the GAT1, marketed by Plessey Optoelectronics and Microwave Ltd. The development of electron beam lithography allowed the first 1 µm device to be produced in 1971, giving gain at 10 GHz, and this was later marketed as the GAT3. In these first ten years, interest in the GaAs MESFET was confined to Caswell and IBM in Zurich.

Along with this activity to produce high performance microwave transistors, the concept of integrating semiconductor devices with microstrip circuitry was already being investigated. The first GaAs MMICs were simple circuits incorporating diodes and microstrip lines, and results were reported by Texas Instruments in 1968: Mao, Jones, and Vendelin [4] presented results for a single-ended 94 GHz mixer employing Schottky barrier diodes, and the design (but not measured performance) of a monolithic Gunn diode oscillator and frequency multiplier were described. In the same issue, Mehal and Wacker [5] presented the design of a 94 GHz balanced mixer and a 30 GHz Gunn oscillator. However, the transistor-based MMIC was probably

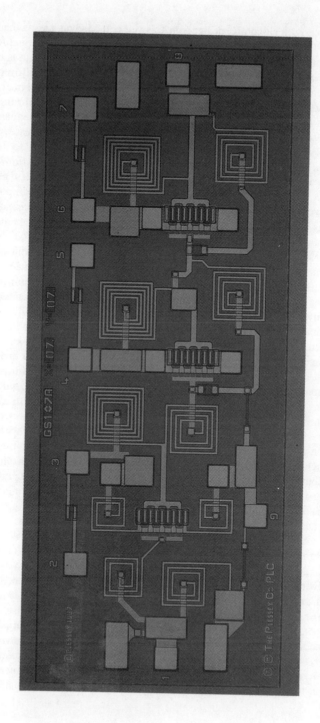

Fig. 1.1 A three-stage low-noise MMIC amplifier developed at Caswell for a satellite receiver
(Photograph courtesy of GEC Marconi Materials Technology Ltd.)

first conceived by a silicon IC designer at Caswell call Michael Gay, who created a layout of a 5 GHz receiver chip employing MESFETs with capacitors, inductors, and resistors. Whilst it was never realised, the idea led to the fabrication at Caswell of the world's first transistor-based MMIC, reported by Pengelly and Turner [6]: this was a single-stage X-band amplifier based on the GAT3 device, but with integrated loop inductors and interdigital capacitors. Later, Joshi et al. [7] presented results of the first FET oscillator MMIC, operating in J-band.

After these early successes, interest grew tremendously around the world and by 1979 the IEEE had established a symposium dedicated to GaAs IC developments. Thousands of papers have been published since then, many resulting from advances in the device technology which have yielded higher frequencies and better performance. However, there are a number of papers in which really important new techniques were presented; often these papers attracted world-wide interest and their contributions became real milestones in the history of MMIC development. Some of these particularly noteworthy papers are introduced now, with apologies in advance to those whose work may have been missed. In 1981, Hornbuckle and Van Tuyl [8] presented results for direct-coupled amplifiers using transistors and level-shifting diodes exclusively. In 1982, Jamison et al. [9] presented the design of MMIC amplifiers using planar spiral transformers, and later the technique was extended to realise receiver chips with exceptionally high packing density [10]. For MMIC amplifier design the parallel [11] and series feedback [12] techniques were demonstrated, and both techniques showed the advantage of the well controlled parasitics inherent in MMICs. In power amplifier design the thermal and matching problems encountered with MMICs were largely overcome with the introduction of the cluster matching technique described by Pavlidis [13]. For broad-band amplifiers the travelling-wave (or distributed) amplifier technique was re-visited, making full use of the high performance of the MESFET combined with the low parasitics of MMICs: in 1982, Strid and Gleeson [14] presented a particularly important paper describing a DC-12 GHz distributed amplifier along with some of the first ever results for RF-on-wafer probing. For many years various companies have competed in the distributed amplifier performance stakes. However, the small size of the circuit reported by Kennan et al. [15] is still remarkable, with the coupling within the inductors actually being used to improve the performance. For high power the distributed amplifier has always been at a disadvantage, but the capacitively-coupled distributed amplifier technique presented by Ayasli et al. [16] in 1984 showed that the power limitations were not insurmountable. Distributed amplifiers entered a new era when HEMTs became available, and the standards were re-defined by the 5-100 GHz bandwidth reported in 1990 by Majidy-Ahy et al. [17] for an InP MMIC distributed amplifier using coplanar waveguides. The distributed amplification technique received widespread interest for other circuit functions, and the broad-band mixers and active baluns demonstrated by Pavio [18, 19] received considerable acclaim. In Europe there was tremendous interest in satellite TV receivers, and the first 12 GHz MMIC receiver chips [20, 21] were important milestones. On-chip antennas were first demonstrated in diode mixer receivers [22], since their size dictates that they are only used in the millimetre-wave band. Now that millimetre-wave HEMT technology has matured, on-chip antenna elements are of even greater interest, with wafer-scale active phased-arrays having been proposed. In order to miniaturise circuits, active techniques have always been of interest; the active filter employing FETs in a transversal distributed amplifier topology was presented by Schindler *et al.* [23] and the active inductor was presented by Hara *et al.* [24], both in 1989. Finally, Hirota *et al.* [25] demonstrated that 'uniplanar' MMICs employing coplanar waveguide and slotline offered scope for a huge range of novel circuits, and Hiraoka *et al.* [26] demonstrated one of the first multi-layer circuits employing the thin-film microstrip technique.

As far as new circuit design techniques are concerned, the 1980s must surely represent the heyday for the MMIC designer. In the 1990s there has been very rapid development of new devices, and an increasing emphasis on the civil applications of MMIC technology in communications. Hence, the designer is often employing established circuit ideas but is making far more use of CAD tools in order to miniaturise the circuit and make sure it works first time.

1.2 Advantages and disadvantages of MMICs

The advantages and disadvantages of MMICs when compared with hybrid MICs are listed in Table 1.1.

Table 1.1 Advantages and disadvantages of MMICs

MMICs	Hybrid MICs
Cheap in large quantities	Simple circuits can be cheaper; automatic assembly is possible
Very good reproducibility	Poor reproducibility due to device placement and bond-wires
Small and light	Not very large anyway
Reliable	Hybrids are mostly 'glued' together and so reliability suffers
Less parasitics – more bandwidth and higher frequencies	The best transistors are always available for LNAs and HPAs
Space is at a premium; the circuit must be made as small as possible	Substrate is cheap, which allows microstrip to be used abundantly
Very limited choice of component	A vast selection of devices and components is available
Long turn around time (3 months)	Can be very fast (1 week), making redesigns easy
Very expensive to start up	Very little capital equipment is required

1.2.1 Cost

The low cost advantage stems from the simple fact that a single wafer can produce upwards of 1000 working amplifiers (for example), each with very similar performance and requiring no hand tuning and minimal assembly work. This argument holds as long as the circuit is fairly complicated and requires a large number of components. However, in many cases it is possible to use packaged transistors soldered onto a cheap substrate. Then the most expensive components are the transistors; the passive components such as filters and matching networks cost next to nothing, and the assembly is straightforward with packaged transistors. Even with bare chip devices, automated assembly can be used to good effect. For a circuit using a few transistors with a large amount of passive circuitry it would be quite difficult to reduce the cost of the circuit by using MMICs: by including the passive circuitry on the GaAs the cost is almost certain to increase. Table 1.2 shows the approximate chip fabrication cost against chip size for a high yield MESFET process using ion-implantation. This represents 1995 prices for high volume manufacture with 3" wafers, calculated from my own estimates after discussion with various foundries. As the chip size increases you get less chips from the wafer *and* lower yield; the cost rises rapidly as a result. However, the yield decrease would not be as bad if the circuit had a lot of passive components, so it is impossible to generalise accurately. What the table does show is that for a high volume application such as DBS (satellite TV distribution), an MMIC must have high packing density to be competitive. A chip size of 2 mm² is really the maximum that should be considered for this function. At the other end of the scale, if you could design a single chip phased-array radar module in a 10 mm × 10 mm chip, you would still get some customers even though each chip cost $440. In highly competitive commercial applications, such as personal communications and wireless LANs, there is tremendous pressure for chip costs to be reduced to less than $1 per square millimetre, and this makes the use of 4" wafers (or larger) advantageous if the necessary capital investment can be justified.

Table 1.2 Chip cost against size

Chip size (mm)	Yield (%)	Working circuits per 3" wafer	Bare chip cost at $4k per wafer
1 × 1	80	3600	1.1
2 × 2	70	800	5
5 × 5	45	80	50
7 × 7	30	25	160
10 × 10	20	9	440

1.2.2 Performance

Most MMIC devices have to be tailored to volume production and tend not to give state-of-the-art performance. This can be a serious problem for low-noise and high-power amplifier design, where performance is of prime concern. For the very best noise figure and power efficiency it is often necessary to use discrete transistors before and after the MMIC parts. With a hybrid MIC the designer can choose the best transistor for the job, and the transistors may be from different manufacturers. Discrete transistors can use the shortest gate length and optimum active layer structure for their application, regardless of other requirements, because yield is much less important. Using the same fabrication techniques for a complex MMIC would result in a very poor yield because of the multiplicative effect of the yields of individual transistors on the same chip. For low-density circuits this is not so much of a problem, and so millimetre-wave circuits using HEMTs should give good performance and yield.

Special devices, such as Gunn diodes, PIN switches, and hyperabrupt varactor diodes, are rarely incorporated into MMIC processes and this can introduce further compromises for the MMIC designer: the FET switch is a poor substitute for the PIN diode, and the HEMT millimetre-wave oscillator will have a low output power compared with a Gunn diode. Most of these compromises can be absorbed into the specifications of the system design, and good communications between the circuit designer and systems designer are very beneficial to the final product.

1.2.3 Investment required

Setting up an MMIC fabrication facility for mass production is prohibitively expensive. There was a time when most major microwave companies had their own facilities as a matter of course, in order to demonstrate their capability. When the profitable applications were not forthcoming, there was a period of regrouping and consolidation. There are still over 25 major foundries/suppliers of GaAs MMICs [27], and there is a rapidly growing number of companies in the silicon MMIC business. The major manufacturers are listed in Table 1.3. Given that foundries are now widely available for MMIC fabrication, what is to prevent the entrepreneur from starting a profitable business? The major investment required nowadays is that of CAD facilities, since foundries can offer RF-on-wafer (RFOW) measurements. The task of designing a competitively priced circuit should not be underestimated, and extensive CAD facilities along with experienced designers are essential. The 'wireless revolution' offers considerable rewards to the company that succeeds but the competition is fierce, not only between companies but also between different technologies within the same company.

In contrast, developing hybrid circuits using packaged transistors can be carried out with a PC-based packaged like Touchstone™ and with little more than an etching tank and soldering iron. A simple hybrid MIC could be laid out, constructed, and tested in one day. Because the substrate is cheap and plentiful the designer doesn't have to spend time squeezing the circuit into the smallest possible space. The prototype can be tweaked with a scalpel and conducting paint, and then the final design completed. The capital investment required and development times are both considerably less for MICs, with obvious benefits.

Table 1.3 Major MMIC manufacturers

GaAs MMIC	GaAs digital	Silicon MMIC
Alenia	AT&T	Daimler Benz
Alcatel-Telettra	Fujitsu	GEC Plessey
Alpha Industries	Motorola	HP-Avantek
Anadigics	Philips	IBM
Avantek	Rockwell	M/A-COM
Celeritek	Thomson	Mini Circuits
COMSAT	Triquint	Motorola
Daimler Benz	Vitesse	National Semiconductor
Fujitsu		NEC
GEC Marconi		SGS-Thomson
HP		Siemens
Hughes		Watkins Johnson
ITT		
Litton		
Lockheed		
M/A-COM		
Motorola		
NEC		
Pacific Monolithics		
Philips		
Raytheon		
Rockwell		
Samsung		
Siemens		
Texas Instruments		
Thomson		
Triquint		
TRW		
Varian		
Watkins-Johnson		

1.2.4 Reproducibility

Reproducibility is excellent for MMICs because the active and passive components are produced by the same well-controlled fabrication steps using the same photolithographic masks. Furthermore, since the active and passive component variations are related to the same physical parameters (such as the sheet resistance of the active layer), there is considerable scope for employing design centring methods for maximising reproducibility and yield. In comparison, the hybrid MIC suffers from device placement and wire-bonding variations from circuit to circuit. Most hybrid circuits will require some form of hand tuning of the response, and this is very uneconomical for large volume manufacture.

In millimetre-wave circuits the ability to control interconnection parasitics is a major advantage. It is fair to say that the ease with which 30 GHz-plus monolithic circuits can be manufactured will make these frequencies viable for widespread commercial use, whereas this frequency resource has been limited previously to expensive military and research activities. Most conventional millimetre-wave parts are waveguide components which are assembled and tuned individually by hand, and they are very expensive as a result. Microstrip-based MICs are exceedingly difficult to construct above 30 GHz; finding a suitable substrate and making a good launch to microstrip at 30 GHz are difficult enough, and then you have to overcome the huge parasitics of components such as chip capacitors and resistors, as well as the large reactances presented by the bond-wires. With experience and careful design, one can overcome these parasitics to construct a circuit in the research lab, but in large volume manufacture the parasitics would be very hard to control.

1.2.5 Reliability

MMICs are more reliable than hybrid circuits, as long as the fabrication process is carefully controlled and qualified: the GEC-Marconi F20 MESFET process has attained space qualification from the European Space Agency. In contrast, the mechanical limitations of chip attachment and wire-bonding become all too apparent when a hybrid circuit is subjected to temperature cycling, shock, and vibration. Hewlett-Packard have already updated some instruments to incorporate GaAs MMICs. This has enabled a direct comparison to be made between the new and old circuits in the field, and so far the MMICs have proved significantly better [28, 29].

1.2.6 Size and mass

MMICs are very compact, and this can be a significant advantage. In many commercial applications there is an increasing need to miniaturise the microwave circuits, with smart cards and radio transceivers for PCMCIA slots (in notebook computers) being good examples. The active phased-array antenna is the primary application where size is very important: since each face of an array may have many hundreds of modules, MMICs are essential in order to minimise the antenna size and mass.

1.3 MMIC technologies

MMICs predominantly use GaAs for two key reasons: (1) GaAs has higher saturated electron velocity and low-field mobility than silicon, resulting in faster devices; (2) GaAs can readily be made with high resistivity, making it a suitable substrate for microwave passive components. As a result, GaAs completely dominated the first 15 years of MMIC development and even now the vast majority of MMICs are GaAs based. Nevertheless, improvements to the silicon bipolar device such as narrow emitters and shallow bases have enabled conventional silicon bipolar devices to achieve f_ts of over 20 GHz, and silicon has captured a large market for general purpose MMIC amplifiers operating up to 5 GHz. The competition between GaAs and silicon has become more fierce as a result of important developments in heterojunction devices with both materials. These developments in active devices are described in Chapter 2. The different technologies which are now used for MMICs are as follows:-

>GaAs MESFET
>GaAs HEMT (conventional and pseudomorphic)
>GaAs HBT
>Silicon bipolar
>Silicon-germanium HBT
>Indium phosphide HEMT
>Indium phosphide HBT
>IMPATT, Gunn and Schottky diodes

The GaAs MESFET was the first microwave transistor and remains an important workhorse for many MMIC designs. It is easily fabricated using ion-implantation for high volume applications and has good noise figure and output power performance. Most GaAs foundries offer 0.5 µm gate length MESFET processes which are useful for circuits operating up to 20 GHz, and circuits operating to over 30 GHz have been reported with shorter gate lengths. Foundries generally offer a selection of different device types including general purpose, low DC power, high RF power, switching, and enhancement-mode/depletion-mode devices. Using selective ion-implantation it is often possible to use different types in the same circuit.

The GaAs-based pseudomorphic HEMT (high electron mobility transistor) offers a considerable increase in transconductance over the MESFET and, with gate lengths as short as 0.1µm, circuits operating to over 100 GHz have been reported [30]. However, these high performance devices are less suited to high volume manufacture than the MESFET: the intricate material layers must be fabricated with either molecular beam epitaxy (MBE) or metal organic chemical vapour deposition (MOCVD), both of which are time consuming. The gates must be defined with electron beam lithography, which is rather slow, and for good noise figures the mushroom gate is needed, which must be produced using a precise multi-layer photoresist technique. HEMTs have also been reported based on indium phosphide, and these have produced the absolute best high frequency performance, with noise figures of 0.8 dB at 60 GHz and 1.2 dB at 94 GHz having been achieved [31]. Monolithic amplifiers operating at over 100 GHz have been reported [32] using InP-based HEMTs.

The GaAs heterojunction bipolar transistor (HBT) has some advantages over the HEMT because it is a vertical structure. The critical device dimensions are defined by the material growth and doping rather than by lithography. An HBT with an emitter width of several microns can offer good microwave performance. The HBT is intrinsically a very high-gain device but tends to suffer from large parasitic resistances and capacitances; it can offer exceedingly high power-density and efficiency [33], but suffers from thermal limitations which have made it most suitable for pulsed applications so far. The high base resistance of HBTs means that their noise figure is considerably higher than that of HEMTs, but the device is continually being improved. In the US several manufacturers are offering HBT foundry processes, and already some very competitive products are on offer, particularly low supply voltage components for mobile handsets.

Silicon bipolar technology has advanced tremendously in recent years and the homojunction device has already achieved f_ts of over 20 GHz. In addition, the heterojunction bipolar transistor based on silicon-germanium [34] has been reported to have f_ts of over 100 GHz [35] and noise figures of less than 1dB at 10 GHz have been achieved [36]. These technologies offer a major threat to the dominance of GaAs in MMIC development. Their major advantages are that 8" silicon wafers are readily available, whilst GaAs MMIC fabrication is limited to 4" at the moment, and that silicon bipolar devices can be integrated readily with conventional VLSI circuitry based on CMOS. The combined Bi-CMOS processes are expensive, but offer the ideal solution of true single chip integration in consumer products such as mobile telephones and intelligent vehicular systems. GaAs technology still has a major advantage in areas where high performance is paramount, and where the superior opto-electronic properties of GaAs are needed. However, if GaAs and silicon vendors are offering a chip with the same functionality, then it is the lowest price that determines the choice made by the customer. Ultimately, since 8" silicon wafers are four times the area of 4" GaAs wafers then, if the processing effort is roughly the same, the GaAs manufacturer has to contend with four times the raw chip manufacturing cost.

For millimetre-wave applications the picture today is very attractive for the GaAs industry, with the pseudomorphic HEMT having no real equal in performance. Nevertheless, there remains the fact that simple transmitter-receivers can be realised [37, 38] using IMPATT or Gunn diode oscillators and Schottky barrier diode receivers. Although they have very limited performance (particularly noise figure) and limited functionality, these technologies may steal some of the market from HEMT manufacturers.

1.4 Applications

The major applications of MMICs are shown in Table 1.4. Military and space applications have been a major driving force behind MMIC technology, and the MIMIC programme in the US is a clear example of this. The adaptive phased-array antenna is probably the single most important application for MMICs in these applications. With hybrid technology the size and mass of the individual transmitter-receiver modules would simply make a large adaptive phased-array antenna

Table 1.4 Applications of MMICs

Military	Space	Civil
Phased-array radar	Communications satellites	Satellite TVRO
	Remote sensing	VSAT earth terminals
Electronic warfare	Synthetic aperture radar	Mobile phones
		LOS communications
Smart munitions	Radiometers	Wireless LANs
Remote sensing		Fibre-optic systems
Synthetic aperture radar	Astronomy	Global positioning (GPS)
		Smart cards / tagging
Decoys	Low earth orbit satellites (IRIDIUM)	Search and rescue transponders
		M$^{(3)}$VDS and the wireless local loop
Altitude meters	Steerable phased-array antennas	Anti-collision radar
		Automatic tolling
Instrumentation		Medical systems

impractical. Since the requirements of adaptive phased arrays have been so beneficial to MMIC development, it is important to understand them: Fig. 1.2 compares a conventional satellite transmitter with a solid-state transmitter using adaptive beam-forming. The conventional transmitter uses a single high-power travelling-wave tube amplifier (TWT) transmitting through a beam-forming network into the individual antenna feeds. This system has been used successfully on all the major communications satellites. However, the TWT is a vacuum tube device and as such has limited reliability and lifetime, requires high voltage supplies, and is rather non-linear. The beam-forming network is required to shape the transmitted beam so that coverage of the ground can be carefully designed to make optimum use of the satellite's precious transmitter power. However, since the beam-forming network may

Introduction 13

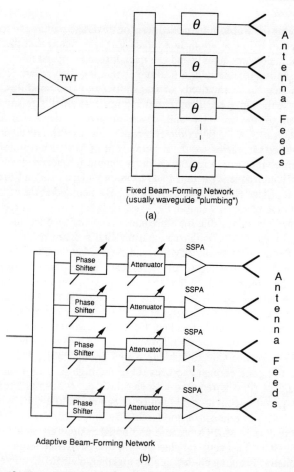

Fig. 1.2 Phased-array antennas: (a) Conventional, using a travelling-wave tube amplifier; (b) Adaptive, using solid-state power amplifiers

be handling many hundreds of watts, is it impractical to employ electronic control of the beam-forming. The beam-forming network is usually a mass of carefully designed waveguide plumbing.

In the solid-state transmitter the power amplification is distributed around the antenna. Hence the individual power amplifiers are only 20 W or so, and solid-state amplifiers can readily be used. This means an improved lifetime and reliability, low voltage supplies, and excellent linearity. In addition, there is clearly a large degree of redundancy in the arrangement. Since each antenna feed is now fed from a separate amplifier, the phase and amplitude of the signal can be individually controlled with electronic attenuators and phase-shifters placed before the power amplifier. This means that the antenna beam can be steered electronically. In a satellite payload this

14 *Introduction*

gives the operator the freedom to reconfigure the coverage patterns to match demand whilst the satellite is in operation. In a radar system it means that the beam can be steered electronically very quickly and can track fast-moving and multiple targets in a fashion which is impossible with a traditional rotating-dish radar. Since a phased-array antenna may have hundreds of transmitter-receiver modules, MMICs are essential because they give excellent reproducibility with small size and mass.

In addition to the military and space industries' drive for phased-array components, the satellite TV (DBS-direct broadcast by satellite) receiver (TVRO) has provided an important 'carrot' for the development of MMIC technology. This high volume consumer application requires a fairly simple frequency downconverter. In the US the received frequency is in C-band at around 4 GHz, and in Europe it is in X-band at 11 to 12 GHz. The 'low noise block' (LNB) mounted at the antenna feed has to downconvert the received signal to the 1 to 2 GHz region, so that the signal can be fed through cheap coaxial cable into the indoor receiver/decoder unit. It was rather disappointing for the MMIC enthusiast that the first systems designed for the mass consumer market did not use any MMICs, and even today the ones that do employ an MMIC are far from being the totally integrated single-chip receivers that were dreamed of in the pioneering days. To see why this is so, consider the simplified LNB block diagram shown in Fig. 1.3. This is a very straightforward frequency downconverter. However, the noise figure of the LNA (low noise amplifier) has to be as low as possible, and certainly less than 1 dB. The lower the noise figure, the better the TV picture or the smaller the dish. However, the first MMIC satellite TV receiver chips [20, 21] were not able to achieve a low enough noise figure because the MESFET was not good enough and because of the high on-chip matching network loss. The bandpass filter is required to give 50 dB or so of image rejection, and a filter of the required selectivity could not be integrated onto a chip by using lumped elements (too lossy) or microstrip (too large). The mixer has always been a difficult component to realise on MMICs because traditional designs are so incompatible with the planar MMIC. The oscillator design is quite easy, but requires an off-chip dielectric resonator. Bringing all these points together, the MMIC single-chip satellite TV receiver actually needs an off-chip discrete LNA stage, an off-chip image filter, and an off-chip dielectric resonator, and on top of this the IF stages are cheaper in silicon. Hence, the MMIC has found it surprisingly hard to make inroads into this market, although eventually it did to good effect. Nevertheless, the discrete HEMT front end stage feeding a MESFET-based MMIC appears to be preferred to a complete HEMT-based MMIC with low noise figure. This reflects the high costs of HEMT manufacture compared with ion-implanted MESFETs.

In the communications field, MMICs are applicable to fibre optic systems and to satellite, fixed-terrestrial, and mobile radio systems. MMICs are an important part of the 'wireless revolution'. In the UK many applications have had frequency allocations 'pencilled in' for frequencies as high as 300 GHz [39], under the assumption that MMIC technology will one day make radio systems at these frequencies affordable. At present, very small aperture terminal (VSAT) systems provide a substantial market opportunity in satellite communications for MMIC manufacturers, but mobile communications is really the most important application.

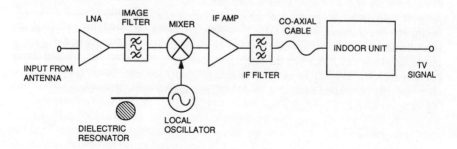

Fig. 1.3 Block diagram of a satellite TV receiver's low-noise block (LNB)

Many countries have already introduced cellular mobile telephone systems operating in the 1–2 GHz region, and major new systems are being developed using low earth orbit satellites [40]. Also, as urban traffic builds up, and as new high data-rate services are envisaged, there are already major projects underway in Europe to develop MMICs for microcellular systems operating at 62.5/63.5 GHz [41]. In addition, wireless data communications is seen as a major growth area as personal digital assistants (PDAs), such as the Apple Newton, and notebook computers become commonplace, and there is a large market for MMIC transceivers. In-building wireless LANs (local area networks) are increasingly being promoted [42, 43], with frequencies of 2.4, 5.2, 17, 18 and even 60 GHz being proposed for new systems. The rationale behind this new application of microwaves is that the transmitter-receivers are so much cheaper nowadays that the wireless system is an economical alternative to the extensive routing and laying of coaxial or fibre optic cables (although these are still used as the backbone of the network in most cases).

The microwave or millimetre-wave video distribution system (MVDS or M^3VDS) is a combined fibre optic and radio TV distribution system, in which the enormous expense and effort required to connect homes with fibre optic cable is overcome by distributing the many TV channels over the last mile or so by a short radio link, as shown in Fig. 1.4. This system was extensively developed by British Telecom, who ran a 29 GHz trial service at Saxmundham in the UK [44]. The combined cable and radio system is ideal for rural areas which are not economically viable for purely cable systems. The work at BT was extensively concerned with the development of an MMIC receiver for a low cost subscriber unit [45]. Anyone who has purchased millimetre-wave waveguide components will appreciate that since a single component such as a mixer can cost £1500, an MMIC receiver is rather essential for a consumer application.

In the millimetre-wave frequency range the proposal for intelligent-vehicle highway systems (IVHS) offers a tremendous opportunity for MMIC manufacturers

16 Introduction

[46, 47]. This idea, which is being promoted around the world under a number of schemes, requires that vehicles have anti-collision radar systems on board, as well as equipment for communicating with roadside units, primarily for automatic tolling purposes. In order to give acceptable resolution, the radar must operate in the millimetre-wave band, and 77 GHz is rapidly becoming a standard. Many motor manufacturers are already introducing anti-collision devices, particularly for parking purposes. However, it is when the intelligent motorway becomes a reality and every car has to have a 77 GHz unit on board, so that they can travel together with electronically controlled gaps, that the MMIC manufacturers will be rewarded.

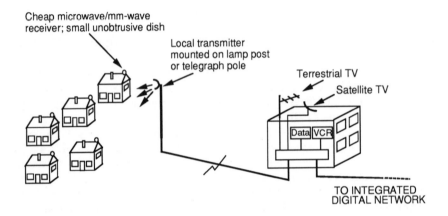

Fig. 1.4 The millimetre-wave video distribution system (M^3VDS)

1.5 Types of MMIC

As well as the different device technologies, in general there are three quite distinct circuit design techniques which are used in MMICs, the choice of which largely depends on the operating frequency of the circuit. The three techniques can be classified as 'all-transistor' techniques, lumped-element techniques, and distributed techniques. There is inevitably some overlap of each approach's useful frequency range of application, and the techniques may often be blended together in the same design. Fig. 1.5 shows the approximate frequency range to which each approach is applicable.

Circuits using entirely active techniques [8] tend to use small device peripheries so that the resulting small input and output capacitances do not unduly affect the performance of circuits such as operational amplifiers. This approach is usable up to approximately 5 GHz, and such high frequency of operation is achieved largely because of the low capacitance of a microwave transistor, rather than through

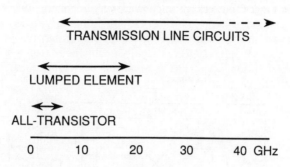

Fig. 1.5 Typical frequency range of operation for the three approaches to MMIC design

the use of classical microwave design techniques. Nevertheless, the design of these 'conventional' circuits at such high frequencies does require tremendous design ability and experience. In the silicon industry this sort of all-transistor design experience is widely available, but the GaAs industry has naturally tended to have microwave design engineers. It is partly because of this that, until recently, entirely active techniques have been used relatively little in MMICs. In addition, most of the original interest in MMICs has been for space and military applications where DC power consumption has been of considerable importance, making active techniques less attractive. However, the major advantage of active techniques is their high packing density, which leads to competitively priced products. Hence, in recent years there has been far more interest in these techniques for microwave applications such as mobile telephones operating at 1.8 GHz and wireless LANs operating at 2.4 GHz. Part of the reason for this has been the availability of high performance silicon bipolar and GaAs HBT technology. These bipolar devices seem much more amenable to complex active circuits than MESFETs and HEMTs, and part of the reason for this must be the relatively poor control of the FET's DC characteristics, which has also limited their application to digital circuits. Fig 1.6 shows an example of an all-transistor MMIC: this is a 5 GHz isolator developed at University College London [48].

For higher operating frequency the transistor input and output capacitance must be accounted for: on MMICs, lumped-element matching networks (using spiral inductors and overlay capacitors) provide the best solution at frequencies below 20 GHz. Fig. 1.7 shows an example of an MMIC employing lumped-elements. This could be considered to be the classical type of MMIC, with the spiral inductors being the visually striking 'trade-mark' of MMIC technology. The chip is a 1–2 GHz single-stage amplifier developed at King's College. The principle advantage of this 'low packing density' circuit is that it has high performance in terms of noise figure and power handling when compared to an all-transistor design. Hence it is well suited to

18 *Introduction*

applications such as satellite payloads and phased-array antennas where high performance and low DC power consumption are very important [49].

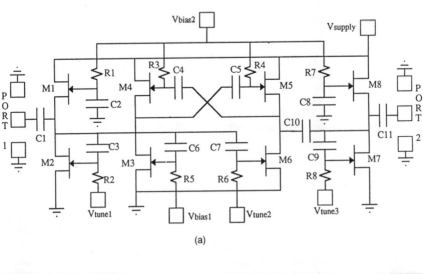

(a)

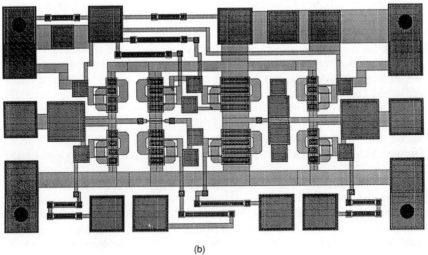

(b)

Fig. 1.6 An 'all-transistor' MMIC active isolator: (a) Circuit diagram; (b) Layout (courtesy of David Haigh, University College London)

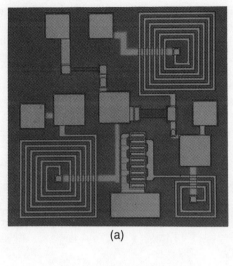

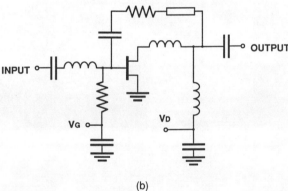

Fig. 1.7 A lumped-element 1 to 2 GHz MMIC amplifier: (a) Photograph; (b) Circuit diagram

At frequencies higher than 20 GHz the spiral inductors will be beyond their useful frequency range because of their own self-resonance, and so distributed elements are used. MMICs operating at over 20 GHz will tend to use stubs and transformers for matching to the device input and output impedance. The distributed elements can be realised in a number of transmission-line media, with microstrip and coplanar waveguide (CWP) being by far the most common. The coplanar waveguide has been extensively used for millimetre-wave circuits, its chief advantages being low dispersion and the ability to provide low inductance grounding of devices. With distributed matching techniques the highest operating frequency of an MMIC is

20   *Introduction*

limited only by the maximum frequency at which the active devices still have usable available gain. The *lowest* frequency of operation for distributed circuits is determined by the chip size, since the physical length of matching elements is too great at frequencies below approximately 5 GHz. Fig. 1.8 shows an example of a circuit employing distributed matching elements: it is a 14–20 GHz two-stage amplifier, developed at King's College. High impedance microstrip lines are used throughout as shunt stubs and series lines for the matching networks.

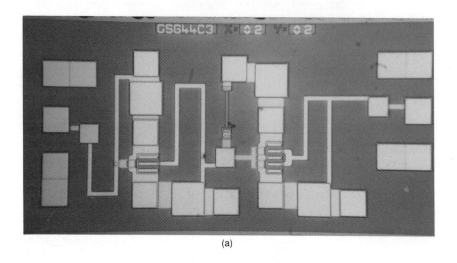

(a)

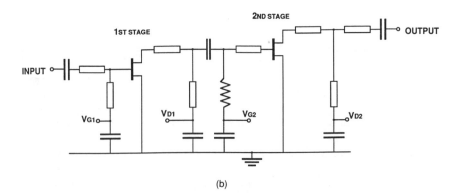

(b)

Fig. 1.8   MMIC 14–20 GHz two-stage amplifier using transmission-line matching elements: (a) Photograph; (b) Circuit diagram

1.6 A guided tour through the book

Chapter 2 describes the active devices and gives an introduction to the fabrication technology. Every MMIC designer should have some knowledge of the processing procedures used for the fabrication of his circuits. A little knowledge goes a long way in helping the designer to understand and obey design rules and have sympathy with the technologists' concerns. Often, advanced new components and techniques can be devised by making full use of the capability of a particular process.

Chapter 3 describes the passive components used in MMIC design. It is the extensive use of passive components that makes MMIC design so unique and enjoyable. Often a circuit using spiral inductors and meandered microstrips can be really appealing to the eye; a true work of engineering art. Whilst a VLSI chip might have one million transistors as opposed to the MMIC's ten or twenty, the MMIC will always look more interesting. The passive components used include resistors, spiral inductors and overlay capacitors, as well as microstrip lines and components such as couplers and power dividers.

Chapter 4 describes the modelling and computer-aided design techniques used for MMIC design. The approaches that can be used range from simple lumped-element models, with parameter values taken from design graphs and equations, up to sophisticated electromagnetic modelling using powerful workstations. It has always been the dream of the MMIC designer to be able to simulate a layout directly; this is now extremely close to reality, as packages are available that can import arbitrary component layouts in GDSII format and simulate them directly. This approach is very demanding of computer power, and certainly can not model a complete chip layout as yet, but computer technology moves at a fast pace and speed is said to double every 18 months to two years.

Chapter 5 describes a wide range of different amplifier design techniques. The amplifier is the most common application of MMIC technology and amplifier performance is often used as the performance benchmark for new devices and technologies. The chapter describes the classical designs such as the feedback amplifier and distributed amplifier, as well as describing the numerous improvements that have been reported in the literature.

Chapter 6 describes the state-of-the-art in the design of MMIC mixers. Conventional mixers use diodes and wire-wound or suspended-microstrip baluns. Monolithic diodes, usually being planar in nature, have been found wanting in performance, and baluns have been extremely hard to realise on MMICs. Nevertheless, advanced new techniques have been applied to mixer design, including active couplers and baluns. FET mixers have been used widely as an alternative to diode mixers, and a number of active mixer design techniques are described.

Chapter 7 describes design techniques for phase shifters. Phase shifters are of particular importance for the adaptive beam-forming in phased-array antennas. Both digital and analogue phase shifter designs are described, and the detailed design principles for reflection type, switched-line, loaded-line and switched-filter topologies are described. The insertion loss variation introduced by the phase shifter is particularly important for beam-forming applications, and state-of-the-art designs are presented for maintaining constant insertion loss.

Chapter 8 describes techniques for the design of switches and gain control circuits. Switches (SPDT, SPST, etc.) find many applications in MMICs, including transmit-receive switches in radar modules, antenna selective diversity switches in communications transceivers, and redundancy switches for satellite payload circuitry. Gain control circuits include variable and switched attenuators and variable gain amplifiers. These find application in automatic gain control (AGC) loops and in phased-array systems.

Chapter 9 describes the design and performance of negative resistance and feedback oscillators. The theoretical and practical noise performance of FET and bipolar transistor oscillators is discussed. The different circuit topologies and types of resonator are described. A design example is given of an X-band negative-resistance varactor-tuned VCO.

Chapter 10 describes the technology and design of silicon millimetre-wave monolithic circuits based on high resistivity substrates. In this frequency range silicon can be used to realise IMPATT and Schottky diodes for transmitter-receivers, and these have been integrated with microstrip antenna arrays. In addition, the Si-Ge HBT device has achieved millimetre-wave performance and offers a serious challenge to the total monopoly that the III-V transistors have had in millimetre-wave applications.

Chapter 11 describes the measurement techniques used for MMICs. On-wafer microwave probing is the most important method for characterising devices and components, as well as for testing final circuits prior to packaging and shipping. Nevertheless, test fixture measurements still provide an important means of testing circuits, particularly when additional off-chip components are necessary for the operation of the circuit. Specially developed adaptable test jigs using in-fixture calibration methods are described, as well as the more conventional test fixtures and de-embedding methods. In addition to these methods of 'black box' S-parameter characterisation, a number of techniques are described for the field probing of circuits; these methods allow the non-invasive measurement of signals *within* the circuit and are particularly useful for troubleshooting purposes.

Chapter 12 describes a wide range of advanced techniques which as yet are outside the mainstream of everyday MMIC design, but which are likely to be used extensively in the future. These techniques include active filters, which have approached waveguide filters for selectivity if not for power handling and linearity. Coplanar waveguide and multi-layer techniques are described, and a number of example circuits are shown. Monolithic active antennas are described, and are seen to be an important component for millimetre-wave applications. In the communications area, direct microwave implementations of modulators and demodulators are of increasing interest for minimising the RF hardware requirements, and a number of techniques are described for achieving this. The role of MMICs in opto-electronic systems is briefly discussed, and a wide range of new components and technologies are described.

1.7 References

1. METZ, E. P. A., MILLER, R. C., and MAZELSKY, R.: 'A technique for pulling single crystals of volatile materials', *Journal of Applied Physics*, Vol. 33, no. 6, 1962, pp. 2016-2017
2. MULLIN, J. B., STRAUGHAN, B. W., and BRICKELL, W. S.: 'Liquid encapsulation techniques: The use of an inert liquid in suppressing dissociation during the melt-growth of InAs and GaAs crystals', *Journal Phys. Chem. Solids*, Vol. 26, 1965, pp. 782-784
3. TURNER, J. A.: 'History of the GaAs FET at Caswell (1964-1985)', IEE Colloquium on Modelling, Design, and Application of MMICs, 17th June 1994, Savoy Place, London, digest no. 1994/092, pp. 1/1-1/3
4. MAO, S., JONES, S., and VENDELIN, G. D.: 'Millimeter-wave integrated circuits', *IEEE Journal of Solid State Circuits*, SC-3, no. 2, June 1968, pp. 117-123
5. MEHAL, E. W., and WACKER, R. W.: 'GaAs integrated microwave circuits', *IEEE Journal of Solid State Circuits*, SC-3, no. 2, June 1968, pp. 113-116
6. PENGELLY, R. S., and TURNER, J. A.: 'Monolithic broadband GaAs FET amplifiers', *Electron. Lett.*, Vol. 12, no. 10, May 1976, pp. 251-252
7. JOSHI, J. S., COCKRILL, J. R., and TURNER, J. A.: 'Monolithic microwave GaAs FET oscilllators', IEEE GaAs IC Symposium, Lake Tahoe, Nevada, Sept. 1979
8. HORNBUCKLE, D. P. and VAN TUYL, R. L.: 'Monolithic GaAs direct-coupled amplifiers', *IEEE Trans.*, ED-28, no. 2, Feb. 1981, pp.175-182
9. JAMISON, S. A., PODELL, A., HELIX, M., NG, P., and CHAO, C.: 'Inductively coupled push-pull amplifiers for low cost monolithic microwave ICs', IEEE GaAs IC Symposium Digest, 1982, pp. 91-93
10. FERGUSON, D., BAUHAHN, P., KEUPER, J., LOKKEN, R., CULP, J., CHAO, C., and PODELL, A.: 'Transformer coupled high-density circuit technique for MMIC', IEEE Int. Microwave Symp. Dig., 1984, pp. 34-36
11. RIGBY, P. N., SUFFOLK, J. R., and PENGELLY, R. S.: 'Broadband monolithic low noise feedback amplifiers', IEEE Microwave and Millimeter-Wave Monolithic Circuits Symposium Digest, June 1983, pp. 71-75
12. LEHMANN, R. E., and HESTON, D. D.: 'X-Band monolithic series feedback LNA', IEEE Microwave and Millimeter-Wave Monolithic Circuits Symposium Digest, June 1985, pp. 54-57
13. PAVLIDIS, D., ARCHAMBAULT, Y., EFTHIMEROU, M., KAMINSKY, D., BERT, A., and MAGARSHACK, J.: 'A new specifically monolithic approach to microwave power amplifier design', IEEE microwave and millimeter-Wave Monolithic Circuits Symposium Digest, June 1983, pp. 54-58
14. STRID, E. W., and GLEESON, K. R.: 'A DC-12 GHz monolithic GaAs FET distributed amplifier', *IEEE Trans. Microwave Theory Tech.*, MTT-30, no. 7, July 1982, pp. 969-975

15. KENNAN, W., ANDRADE, T., and HUANG, C.: 'A miniature 2 to 18 GHz monolithic GaAs distributed amplifier', IEEE Microwave and Millimeter-Wave Monolithic Circuits Symposium Digest, May 1984, pp. 41-44
16. AYASLI, Y., MILLER, S. W., MOZZI, R., and HAINES, L. K.: 'Capacitively coupled travelling-wave power amplifier', IEEE Microwave and Millimeter-Wave Monolithic Circuits Symposium Digest, May 1984, pp. 52-54
17. MAJIDI-AHY, R., NISHIMOTO, C. K., RIAZIAT, M., GLEEN, M., SILVERMAN, S., WENG, S. L., PAO, Y. C., ZDASIUK, G. A., BANDY, S. G., and TAN, Z. C. H.: '5-100 GHz InP coplanar waveguide MMIC distributed amplifier,' *IEEE Trans. Microwave Theory Tech.*, MTT-38, Dec. 1990, pp. 1986-1993
18. HOWARD, T. S., and PAVIO, A. M.: 'A dual-gate 2-18 GHz monolithic FET distributed mixer', IEEE Microwave and Millimeter-Wave Monolithic Circuits Symposium Digest, June 1987, pp. 27-30
19. PAVIO, A. M., *et al*.: 'Double balanced mixers using active and passive techniques', *IEEE Trans. Microwave Theory Tech.*, MTT-36, no. 12, December 1988, pp. 1948-1957
20. HARROP, P., LESARTRE, P., and COLLET, A.: 'GaAs integrated all-FET front end at 12 GHz', IEEE GaAs IC Symp. Dig., Nov. 1980, paper no.28
21. KERMARREC, C., *et al*.: 'The first GaAs fully integrated microwave receiver for DBS applications at 12 GHz', 14th European Microwave Conference Proceedings, 1984, pp. 749-754
22. NIGHTINGALE, S. J., *et al*.: 'A 30 GHz monolithic single balanced mixer with integrated dipole receiving element', IEEE Microwave and Millimeter-Wave Monolithic Circuits Symposium Digest, June 1985, pp. 74-77
23. SCHINDLER, M. J., and TAJIMA, Y.: 'A novel MMIC active filter with lumped and transversal elements', *IEEE Trans. Microwave Theory Tech.*, MTT-37, no. 12, December 1989, pp. 2148-2153
24. HARA, S., TOKUMITSU, T., and AIKAWA, M.: 'Lossless, broadband monolithic microwave active inductors', IEEE Int. Microwave Symp. Dig., June 1989, pp. 955-958
25. HIROTA, T., TARUSAWA, Y., and OGAWA, H.: 'Uniplanar MMIC hybrids—a proposed new MMIC structure,' *IEEE Trans. Microwave Theory Tech.*, MTT-35, June 1987, pp. 576-581
26. HIRAOKA, T., TOKUMITSU, T., and AKAIKE, M.: 'Very small wide-band MMIC magic-T's using microstrip lines on a thin dielectric film', *IEEE Trans. Microwave Theory Tech.*, MTT-10, Oct. 1989, pp. 1569-1575
27. Gallium Arsenide IC User Survey, BIS Strategic Decisions Limited, August 1993
28. ESTREICH, D. B.: 'A monolithic wide-band GaAs IC amplifier', *IEEE Journal of Solid-State Circuits*, SC-17, no. 6, December 1982, pp. 1166-1173
29. ESTREICH, D. B., (Hewlett-Packard), Proceedings of the King's College MMIC Design Course, 1987
30. WANG, H., *et al*.: 'A high gain low noise 110 GHz monolithic two-stage amplifier', IEEE Int. Microwave Symp. Dig., June 1993, pp. 783-785

31. DUH, K. H. G., *et al.*: 'A super low-noise 0.1μm T-Gate InAlAs-InGaAs-InP HEMT', *IEEE Microwave and Guided Wave Lett.*, Vol. 1, no. 5, May 1991, pp. 114-116
32. WANG, H., *et al.*: 'A monolithic W-band three-stage LNA using 0.1μm InAlAs/InGaAs/InP HEMT technology', IEEE Int. Microwave Symp. Dig., June 1993, pp. 519-522
33. SHIMURA, T., *et al.*: '1W Ku-band AlGaAs/GaAs power HBTs with 72% peak power added efficiency', IEEE Int. Microwave Symp. Dig., May 1994, pp. 687-690
34. PATTON, G. L., *et al.*: '75 GHz f_T SiGe-Base HBT's', *IEEE Electron Device Lett.*, Vol. 11, 1990, pp. 171
35. LUY, J. F., STROHM, K. M., and SASSE, E.: 'Si/SiGe MMIC technology', IEEE Int. Microwave Symp. Dig., May 1994, pp. 1755-1757
36. SCHUMACHER, H., ERBEN, U., and GRUHLE, A.: 'Low-noise performance of SiGe heterojunction bipolar transistors', IEEE Int. Microwave Symp. Dig., May 1994, pp. 1167-1170
37. RUSSER, P.: 'Silicon monolithic millimetre-wave integrated circuits', 21st European Microwave Conference Proceedings, 1991, pp. 55-71
38. BÜCHLER, J., KASPER, E., LUY, J. F., RUSSER, P., and STROHM, K. M.: 'Silicon millimeter-wave circuits for receivers and transmitters', IEEE Microwave and Millimeter-Wave Monolithic Circuits Symposium Digest, 1988, pp. 67-70
39. 'The use of the radio frequency spectrum above 30 GHz', DTI Radiocommunications Division, 1988, ISBN 1-870837-00-46
40. LEOPOLD, R. J., and MILLER, A.: 'The IRIDIUM communications system', IEEE Int. Microwave Symp. Dig., June 1993, pp. 575-578
41. CHELOUCHE, M., and PLATTNER, A.: 'Mobile broadband system (MBS): trends and impact on 60 GHz band MMIC development', *IEE Electronics & Communication Engineering Journal*, June 1993, pp. 187-197
42. DEVLIN, L. M., BUCK, B. J, CLIFTON, J. C., DEARN, A. W., and LONG, A. P.: 'A 2.4 GHz single chip transceiver', IEEE Microwave and Millimeter-Wave Monolithic Circuits Symposium Digest, June 1993, pp. 23-26
43. WILLIAMS, D. A.: 'A frequency hopping microwave radio system for local area network communication', IEEE Int. Microwave Symp. Dig., June 1993, pp. 685-690
44. PILGRIM, M., and SEARLE, R. P.: 'MM-wave direct-to-home multichannel TV delivery system', IEEE Int. Microwave Symp. Dig., June 1989, pp. 1095-1098
45. WILSON, P. G., and BARNES, B. C.: 'Millimetre-wave downconverter using monolithic technology for high volume application', IEEE Int. Microwave Symp. Dig., June 1989, pp. 1099-1102
46. SUSSMAN, J. N.: 'Intelligent vehicle highway systems: challenge for the future', IEEE Int. Microwave Symp. Dig., June 1993, pp. 101-104

47. RAFFAELLI, L., STEWART, E., QUIMBY, R., BORELLI, J., GESSBERGER, A., and PALMIERI, D.: 'A low cost 77 GHz monolithic transmitter for automotive collision avoidance systems', IEEE Microwave and Millimeter-Wave Monolithic Circuits Symposium Digest, June 1993, pp. 63-66
48. HAIGH, D.: 'Circuit techniques for efficient linearised GaAs MMICs', IEEE Int. Microwave Symp. Dig., June 1992, pp. 1035-1038
49. BAYAR, E., ROBERTSON, I. D., and AGHVAMI, A. H.: 'L-Band GaAs MMIC amplifier design and test', ESA Workshop on MMICs for Space Applications, ESTEC, Noordwijk, Netherlands, March 1990.

Chapter 2

Active devices and fabrication technology

A. A. Rezazadeh and C. Sansom

2.1 Introduction

Since the first GaAs metal-semiconductor field effect transistor (MESFET) was reported [1] there has been a growing interest in GaAs-based transistors as discrete devices and in analogue and digital integrated circuits. This has principally been due to the high electron mobility and the low interconnection capacitance associated with the high resistivity of semi-insulating GaAs substrates compared with silicon. For both hybrid-MICs and MMICs the GaAs MESFET has been an important workhorse and has continued to outperform world-wide the silicon-based counterparts. In recent years, MESFET technology has matured sufficiently to enable its use as the principle active device in commercial hybrid and monolithic integrated circuits and as a standard 'foundry' device.

The demand for more control and sophistication in simple GaAs MESFETs has stimulated much advanced materials growth activity. Techniques such as molecular beam epitaxy (MBE) and metal-organic chemical vapour deposition (MOCVD) have proved invaluable in improving devices and circuits. The evolution of MBE and MOCVD to today's level of sophistication has made possible a generation of new devices which utilise heterojunctions between the alloy III-V aluminium gallium arsenide (AlGaAs) and the binary III-V compound, GaAs. The band gap increases with aluminium concentration which leads to a difference (discontinuity) in the band-gap energies at GaAs/AlGaAs heterojunctions. The ability to so engineer the band structure of transistors has opened up new and exciting possibilities in transistor material structures, which has subsequently led to the inception, development and rapid demonstration and exploitation of new heterostructure devices.

The most notable of these new devices are two generically different operational types: the high electron mobility transistor (HEMT or MODFET) and the heterojunction bipolar transistor (HBT). The basis of the HEMT is a sophisticated MESFET (analogous to the Si MOSFET) depending on lateral unipolar transport. On the other hand, the HBT is a III-V analogue of the silicon bipolar transistor.

Of these two main heterostructure devices, the HEMT has received most attention due to its comparative ease of material preparation and its potential as a low-noise analogue or low-power digital device. The performance advantage of HEMTs compared with MESFETs is derived from the separation of the carriers (electrons) from their parent donor atoms across the AlGaAs/GaAs heterojunction, which virtually eliminates ionised impurities as a carrier scattering mechanism. The HBT concept, long recognised for its potential in high-speed applications, has recently stimulated an increased level of research activity, particularly in the GaAs/AlGaAs material system.

Analogue applications for HEMT and HBT devices are arguably more immediate. In particular, HEMT devices are ideally suited to the critical front-end amplification stage of a low-noise microwave receiver system. Such amplifiers could be hybrid circuits based on discrete HEMT devices or more advanced monolithic microwave integrated circuits (MMICs). Already, HEMT devices have achieved lower noise figures than any other transistor types for comparable frequencies.

2.2 The GaAs MESFET

All GaAs MESFETs look alike at first glance, and are deceptively simple in their construction. As shown in Fig. 2.1, there are comparatively few elements to consider. The FET is built on a semi-insulating GaAs substrate, with an active layer produced by either ion implantation or epitaxial growth. Implantation is preferred for low cost, high volume processes. Epitaxial layer growth is a more expensive option, even for simple FET structures, and is usually limited to the realisation of profiles that are difficult to achieve by implantation. The exact doping profile is critical to the performance of the FET, as is the presence of deep levels and the density of defects and impurities in the substrate. The active regions need to be isolated from each other, this being achieved by wet chemical etching to define mesa structures or by implant isolation which opens up the option of a planar device.

The order in which the MESFET gate and ohmic contacts are fabricated depends on many factors. For low cost, high density digital circuits it is common to fabricate a high-temperature stable refractory-metal gate initially. The ohmic regions can then be self-aligned to the gate, and fired in to produce the source and drain contacts. We will confine ourselves to typical analogue MMIC devices here, thus considering a process sequence that fabricates the ohmic regions first, with a recessed gate thereafter.

The source and drain contacts are usually produced by standard photolithography and float-off techniques. Alloyed contacts of Au/Ge are most common, although high-temperature contacts are used for specific applications. Metals may be deposited by evaporation or, in the case of high melting point materials, sputtering. Alloyed contacts are heated to around 400-450 °C in either a conventional tube furnace or by optical means in a rapid thermal annealing (RTA) furnace. RTA gives better surface morphology and sharper contacts owing to the rapid temperature ramps that the lamps are able to produce.

Gate processing is the key stage in the fabrication of the GaAs MESFET. Since the gate metal will be defined by float-off techniques once again, the first task is to define a narrow opening in a layer of resist with an appropriate cross-section for float-off to occur (i.e. an overhang). For gate lengths down to about 0.5 μm, this opening can be

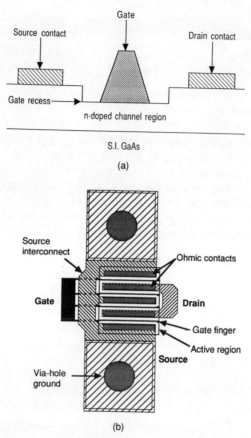

Fig. 2.1 The GaAs MESFET: (a) Cross-section; (b) Typical layout (4 × 75μm device)

produced by conventional photolithographic techniques. At dimensions less than 0.5 μm, optical lithography is stretched to its limit, and strategies based on electron beam lithography or X-ray lithography are common. It is also worth noting at this stage that the silicon IC industry is driving direct step on wafer ('stepper') technology below 0.5 μm, providing the GaAs process engineer with a new technology for GaAs MMIC fabrication.

Depending on the crystal orientation and etch selectivity, different recess shapes can be produced, affecting the performance of the finished device. Either wet etching or dry etching is used in practice, and the etching is carefully controlled and monitored to achieve the final device current required. The gate metals deposited thereafter are often a complex layer of materials: a base layer to form a Schottky contact to the GaAs surface, with good adhesion properties to GaAs; a diffusion barrier of platinum or palladium to produce a high reliability contact; and a top layer of contact gold. It is advisable to passivate the surface as soon as possible after completion of the active device in order to stabilise the exposed GaAs surfaces.

2.2.1 Principles of operation

A simplified expression for the source-drain current, I_{ds}, of a MESFET can be obtained using Poisson's one-dimensional equation together with the current continuity equation given below:

$$I_{ds} = qN v_{sat}(y-d)Z_g \qquad (2.1)$$

where y and d are the depth of the n-type channel and the depletion region, at the point of velocity saturation, v_{sat}, respectively, Z_g is the gate width, q is the unit charge and N is the electron density in the channel. A crude approximation can be obtained by ignoring the potential drop between the source and the point where velocity saturation occurs in the channel. Thus the voltage drop across the depletion layer at this point is approximated to the gate-to-source potential, V_{gc}. Using the depletion approximation for the depletion depth, d, the drain current becomes:

$$I_{ds} = qN v_{sat}[y - (\frac{2\varepsilon V_{gc}}{qN})^{\frac{1}{2}}]Z_g \qquad (2.2)$$

where ε is the semiconductor dielectric constant.

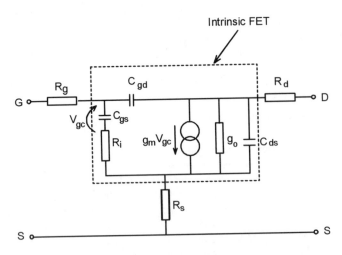

Fig. 2.2 Small-signal equivalent circuit of a MESFET

The small-signal equivalent circuit of a microwave FET, biased in the saturation region, is shown in Fig. 2.2. The gain mechanism of a FET is embodied in the transconductance, g_m, which models the current controlling capability of the gate electrode. The transconductance is defined as the ratio of the small change in drain current produced by a small change in gate-to-channel voltage, V_{gc}, when the drain-to-source voltage is kept constant. The intrinsic transconductance, g_{mo}, is given by:

$$g_{mo} = \left. \frac{dI_{ds}}{dV_{gc}} \right|_{V_{ds}=\text{const}} = \frac{\varepsilon v_{sat} Z_g}{d} \quad (2.3)$$

Although the above equation has been derived for a MESFET, it can also be used to provide useful information about the operation of a HEMT. To do so, however, certain approximations have to be made; for example, about the depth of the depletion region underneath the gate. In MESFETs, the depletion depth, d, increases with the application of negative source-to-gate voltage until the channel is pinched-off. In HEMTs, to utilise their full potential as low-noise devices, it is essential that the AlGaAs supply layer is fully depleted. This means that d in eqn. 2.3 becomes the total thickness of the AlGaAs plus the thickness of the undoped AlGaAs layer.

The extrinsic transconductance, g_m, measured from the device terminals, is influenced by the source resistance, R_s. The extrinsic transconductance can be obtained by considering the voltage drop across the source resistance:

$$V_{gs} = V_{gc} + I_{ds} R_s \quad (2.4)$$

Differentiating this equation with respect to the drain current, g_m can be obtained as:

$$g_m = \frac{g_{mo}}{(1 + g_{mo} R_s)} \quad (2.5)$$

2.2.2 High frequency capability

The cut-off or transition frequency, f_t, is defined as the frequency at which the short circuit current gain of the device drops to unity. Using a simplified model and considering the equivalent circuit given in Fig. 2.2, the input and output currents are given by:

$$I_o = g_m V_{gc} \quad (2.6)$$

$$I_{in} = V_{gc} j\omega C_{gs} \quad (2.7)$$

Thus the current gain, h_{21}, of the device is given by:

$$|h_{21}| = \frac{g_m}{\omega C_{gs}} \quad (2.8)$$

From eqn. 2.8 the cut-off frequency can be calculated as:

$$f_t = \frac{g_m}{2\pi C_{gs}} \quad (2.9)$$

Another high frequency figure of merit is the maximum oscillation frequency, f_{max}, which is defined as the frequency when the unilateral power gain is unity. The value of f_{max} can be estimated as follows:

Considering the equivalent circuit of the FET given in Fig. 2.2, with its input and output matched for maximum power transfer, we have:

$$\text{Power gain} = \left(\frac{I_o}{I_{in}}\right)^2 \times \frac{R_L}{4R_{in}} = \left(\frac{f_t}{f}\right)^2 \times \frac{R_L}{4R_{in}} \qquad (2.10)$$

Therefore the f_{max} is given by:

$$f_{max} = \frac{f_t}{2} \times \left(\frac{R_L}{R_{in}}\right)^{1/2} \qquad (2.11)$$

where R_L and R_{in} are the load and input resistances of the device.

2.2.3 Noise analysis

Noise in microwave FETs is produced by both 'intrinsic sources' of the device and by 'thermal sources' associated with the parasitic resistances. The intrinsic noise source arises from two mechanisms [2]:

(1) thermal (Johnson) noise which is produced in the ohmic section of the device channel
(2) diffusion noise which is produced in the velocity-saturated section of the channel and can be dominant for short gate-length devices.

The minimum noise figure in FETs is related to the transistor small-signal equivalent circuit parameters and can be approximated to [3]:

$$NF_{min} = 1 + k_f \frac{C_{gs}}{g_{mo}} \omega \sqrt{g_{mo}(R_g + R_s)} \qquad (2.12)$$

where k_f is a fitting factor representing the quality of the channel. The magnitude of k_f has been evaluated approximately using a theoretical analysis of the noise theory [4]:

$$k_f \approx 2 \sqrt{\frac{I_{opt}}{E_c L_g g_{mo}}} \qquad (2.13)$$

where I_{opt} is the optimum drain-to-source current at minimum noise figure, E_c is the critical electric field in the channel and L_g is the gate length of the device. The typical value of k_f is 4-6 for GaAs MESFETs, while this is much smaller for HEMT devices due to the higher quality of the electron channel. The value of F_{min} is usually quoted in decibels and this is given by:

$$F_{min}(\text{dB}) = 10 \log F_{min} \qquad (2.14)$$

Active devices and fabrication technology 33

2.3 The high electron mobility transistor (HEMT)

The simplest form of HEMT device is shown in Fig. 2.3. Advanced techniques of materials growth, such as molecular beam epitaxy, are required to produce the layers with an atomically smooth heterojunction interface between them. The electrons contributed by the n-AlGaAs are free to move through the entire crystal until such time as they fall into the lowest energy states allowed to them. In the HEMT, these lowest energy states are to be found just to the GaAs side of the heterojunction interface, with the outcome that all the electrons accumulate there in a thin sheet. Subsequently, they are free to move only in the two-dimensional plane of the interface, and the term 'two-dimensional electron gas' (2DEG) is used to describe the conduction electrons' state or condition. The spatial separation of the conduction electrons from their donor impurities reduces ionised impurity scattering which, coupled with low defect material, produces very high electron mobilities. The ultimate speed of the transistor is therefore greatly enhanced.

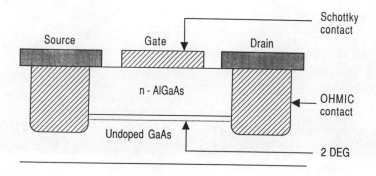

Fig. 2.3 Conventional HEMT structure

From a fabrication point of view, the difference between the MESFET and the HEMT is largely in the starting material. A typical sequence of layers for a simple HEMT may consist of 1 μm of undoped GaAs grown directly onto the GaAs substrate, followed by 400 Å of AlGaAs, of which the first 50 Å is left undoped and the rest is silicon doped to around 1×10^{18} cm^{-3}, followed by an n$^+$ capping layer of several hundred angstroms thickness. The capping layer prevents oxidation of the AlGaAs layer, absorbs the potential at the semiconductor free-surface, and allows low resistance ohmic contacts to be made to the structure. Device fabrication after layer growth is similar to that of conventional MESFETs. An isolation stage is performed, ohmic contacts are fabricated, and the gate is defined. Gate recess etching can bring extra difficulties however, owing to the difference in etch rates between the different hetero-epitaxial layers. For low noise, high frequency operation, mushroom or T-gate cross-sections are required, necessitating multi-level resist strategies and electron beam lithography. Fig 2.4 shows a sub-0.25 µm mushroom gate HEMT device, fabricated by GEC-Marconi Materials Technology.

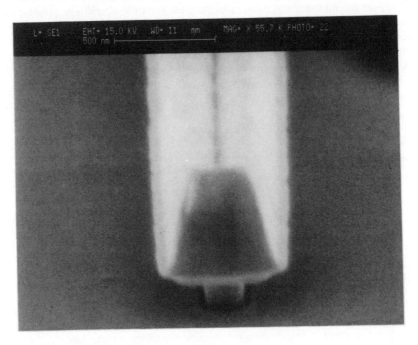

Fig. 2.4 Mushroom gate HEMT technology

HEMT transistors based on the GaAs/AlGaAs heterojunction have serious limitations. The presence of a deep donor level in AlGaAs limits the aluminium composition to around 25-30% [5]. Also, the GaAs/AlGaAs heterojunction has only a small conduction band offset, and hence a small usable 2DEG sheet density. This limits the transistor current drive and high-frequency performance at room temperature. As a consequence, conventional HEMTs have been superseded by :

(i) a lattice matched HEMT of AlInAs/InGaAs grown on InP

(ii) a pseudomorphic HEMT of AlGaAs/InGaAs grown on GaAs. In fact the HEMT illustrated in the micrograph of Fig. 2.4 is a pseudomorphic HEMT.

The principle advantage of the pseudomorphic HEMT is that it achieves better confinement of carriers in the channel. This has important consequences for the performance of the device. Fig. 2.5 compares the noise figure and gain of a conventional HEMT with a pseudomorphic HEMT, both with a 0.5 μm × 120 μm geometry. The important difference to note is that the low noise figure and high gain of the pseudomorphic device is maintained over a broad range of operating current, due to the better confinement of carriers. Table 2.1 compares the equivalent circuit parameters of a MESFET, C-HEMT and P-HEMT, all with the same 0.5 μm × 120 μm geometry. The P-HEMT achieves virtually double the f_t of the MESFET, and of course even higher f_t is achieved with shorter gate-lengths. The noise figures at 10 GHz are similar for the C-HEMT and P-HEMT, and both are considerbly lower than that of the MESFET.

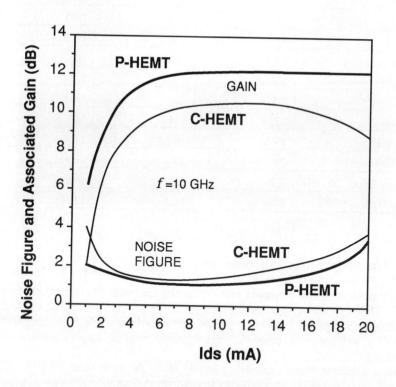

Fig. 2.5 Comparison of the gain and noise figure against drain current for a conventional and pseudomorphic HEMT with 0.5 μm × 120 μm geometry [6, 7]

Table 2.1 Comparison of equivalent circuit element values for MESFET, C-HEMT and P-HEMT of identical layout geometry (0.5 µm × 120 µm) [6, 7]

Equivalent circuit parameters	MESFET	C-HEMT	P-HEMT
V_{ds} (V)	1.8	1.7	1.7
V_{gs} (V)	-1.8	-0.2	-0.2
I_{ds} (mA)	7.0	10.0	11.0
I_{dss} (mA)	34.0	17.0	26.0
g_{mo} (mS)	15.0	32.29	41.3
C_{gs} (pF)	0.081	0.144	0.198
C_{gd} (pF)	0.015	0.028	0.023
C_{ds} (pF)	0.029	0.040	0.027
R_g (Ω)	1.13	1.13	1.13
R_{gs} (Ω)	92.1	1.9	12.7
R_i (Ω)	12.4	14.7	10.0
R_s (Ω)	5.3	1.5	1.3
R_d (Ω)	7.8	6.4	5.6
R_{ds} (Ω)	535	514	567
R_{sum} (Ω)	19.5	18.04	12.4
f_t (GHz)	21	37	40
NF_{min} (dB) @ 10 GHz	2.2	1.3	1.1
G_{as} (dB) @ 10 GHz	8.2	11.5	11.0

2.4 The heterojunction bipolar transistor (HBT)

Bipolar transistors have numerous advantages compared with other transistors such as FETs. The main advantages include a high current gain cut-off frequency, f_t, high current drive capability per unit chip area, high transconductance, high voltage handling capability, well controlled threshold voltage for turn-on of the output current, and low $1/f$ noise.

Heterojunction bipolar transistors largely retain the advantages of their silicon predecessors, but extend them to higher frequencies. A typical device layer structure is shown in Fig. 2.6. Base doping in the range 10^{19} to 10^{20} cm^{-3} is used, with an emitter doping of 5×10^{17} cm^{-3}.

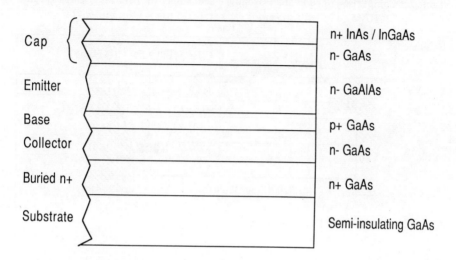

Fig. 2.6 Typical HBT structure

Pseudomorphic InGaAs can be used to form emitter contacts. Base thicknesses are in the range 400-1000 Å. These are complex structures, requiring such growth techniques as MBE or MOCVD. Other material systems are receiving considerable attention currently. These include:

(i) InGaAs/InP : InP-based HBTs
(ii) GaInP/GaAs : lattice-matched HBT
(iii) InAs- and GaSb-based HBTs
(iv) GaP/AlGaP : high bandgap HBTs

Turning now to the fabrication of III-V HBTs, clearly the key structural features of the device are produced by the epitaxial growth process. The objectives of the remaining processing steps are to make contact to the different device layers, and to isolate individual devices. There is a great deal of commonality between the process sequence employed for GaAs FET-based circuits and that for HBT fabrication, and usually a single processing facility can accomplish both tasks.

The key dimensions of HBTs must be reduced to minimise parasitics, particularly base resistance and base-collector capacitance. Device size reduction for digital circuits is most conveniently achieved using self-aligned processing techniques. In an HBT, three patterns must be critically aligned with each other. These are the active emitter area, the emitter contact, and the base contact. Self-aligned techniques are used to eliminate one or more alignments, or at least render the alignment non-critical. A typical self-aligned HBT is shown schematically in Fig. 2.7.

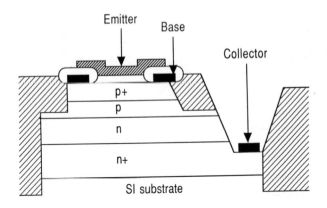

Fig. 2.7 Self-aligned HBT cross-section

In all HBT structures it is the need to access the base which can be the most difficult process to control accurately. The required control is just possible using standard etching techniques, but the task can be greatly simplified by using compositionally selective etching. The use of so-called "etch stops" is possible in both wet or dry etch strategies to cease etching within about 100 Å of the top of the base.

2.4.1 Principles of operation

Fig. 2.8 shows a schematic diagram of an n-p-n homojunction bipolar transistor in common-emitter configuration operating in the active region with the emitter/base forward biased and the base/collector reverse biased. The forward biased emitter/base junction causes electrons to be injected into the base and holes injected from the base to the emitter. Some of the electrons being injected into the base will recombine with holes in the emitter/base depletion region, giving rise to a recombination current, I_{scr}. The remaining electrons are injected into the base giving rise to an electron current I_{neb}. Further electrons recombine with holes in the bulk of the base, giving a base bulk recombination current, I_{bulk}. The remaining electrons reach the edge of the base/collector depletion region and are swept to the collector under the influence of the high electric field at the junction.

The holes injected from the base to the emitter produce a reverse injection current I_{pbe}, which is also shown in Fig. 2.8. Furthermore, there is an additional current component, I_{CBO}, which arises from the base/collector reverse bias junction due to the thermal generation of carriers at this junction. So far we have ignored the surface leakage current. The DC current gain of a transistor is defined as:

$$\beta = \frac{I_c}{I_b} \qquad (2.15)$$

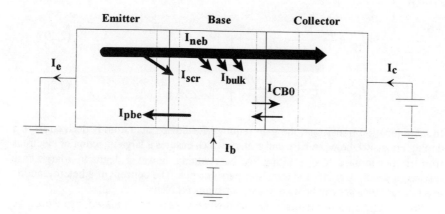

Fig. 2.8 Current components in an n-p-n bipolar transistor biased in common-emitter configuration

Assuming that the base thickness is much smaller than the electron diffusion length in the base then $I_{bulk} = 0$. If it is also assumed that I_{CB0} and I_{scr} are small in comparison with I_{pbe} then the current gain, β, becomes:

$$\beta = \frac{I_{neb}}{I_{pbe}} = \frac{J_{neb}}{J_{pbe}} \tag{2.16}$$

where J_{neb} and J_{pbe} are the electron and hole current densities, respectively.

Now, J_{neb} and J_{pbe} can be expressed in terms of the emitter/base voltage V_{be} and the built-in potential barrier to electrons in the emitter, V_{bie}, and to holes in the base, V_{bib}:

$$J_{neb} = \frac{qD_{nb}N_{DE}}{L_{nb}}\left[\exp\left(\frac{-qV_{be}}{kT}\right) - 1\right]\exp\left(\frac{-qV_{bie}}{kT}\right) \tag{2.17}$$

$$J_{pbe} = \frac{qD_{pe}N_{AB}}{L_{pe}}\left[\exp\left(\frac{-qV_{be}}{kT}\right) - 1\right]\exp\left(\frac{-qV_{bib}}{kT}\right) \tag{2.18}$$

where D_{nb}, D_{pe}, L_{nb} and L_{pe} are the minority carrier diffusion coefficients and minority carrier diffusion lengths for electron and holes in the base and emitter regions, respectively. N_{AB} and N_{DE} are the base acceptor and emitter donor concentrations, respectively. Here we have assumed ideal diodes; that is $n_c = n_b = 1$. Assuming forward bias operation, we can write $V_{be}\text{-}V_{bi} = V_n$ and $V_{be}\text{-}V_{bi} = V_p$ as the heights of the potential barriers to electrons and holes between the emitter and base, respectively, where V_{bi} is the built-in potential for a homojunction. Eqns. 2.17 and 2.18 now simplify to:

$$J_{neb} = \frac{qD_{nb}N_{DE}}{L_{nb}} \exp\left(\frac{-qV_n}{kT}\right) \tag{2.19}$$

$$J_{pbe} = \frac{qD_{pe}N_{AB}}{L_{pe}} \exp\left(\frac{-qV_p}{kT}\right) \tag{2.20}$$

In homojunction transistors, the gain required for device operation is derived from a doping imbalance between base and emitter which ensures a large injection of electrons from emitter to base. The resulting low base doping, however, leads to a large base resistance which degrades the transistor performance. The concept of a heterojunction in a bipolar transistor can be used to overcome this problem.

Some interesting behaviour is revealed when a transistor is biased into the active region and the base and collector currents are measured as functions of the base-emitter voltage, V_{be}, for a particular base-collector voltage ($V_{bc} = 0$ V) on a semilogarithmic scale with the V_{be} on the linear scale (known as the Gummel plot). The magnitude of the collector current density can be determined from the diffusion process of minority carriers in the base and can easily be shown to be given by:

$$J_C = \frac{qD_n n_{ib}^2}{P_b W_b} \exp(qV_{BE}/kT) \tag{2.21}$$

where D_n, P_b, W_b and n_{ib} are the electron diffusion coefficient, hole concentration, thickness and intrinsic carrier density in the base, respectively. The effective value of n_{ib} is slightly higher than in undoped material as a result of bandgap narrowing effects due to the heavily doped p-type base region of the HBT.

From eqn. 2.21 it can be expected that the value of V_{be} required to reach a given J_c is highly reproducible (so that the transistor 'turns on' at a precisely established input voltage). The required turn-on voltage to obtain J_c can be expressed in terms of the bandgap of the base W_{gb} and the conduction and valence band densities of states N_c and N_v from eqn. 2.21:

$$V_{be} = \frac{W_{gb}}{q} - \frac{kT}{q} \ln \frac{qD_n N_c N_v}{J_c P_b W_b} \tag{2.22}$$

V_{be} is dominated by the first term; the typical value of the second term is near 0.25 V, and it changes by only 0.026 V for a 10% change of any of the variables within. The reproducibility of V_{be} is of considerable importance for IC applications.

The transistor transconductance, g_m, of HBTs is given by the behaviour of I_c against V_{be}. The intrinsic g_{mo}, as is well known for homojunction bipolar transistors, has the value :

$$g_{mo} = \left.\frac{dI_c}{dV_{be}}\right|_{V_{ce}=\text{constant}} = \frac{qI_c}{kT} \qquad (2.23)$$

The extrinsic transconductance, measured from the device terminals, is influenced by emitter and base series resistances $R_{ee'}$ and $R_{bb'}$, respectively. Transconductance per unit device width is typically used as a figure of merit of FET performance, with values up to 300 mS/mm; the corresponding value for HBTs is 10^4 mS/mm.

2.4.2 High frequency capability

The cut-off frequency, f_t, corresponds to the frequency at which the incremental short-circuit current gain decreases to unity. For most electronic devices, for a wide range of frequencies below f_t (but not necessarily extending down to DC), the magnitude of the current gain G_i is given by:

$$|G_i| = \frac{f_t}{f} \qquad (2.24)$$

where f_t can be simply related to the transit time of carriers across the device within a simple charge-control model of a transistor. From this model, f_t is found to be equal to:

$$f_t = \frac{1}{2\pi\tau_{ec}} = \frac{1}{2\pi(\tau_e + \tau_b + \tau_d + \tau_c)} \qquad (2.25)$$

where τ_{ec} is the total emitter-to-collector delay time. Here, τ_e corresponds to the hole charge stored at the emitter edge of the base, within the emitter-base space charge layer, or within the emitter itself. This latter contribution is negligible in HBTs due to the wide emitter bandgap (although this is significant for homojunction devices). For the most part, the storage of holes corresponds to an expansion or contraction of the space charge region width, and the magnitude of the associated charge is the value inferred from the base-emitter space charge layer capacitance, C_{je}, multiplied by the change in V_{be} required to change the output current. Thus, the value of τ_e is given by:

$$\tau_e = r_e C_{je} = C_{je} \frac{kT}{qI_c} \qquad (2.26)$$

According to eqn. 2.26, to minimise τ_e, the transistor should be operated at the maximum possible current I_c. Additionally, C_{je} can be decreased by decreasing the emitter doping which is possible in HBT structures. The contribution τ_b corresponds to the charge of electrons traversing the base which can be represented by:

$$\tau_b = \frac{W_b^2}{\eta D_n} \qquad (2.27)$$

where W_b is the base width, D_n is the minority carrier diffusion coefficient and $\eta = 2$ for a uniformly doped base structure.

The contribution τ_d corresponds to the base charge induced by the electrons that are crossing the base-collector space charge layer. Under the simplifying assumption that the electrons travel at a constant saturated drift velocity, v_{sat}, it can be shown that the value of τ_d is related to the saturated drift velocity and the width of the collector-base depletion width, x_d:

$$\tau_d = \frac{x_d}{2v_{sat}} \qquad (2.28)$$

A final contribution to τ_{ec} corresponds to the fact that a change in output current produces a change in base-collector voltage, principally through resistive drops. The change in junction voltage in turn induces a charge in the base, through the depletion layer capacitance. The corresponding τ_c is given by:

$$\tau_c = (R_{ee'} + R_{cc'} + \frac{kT}{qI_c})C_{jc} \qquad (2.29)$$

where $R_{cc'}$ and $R_{ee'}$ are the parasitic collector and emitter series resistances, respectively, and C_{jc} is the base-collector capacitance (including both intrinsic and extrinsic components).

Although the forward transit time, $\tau_f = (\tau_b + \tau_d)$ contributions to τ_{ec} depend solely on the vertical structure of the device, the sum of $(\tau_e + \tau_c)$ depends on the lateral structure of the device. The reduction of this contribution during the course of HBT fabrication process development has been the major factor in the increases reported in the cut-off frequency of HBTs.

Substituting eqns. 2.26 to 2.29 in eqn. 2.25 gives the cut-off frequency:

$$f_t = \frac{1}{2\pi}\left[\frac{kT}{qI_c}C_{je} + \frac{W_b^2}{2D_n} + \frac{x_d}{2v_{sat}} + (R_{ee'} + R_{cc'} + \frac{kT}{qI_c})C_{jc}\right]^{-1} \qquad (2.30)$$

To obtain a high f_t, each time constant contributing to the τ_{ec} in eqn 2.30 must be minimised. The transistor should therefore have a narrow base width, a narrow collector region and should be operated at a high current level. We now discuss how the heterojunction bipolar transistor yields a higher f_t than a homojunction transistor:

(1) Comparing a homojunction bipolar with an HBT of identical dimensions, the HBT can permit a very low emitter-base charging time, τ_e, at low current density due to the lower emitter doping compared with the base. Thus a higher f_t can be obtained for the HBT at low current densities.

(2) The base transit time, τ_b, is another factor in the total transit time, and it varies with the square of the base thickness. Thus, this time can be reduced

by decreasing the base thickness. At an equivalent doping level, GaAs has a diffusion constant, D_n, four times greater than silicon. As a result, with an identical base thickness, τ_b can be reduced by a factor of about 2, while achieving a base resistivity about 5 times smaller than that in silicon transistors.

(3) Finally, the last two terms in τ_{ec} are, respectively, functions of the collector thickness and doping level. Depending on the applications, a compromise must be sought between the base-collector breakdown voltage and the collector doping level which together determine the depletion region thickness and the base-collector transition capacitance. Two points may be noted here: first, the use of very thin high mobility epitaxial collector layers tends to result in very low series collector resistances; secondly, the velocity overshoot in GaAs is higher than in Si. Thus the GaAs/AlGaAs heterojunction transistor allows a considerable overall gain to be achieved.

Another figure of merit in bipolar transistors is the maximum frequency of oscillation, f_{max}, which describes the frequency at which the power gain of the transistor falls to unity. The power gain is defined as:

$$G_p = \frac{\text{power delivered to load}}{\text{power input to device}} \qquad (2.31)$$

Over a significant frequency range, the maximum available power gain of a single-stage transistor amplifier is often found to obey the relation:

$$G_p \approx \frac{f_{max}^2}{f^2} \qquad (2.32)$$

Under optimal input and output impedance matching conditions, G_p of the amplifier may be expressed in terms of the device current gain G_i, input Z_{in} and output Z_{out} impedances of the device as:

$$G_p \approx \frac{|G_i|^2}{4} \frac{\text{Re}(Z_{out})}{\text{Re}(Z_{in})} \qquad (2.33)$$

At sufficiently high frequency, $Z_{in} = R_{bb'}$, the base resistance. The output impedance, (Z_{out}), is established indirectly by feedback through the capacitor C_{jc} of a current that is then amplified by the forward gain of the transistor. The corresponding $\text{Re}(Z_{out})$ is approximately $1/(2\pi C_{jc} f_t)$. From these, an expression for f_{max} can be derived:

$$f_{max} \approx \left[\frac{f_t}{8\pi R_{bb'} C_{jc}}\right]^{\frac{1}{2}} \qquad (2.34)$$

It is clear from eqn. 2.34 that to have high f_{max} it is necessary to minimise the base (access) resistance of the device and decrease the magnitude of its feedback capacitance C_{jc}. In HBTs high values of f_{max} can be obtained due to the lower values of $R_{bb'}$ obtained as a result of the higher base doping structure of this device. An additional, well-established technique is to use an interdigitated structure in the transistor layout, with narrow emitter fingers interspersed with base contacts. In Si bipolar technology 0.35 µm emitters with a 2 µm pitch have been attained. In HBT technology, to date, 1.2 µm emitters with a 2.8 µm pitch are the narrowest structures employed.

The contribution to C_{jc} from the intrinsic device (under the active emitter) can be decreased only by increasing the width of the base-collector depletion region. This, in turn, causes a decrease in f_t due to the resultant increase in the space charge region transit time. For optimal RF performance, design trade-offs must thus be made. The contribution to C_{jc} from the extrinsic base area (outside of the active emitter) is typically also significant. In HBTs there are two possible ways to reduce this capacitance: (i) by making use of ion bombardment to introduce lattice damage into the collector layer under the base, reducing its effective doping and (ii) by using an inverted transistor structure which makes use of a wide-bandgap emitter buried beneath the base and collector, and a collector layer located on the wafer surface.

The small-signal RF characteristics of HBTs can be described in terms of an equivalent circuit model, such as that shown in Fig. 2.9 (known as a π-model). This is similar to the one used most frequently for Si bipolar transistors, although for both cases the models are not unique. In Fig. 2.9 the capacitance C_π and resistance R_π are given by $C_\pi = g_m \tau_f$ and $R_\pi = \beta/g_m$, respectively. A corresponding set of equivalent circuit element values can be obtained by fitting measured S-parameter data at many frequencies. Table 2.2 shows typical equivalent circuit values for a non-self-aligned AlGaAs/GaAs HBT with emitter/base dimensions of 3 × 30 µm² [8]. The element values change with bias conditions, however, and for the most part scale with transistor size, so that care must be exercised in applying the equivalent circuit.

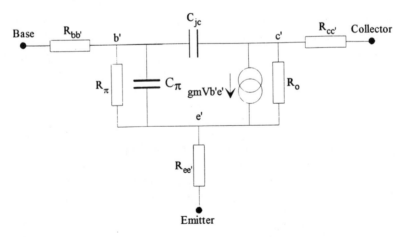

Fig. 2.9 Small-signal equivalent circuit π-model of an HBT

Table 2.2 Small signal equivalent circuit parameters measured for non-self aligned AlGaAs/GaAs HBTs with emitter/base dimensions of 3×30 μm² [8]

Small signal parmaters	Definition	Measured values
V_{ce} (V)	Collector-emitter voltage	3
J_c (A/cm²)	Collector current density	2×10^4
g_{mo} (mS)	Transconductance	720
C_π (pF)	e/b depletion and diffusion capacitances	0.77
C_{bc} (pF)	b/c depletion capacitance	0.066
R_π (Ω)	Equivalent input resistance	31.3
R_{je} (Ω)	Dynamic input resistance	1.4
R_{bc} (Ω)	Base-collector isolation resistance	10^5
$R_{bb'}$ (Ω)	Base series resistance	13
$R_{ee'}$ (Ω)	Emitter series resistance	7.4
$R_{cc'}$ (Ω)	Collector series resistance	2
L_b (nH)	Base inductance	0.15
L_e (nH)	Emitter inductance	0.5
L_c (nH)	Collector inductance	1
f_t (GHz)	Cutt-off frequency	31
f_{max} (GHz)	Maximum oscillation frequency	40

2.4.3 HBT material systems

The AlGaAs/GaAs material system for HBTs has been established and well understood for many years now. There have, however, been studies of many other material systems reported in the literature. Different materials are investigated in order to overcome the difficulties with the AlGaAs/GaAs system. Some of the common recent material systems for HBTs are as follows: AlGaAs/GaAs, InGaP/GaAs, InP/InGaAs, InAlAs/InGaAs, InGaAsP/GaAs and Si/SiGe.

Conventional AlGaAs/GaAs HBTs have demonstrated excellent device and circuit performance. However, there are several disadvantages of using this material system. The main problem with this structure is that the majority of the bandgap discontinuity is in the conduction band, reducing the emitter injection efficiency. This discontinuity can be reduced by grading the Al in the emitter, but this increases the space-charge recombination in the emitter-base junction. The alternative is to use Al-free InGaP/GaAs-based HBTs which has attracted considerable interest in the recent years. HBTs based on InGaP materials have many advantages including the high selectivity of chemical etching between InGaP and GaAs which is very useful for making contacts to a very thin base, providing better device uniformity and yield for the HBT production.

The Gummel plot for a typical InGaP/GaAs HBT is shown in Fig. 2.10. The first noticeable feature from this figure is that the base current is nearly parallel to the collector current, giving a low base ideality factor of 1.09. At low base currents (V_{be}<0.96 V) it is seen that there is a presence of an additional current component which indicates that space-charge recombination starts to dominate at low V_{be}.

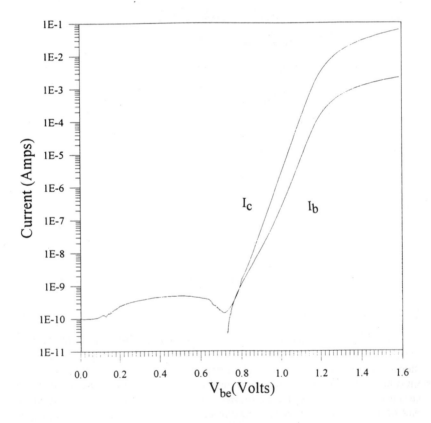

Fig. 2.10 Measured Gummel plot for an InGaP/GaAs HBT [9]

The improvement in emitter-base junction quality with the InGaP emitter can be clearly seen in Fig. 2.11 where the log of current gains for the two HBTs have been plotted against log collector current, I_c. Both characteristics show increasing β with I_c at low I_c levels, and saturated β at high I_c. However, at low I_c, β of the AlGaAs/GaAs HBT starts with a lower value and increases with I_c at a faster rate. In contrast, the current gain of the InGaP emitter HBT remains relatively constant over five decades of current and shows useful gain at collector current levels as low as 0.1 µA. By contrast, the device with the AlGaAs emitter shows no gain at collector currents below ≈20µA. These results show the superior performance of the InGaP/GaAs material system compared with AlGaAs/GaAs.

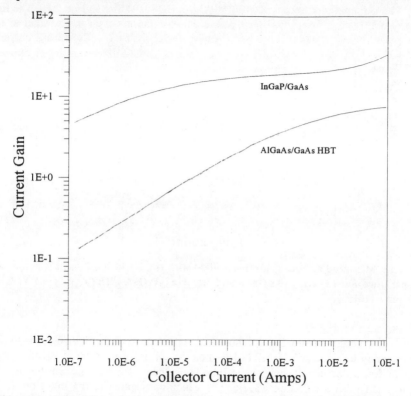

Fig. 2.11 Collector current dependence of current gains for InGaP/GaAs and AlGaAs/GaAs HBTs at $V_{CE} = 3$ V [9]

For power applications it is important that HBTs demonstrate stable current gain over an extended temperature range. This criterion is particularly important since the thermal conductivity of GaAs is about three times lower than Si. The current gain for InGaP/GaAs HBTs is found to remain almost constant with temperature above 280 K, as shown in Fig. 2.12. This is attributed to the high quality of the emitter/base junction and large valence-band energy discontinuity associated with the InGaP/GaAs HBT system.

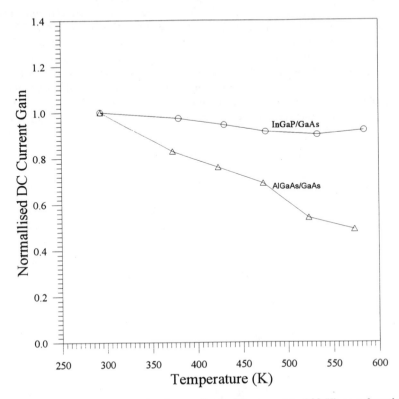

Fig. 2.12 Measured current gains (normalised with respect to 300 K) as a function of temperature for $Al_{0.3}Ga_{0.7}As/GaAs$ and $In_{0.5}Ga_{0.5}P/GaAs$ HBTs ($V_{CE} = 3$ V, $I_c = 100$ mA) [10]

For low power circuits the threshold (turn-on) voltage of the device needs to be considerably lower than 1 V. HBTs based on the InP/InGaAs material structure are ideal for low power circuit applications. In addition, these transistors are sensitive to the 1.3 µm and 1.5 µm radiation wavelengths which correspond to the low fibre loss windows, and thus are used almost exclusively in long-haul optical fibre telecommunications systems [11]. A comparison of the turn-on voltage for various HBTs fabricated using the same technology and device layout is shown in Fig. 2.13 [12]. It is clear from this comparison that InP/InGaAs HBTs exhibit much lower turn-on voltage (0.2 V) compared with GaAs-based HBTs (0.8 V). This difference is caused primarily by the difference in the band gap energies of the InGaAs ($W_g=0.75$ eV) and GaAs ($W_g=1.43$ eV) base materials. The inherent low turn-on voltage of InP-based HBTs is of considerable importance for low power consumption circuits.

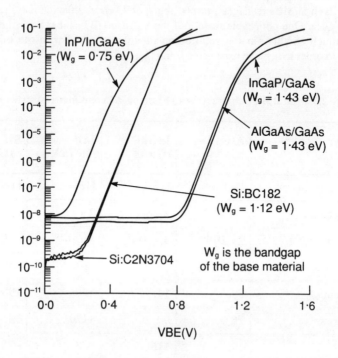

Fig. 2.13 I_c against V_{be} for various HBTs [12]

Another interesting HBT material structure is Si/SiGe. It has often been thought that Si has been optimised to its maximum potential and cannot be improved any further for increased frequency of operation. However, by using the techniques from the III-V heterojunction materials, this technology can be transferred to silicon to provide Si/SiGe HBTs for high speed circuits [13]. Since Si has been firmly established and its material characteristics are well understood, this technology can easily be converted for use with Si/SiGe. Another major advantage is that the cost of Si is very much lower than that of the conventional III-V compounds. There are, however, several disadvantages of SiGe and therefore these materials will not entirely replace the heavily researched III-V compound materials. The growth processing of SiGe is quite complex due to the difference in lattice constants of SiGe and Si. Owing to the narrow bandgaps of SiGe this device is not really suitable for high temperature and high power applications. Their lack of capability to emit and detect light means they are not suitable for optical detectors and receivers. Si/SiGe will, however, be particularly suitable for high speed digital circuits which can replace existing Si homojunction transistor circuits.

Table 2.3 shows typical characteristics obtained for HBTs reported in the literature and compared with AlGaAs/GaAs HBTs. The newer material structures all show comparable or even improved device performance compared with the AlGaAs/GaAs structure. Even Si/SiGe manages a respectable f_t of 73 GHz, although the f_{max} is rather low at 28 GHz. One noticeable feature of the Si/SiGe HBT is that the emitter width must be much smaller (in the sub-micron range) compared with the other HBT structures in order to achieve reasonable f_t and f_{max}.

Table 2.3 Typical HBT characteristics based on different material structures

Parameter	AlGaAs/GaAs	InGaP/GaAs	InP/InGaAs	Si/SiGe
Reference	[14]	[15]	[16]	[17]
Emitter area (μm^2)	1.4×12	2×60	2×5	0.6×19.3
Base doping (cm^{-3})	2×10^{19}	3×10^{19}	5×10^{19}	$\approx 10 \times 10^{19}$
Base thickness (Å)	1000		500	350
Current gain	303	14	≈ 30	290
f_t (GHz)	76	50	≈ 120	73
f_{max} (GHz)	102	116	≈ 73	≈ 28

2.5 Passive components

In addition to the fabrication of the active devices, the MMIC also requires the processing of passive components. The genre is usually taken to mean the processing of resistors, capacitors, inductors and transmission lines. These elements may be either distributed or lumped. Distributed elements are physically large enough that transmission-line characteristics play a significant role in their function. Distributed elements have inductive, capacitive, and resistive aspects, all of which are taken into account by transmission-line treatment. Lumped elements are physically small enough that transmission-line effects do not play a significant role in their function. This generally requires that they are less than 0.1 wavelength in size. Nevertheless, even lumped elements are not purely inductive or resistive or capacitive, but have aspects of all three. This must be borne in mind during MMIC circuit design. We will discuss the fabrication of monolithic capacitors, inductors and resistors in the remainder of this section.

2.5.1 Capacitors

Capacitance can be realised on the MMIC in one of four ways. These are:

(i) a stub in a microstrip transmission-line
(ii) coupled lines, forming an interdigitated capacitor
(iii) a metal-insulator-metal sandwich, the MIM capacitor
(iv) the Schottky diode.

The MIM structure is the most common configuration on modern MMICs, and it is this option that is outlined below. Details of other design options can be found in the literature [18].

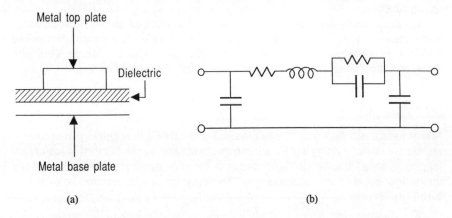

Fig. 2.14 Monolithic MIM capacitor: (a) Simplified cross-section; (b) Model

A schematic diagram of a monolithic MIM capacitor is shown in Fig. 2.14 (a). The simple thin film capacitor comprises two metal plates separated by a dielectric layer. Since this forms a parallel plate capacitor, the capacitance to first order is given by:

$$C = \frac{\varepsilon A}{d} \tag{2.35}$$

where ε is the permittivity of the capacitor dielectric film, A is the plate area and d is the dielectric thickness.

However, these series capacitors also have other components (resistive, inductive, and shunt capacitive) as shown in the equivalent circuit of Fig. 2.14 (b). There are also distributed effects and fringing fields at the plate edges which can affect the measured value of the capacitance [19, 20].

The monolithic MIM capacitor is fabricated in the following way: the base plate of the capacitor is usually formed during the processing of the active components, and might for example be the GaAs FET gate metallisation. This would produce a mainly gold base plate with a smooth surface. This surface quality is important if we are to eliminate the possibility of pinholes occurring in the sandwich dielectric. The dielectric layer is most commonly chosen to be silicon nitride or silicon dioxide. Since these films are used to passivate the active devices immediately after their fabrication, it is natural to use them in capacitor construction as well. Dielectric films are deposited on MMIC wafers by either plasma-assisted deposition or sputtering, and are typically in the thickness range 0.1 to 0.5 µm. The top plate should be at least two skin depths thick to provide an adequate Q-factor (skin depth at 10 GHz is about 0.8 µm). Capacitor top plates are usually formed from thick gold, greater than 3 µm thickness, fabricated by sputtering and ion-beam milling techniques, or by a selective area gold plating technique. This stage usually forms the transmission-line links, inductors and bond pads as well.

MIM capacitor yield is a key issue. A single pinhole can destroy an MIM capacitor and make a complete MMIC useless. Thus it is important to ensure that the processes that form the metal base plate and the dielectric film are extremely well controlled, with the utmost attention given to the reduction of any particulate contamination.

2.5.2 Inductors

MMIC inductors, which may be either lumped or distributed, function as part of a tuned microwave circuit with typical values of between 0.5 and 20 nH. Lumped inductors are lengths of metal line which have mutual inductance by virtue of electromagnetic interactions between the metal segments. The most common layouts are shown in Fig. 2.15 [18].

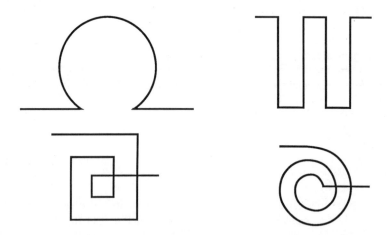

Fig. 2.15 Lumped inductor configurations

Most modern MMIC circuits include rectangular or circular spiral inductors, fabricated from sputtered or plated gold onto insulating or dielectric materials. Lumped inductors are difficult to model accurately, and it is preferable at higher frequencies to form the inductors as transmission lines on the GaAs substrate.

The major characteristics of the transmission line are its length and impedance. For MMIC purposes, transmission lines are either microstrip or coplanar, with microstrip being the most common. These are illustrated in Fig. 2.16.

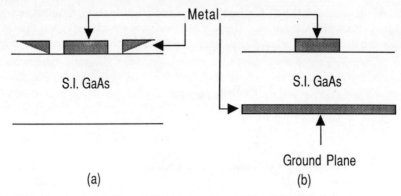

Fig. 2.16 Transmission lines: (a) Coplanar waveguide; (b) Microstrip

Coplanar lines are less common because their lateral ground planes eat up more GaAs real estate on a wafer and a pure coplanar mode is difficult to obtain. However, from a fabrication viewpoint the coplanar configuration is often a considerable advantage, as the back face processing becomes less exacting.

However, microstrip is the most preferred choice in practice. The impedance of the line is determined by the ratio of the conductor width to the substrate thickness (to the first order). An impedance of 50 Ω is generally preferred for input and output transmission lines; this matching the usual impedance of microwave circuits and test configurations. The transmission line must be sufficiently thick to carry the required current without significant ohmic losses. This means that it should be at least two skin depths thick (skin depth at 10 GHz is about 0.8 μm). If the lines are to carry DC bias, the line cross-section must be sufficient to support the current density. The critical current density for gold is around 10^6 Amps/cm^2.

The fabrication of transmission lines in a microstrip configuration requires the following process steps: the lines consist mainly of gold, deposited at the same time as bond pad metal and capacitor top plates on a typical GaAs MMIC. Thus the width of the line is controlled by photolithography, either as a prelude to ion beam milling unwanted gold away, or as part of a selective area plating procedure. Either process will form gold lines of thickness greater than about 2 μm, as skin depth considerations demand. The tolerance placed on substrate thickness depends on the thickness itself. For 100 μm substrates this may be as tight as ± 5 μm. This level of accuracy requires good control as the wafer is thinned following front face processing. Thinning is commonly a chemo-mechanical lapping and polishing process, using precisely aligned jigs to ensure wafer flatness. Finally, the thinned wafer is backface metallised to form the ground plane, again with gold the primary metal used, but with the chip mounting requirements also in mind.

2.5.3 Resistors

Resistors are used in MMICs for feedback, isolation, self-biasing, terminations and voltage dividers in bias networks. GaAs MMIC resistors are of two types: the GaAs resistor and the thin-film resistor. The GaAs resistor is the simplest, as shown in Fig. 2.17.

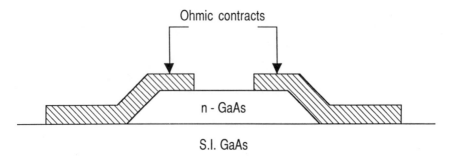

Fig. 2.17 GaAs mesa resistor

GaAs resistors may be mesa or planar, depending on the approach taken to device isolation by the MMIC manufacturer. This is easy to fabricate, as the mesa etching (or implant isolation for a planar structure) is necessarily performed to create the active device islands on the semi-insulating GaAs. The same remarks apply to the ohmic contacts that complete the GaAs resistor. GaAs resistors are, however, of limited use. Current saturation and Gunn diode formation are potential problems to be borne in mind by the circuit designer, but a more serious problem is the large and positive temperature coefficient of resistivity (around 3000 ppm per degree). Many MMICs intended for military use must operate over an extensive temperature range, and compensation techniques may be needed to cater for the large TCR. There are also the potential problems of current handling capacity and thermal dissipation. These issues, coupled to the inflexibility of confining the circuit designer to a range of sheet resistances between 300 - 450 ohms/sq, have led to the development of thin film MMIC resistors.

The need for lower sheet resistances and lower TCR values for increased stability has led to the use of thin film resistors of NiCr (nichrome),TaN (tantalum nitride), and other more exotic materials for specific applications. These produce films with a sheet resistance of 20 - 500 ohms/sq in the case of NiCr, or around 300 ohms/sq for TaN. Both structures produce resistors with low TCR values (and either positive or negative in the case of TaN). NiCr is usually sputtered from a single target, whereas tantalum is usually reactively sputtered in a nitrogen-argon atmosphere to produce the TaN layers. The resistivity of the film and the dimensions of the resistor are the variables that the process engineer needs to have under good control. Sputtered films are deposited from high purity targets at known deposition rates, thus ensuring consistent sheet resistances from run to run. The lateral dimensions of the thin film resistor are determined by the additional masking level that is usually required. A thin film resistor process stage is a plug-in step following the fabrication of active devices. It adds to the overall process sequence and cost, but provides additional freedom in resistor design not achievable using simple GaAs resistors.

2.6 MMIC fabrication

This section will show how the individual components described above are fabricated to form a complete MMIC. No two GaAs manufacturers will fabricate an identical design in quite the same way. Some GaAs foundries utilise air-bridge crossovers whereas others do not. The designer may find through-GaAs vias are not always available, and the allowed wafer thickness may be different. Nevertheless the similarities between the GaAs foundries is such that we can gain much understanding from the consideration of a single process. Thus this section draws heavily on the GEC-Marconi Materials Technology F20 foundry process, fabricated on the 3" process line at Caswell.

The GMMT F20 foundry process is shown schematically in Fig. 2.18. The figure shows how the individual levels are built up to create the MMIC structure. Wafer fabrication begins with a bare semi-insulating GaAs substrate. As GaAs materials technology has advanced, wafer diameters have increased. Currently 3" wafers are the norm, with 4" wafers rapidly coming to the fore in order to increase the output of die out per wafer in. Up to 6" GaAs processing is certainly on the horizon as the explosion in commercial applications for GaAs MMICs begins to influence the major world players.

The choice between implantation and epitaxy for creation of the active surface regions depends on the application. The complex heterostructures that define the HEMT cannot be created by implantation for example. Production of epi-layers is a specialist business, often subcontracted out by the GaAs manufacturers. Techniques such as MBE and MOVPE require considerable capital investment and in-house expertise. Ion implantation has the advantage of low cost, and its inherent guarantee of uniformity and reproducibility can be of great value. It is not usually acceptable to mix the two technologies, although work on selective area epitaxy has been performed.

The first stage of wafer fabrication is usually to define the active devices, such as the transistors (whether MESFET, HEMT or HBT), as described earlier. For high frequency and low noise applications this may call for sub-half-micron gate technology, with mushroom or T-gate cross-sections. These demands are generally beyond the limits of conventional optical lithography, although the silicon IC industry is pushing direct step-on-wafer machines down to 0.5 µm features. Below 0.5 µm there is no option but to abandon optical lithography. Critical structures such as gates are created by electron beam lithography or X-ray techniques. Despite adding complexity to the

56 *Active devices and fabrication technology*

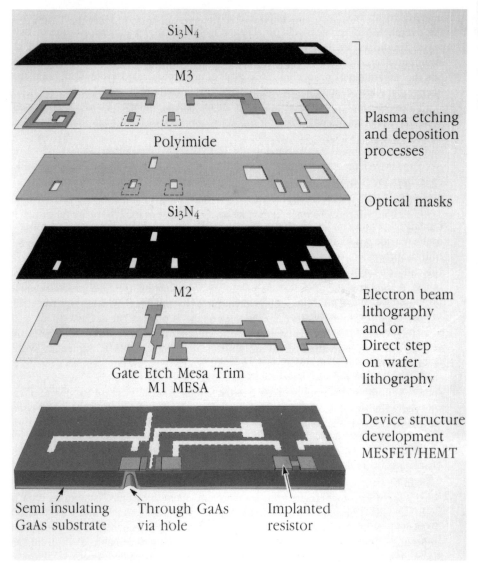

Fig. 2.18 **GMMT F20 GaAs MMIC foundry process**

fabrication processes these approaches have one important advantage: they do increase the yield of critical features, not suffering from the familiar problems of mask run-out or surface contamination which accompanies conventional contact photo-lithography.

Implanted wafers will usually have GaAs resistors defined if required. For many applications thin film resistors are also an additional requirement. These resistors (often made of NiCr, TaN, or WSi) call for a separate mask level, and are fabricated later in the overall MMIC process scheme. Device isolation can be achieved in one of two ways. Simple wet etch techniques can create mesa islands, or proton bombardment in an ion implanter can render exposed material non-conducting. Proton isolation, which has the advantage of producing a planar structure, produces deep level traps in the energy gap which capture the charge carriers, thus preventing current flow.

The metals that complete the active device also act as connecting tracks and the base plate for MIM capacitors. As soon as the active devices have been fabricated it is advisable to passivate the surface. The GMMT F20 process uses silicon nitride to this end, which also acts as the dielectric sandwich in the MIM nitride capacitors.

The GMMT process also uses an additional polyimide layer to produce polyimide capacitors, but the main purpose of the layer is to provide a dielectric crossover. Thus air-bridges are not required and a high yield process is assured. Top level metal (mainly gold) is then deposited to create the top plates of the capacitors, to form the RF transmission lines, and to complete the bond pads for connecting the packaged MMIC to the outside world. Either gold plating techniques or a combined sputtering/ion beam milling process is most commonly used to fabricate this top level metal, which is typically of 2-5 µm thickness in order to facilitate high quality ball or wedge wire bonds. A final silicon nitride coating (or equivalent) passivates the entire structure, and provides mechanical protection. Etching allows only the metal bond pads to be exposed to the environment thereafter.

The amount of backside processing to be performed depends on the circuit design configuration. For a coplanar design the processing is nearly complete. For a microstrip design, with the option of through GaAs via holes to minimise source inductance, much work is still to be done, as described below.

Prior to thinning the GaAs substrate, the front face features must be protected by a sacrificial coating, which must provide mechanical protection but be easily removed at the end of processing. Thinning itself is a chemo-mechanical lapping and polishing process, requiring excellent controls to ensure a finished wafer thickness of between 50 and 200 µm, the exact thickness depending on the application and the power handling characteristics of the circuit. Through GaAs via holes are etched from the back surface to meet the front face features. Vias are extremely important for monolithic ICs because FETs on such an IC may not be near the edge of the chip, but still require a low inductance ground. Most foundries now reactively ion etch via holes, as dry etching gives access to greater process control at a critical stage in the fabrication sequence. A final metallisation forms the ground plane and provides the conducting filling to the etched vias. Cleaning and wafer dicing (either mechanically sawn or scribed) completes the MMIC.

Fig. 2.19 shows a photograph of a completed MMIC wafer; note the five standard process control arrays which contain standard test cells. These DC and RF probeable components are to assure the quality of the finished wafer, measured in comparison with standard library results. A wafer is released for further processing on the basis of the results from these test cells.

58 *Active devices and fabrication technology*

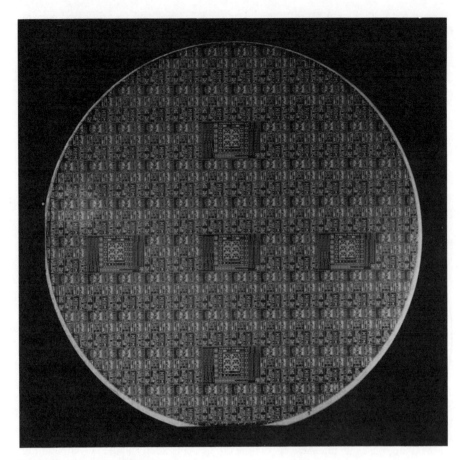

Fig. 2.19 Completed MMIC wafer

The above description has given an overview of the principal stages involved in the fabrication of a GaAs MMIC. It is not within the scope of this chapter to enter into greater detail concerning the individual process steps, but some appreciation of the complexity of the process can be given by consideration of Fig. 2.20. Here, a typical GaAs front face process has been drawn schematically. Each concentric ring represents a process stage (or mask level). A batch of wafers completes a revolution of the innermost circle (metaphorically), proceeding then to the larger adjacent circle, and so on. Hence a batch is routed through the same equipment many times which, allied to the comparatively long cycle times, makes scheduling an art rather than a science. The arrival on the scene of affordable finite capacity scheduling software is likely to be of considerable help in this respect.

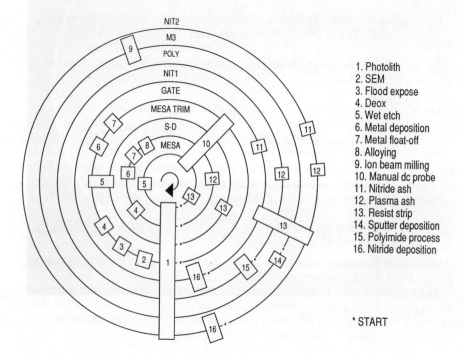

Fig. 2.20 MMIC process representation

The comments on affordable software also apply to computer-aided manufacturing (CAM) techniques. A system such as PROMIS [21] or WORKSTREAM (formerly COMETS) [22] is considered essential to the successful MMIC manufacturer. These computer based systems enable the manufacturer to take an integrated approach to work-in-progress tracking and scheduling, data collection and analysis, equipment status monitoring, process control, and overall systems management.

2.7 Yield and process control

This section concentrates on analogue GaAs MMICs, although many of the arguments and formulae are equally applicable to digital GaAs circuits. The yield of functioning GaAs MMICs can be described in terms of the following individual figures of merit:

(a) Throughput yield: this refers to the loss of whole wafers through, for example, wafer breakage. This yield increases as the process maturity increases, and as techniques such as automated handling are adopted

(b) Component yield: this refers to the loss of individual circuits owing to the catastrophic failure of one or more components

(c) Probe yield: failure of a functioning circuit to meet specification accounts for this yield loss

(d) Visual yield: failure to pass visual inspection criteria results in yield loss

(e) Other: for example, burn-in failure

Increasing circuit complexity has tended to restrict component yields to <70%, and it is this yield metric that is receiving the most attention. It is over 30 years since Hofstein and Heiman [23] published the Poisson yield model for a silicon insulated gate FET:

$$\text{yield } Y = \exp(-NA_g D) \tag{2.36}$$

where A_g is the gate area of each transistor, D is the average defect density (defects per unit area), and N is the total number of transistors.

Defect densities in real semiconductor manufacturing processes appear to vary from wafer to wafer and lot to lot. Therefore, a constant defect density rarely exists. The modelling of such varying defect densities was first proposed by Murphy [24], producing the formula for the mixed Poisson yield model:

$$\text{yield } Y = \int_0^\infty \exp(-AD) f(D)\, dD \tag{2.37}$$

where A is the defect sensitive area, D is the defect density, and $f(D)$ is the defect density distribution.

However, care is needed when applying the Murphy formula. The relationship does not hold when severe defect clustering occurs. This can arise within new or highly complex processes, such as those used to fabricate GaAs transistors. Nevertheless, the formula is freely used in GaAs yield modelling.

Norris and Barratt [25] found no strong evidence of such defect clustering in their analysis of GaAs MMICs. As a result, they appropriately used Poisson statistics to derive the following yield for MMICs produced at Lockheed Sanders:

$$\text{yield } Y \approx A\exp\left[-(D_f F + D_c C)\right] \qquad (2.38)$$

where A is a fixed (process dependent) constant, F is the FET gate periphery, C is the MIM capacitor area, and D_f, D_c are the corresponding defect densities per unit area.

If Murphy's model does hold, it is possible to plot a family of probe yield curves relating percentage die yield to die size for a range of defect densities per unit area, as shown in Fig. 2.21. Superimposed on the plot are corresponding stages in the development of GaAs MMIC manufacturing, from research to high volume production [26].

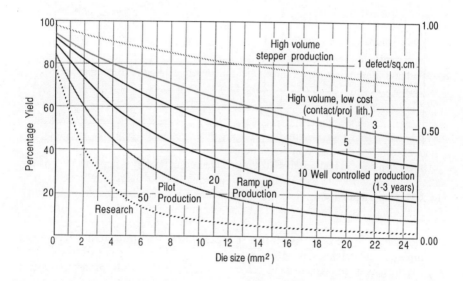

Fig. 2.21 Die yield against die size

Wafer fabrication yields affect production costs. Another insight into the behaviour of production costs over time for a particular product line can be gained from the experience curves. These curves reflect the numerous factors that rely on a competent and effective management team to systematically exploit the cost opportunities that present themselves [27]. The curves include the more familiar 'learning curve' as a subset. For example, the curves shown in Fig. 2.22 demonstrate the relationship between cumulative chip deliveries and relative cost for curves of 70% to 90%. For comparison, a 70% experience curve described the cost of random access memory (RAM) components in the period 1976-84. The curves are particularly useful during production ramp-ups to aid decisions on future pricing strategies. Clearly it is important to develop an accurate cost model in order to achieve meaningful results. Inaccurate apportioning of overhead costs, for example, can lead to misleading conclusions; a fact that has led to the increased use of activity-based costing techniques [28].

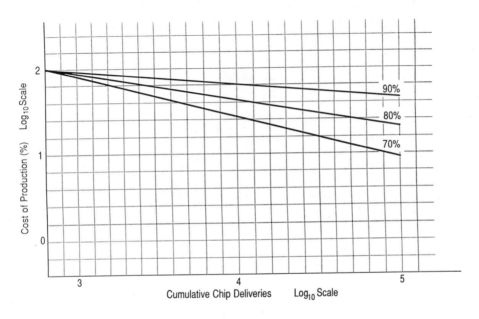

Fig. 2.22 Experience curves

The increasing use of GaAs MMICs in high volume, low cost commercial systems has imposed new demands on MMIC manufacturing. The importance of low cost per chip requires great emphasis to be placed on the quality of the MMIC product, to ensure high yields of chips to specification. This in turn means that the processes that are used to fabricate the MMICs are well-controlled, are individually of intrinsically high yield, and that the equipment on which the processes are run is well-understood and well-maintained. Long established techniques are available to manufacturers of GaAs MMICs that can assist in reaching the goal of high quality production while simultaneously reducing waste and chip costs. Throughout industry in Europe, North America and Japan, statistical process control (SPC) is the primary tool for monitoring and controlling production processes.

The key to SPC is the comparison of measured data collected from a process with the normal or Gaussian distribution. For a fixed and stable process the data measured from run to run will not be identical, but will form a distribution about some central (mean) value. This reflects the natural variation in the process. For example, the small differences occurring in chamber pressure, gas flow or power level during nominally identical dielectric deposition runs will result in slightly different thicknesses from run to run (Fig. 2.23).

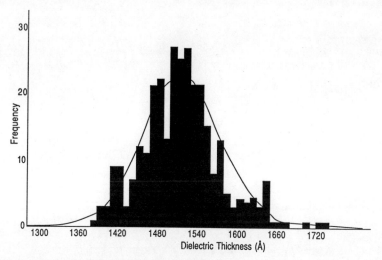

Fig. 2.23 **Natural variation of an MMIC process stage**

The data should approximate to the normal distribution provided that no special causes are present to distort the pattern. In the example, special or 'assignable' causes might be the inconsistent placement of wafers in the equipment, the use of a different machine, or processing by a poorly trained operator.

It is usual for measurements in production to be taken on a sample basis rather than measuring every possible site on every wafer to gain knowledge of the entire population of items in the distribution. However, if all the assignable causes have been eliminated, the central limit theorem can be invoked to infer information about the entire population knowing only the distribution of the sample data.

Thus, a particular process step can easily be characterised by the familiar mean and standard deviation measures, provided that no assignable causes are present. Armed with this knowledge, SPC charts can be constructed based on the mean value of the distribution and the 3 × (standard deviations) limits. On average, we would expect 99.73% of our results to be between these mean ± 3 standard deviation limits. These limits are called control limits, and are warning signals that an assignable cause may be present. If a point plots outside the control limits, action should be taken to investigate the cause. It is important to note that these control limits are fixed by the natural variation of the process. They are not to be confused with specification limits, which are imposed on the process from outside. For a high yield process the process control limits should be well inside the specification limits.

The most common SPC chart is the mean-range (X-R) chart, an example of which is shown in Fig. 2.24. The means and ranges of samples have been plotted on the charts, with the upper (UCL) and lower (LCL) control limits also present.

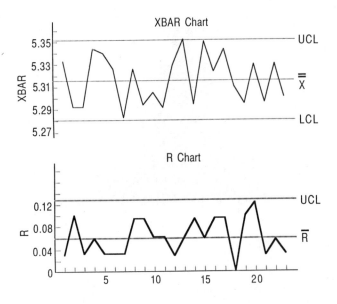

Fig. 2.24 Mean and range for occasional samples

The figure shows a process stage under statistical control, with no control limit violations. Hence there are no assignable causes perturbing the natural variation of the process.

The great value of the technique is that once an assignable cause has been identified the process can be adjusted before any scrap is produced. Thus SPC has become an important feature of yield improvement programmes throughout industry. SPC is also a corner-stone of continuous improvement (or TQM) initiatives. By reducing the natural spreads of processes the control limits will move closer to the mean values, ensuring even fewer process failures. These attributes have given SPC universal business appeal, beyond the manufacturing function. Since every business activity is a process, SPC is equally applicable to support functions, infrastructure and service departments.

2.8 References

1. TURNER, J. A.: 'History of the GaAs FET at Caswell (1964-1985)', IEE Colloquium on Modelling, Design, and Application of MMICs, 17th June 1994, Savoy Place, London, digest no. 1994/092, pp. 1/1-1/3
2. PUCEL, R. A., HAUS, H. A., and STATZ, H.: 'Signal and noise properties of gallium arsenide microwave field-effect transistors', Advances in Electronics and Electron Physics, vol. 38, 1975, New York Academic Press, pp. 195-265
3. FUKUI, H.: 'Optimal noise figure of microwave GaAs MESFETs', *IEEE Trans. Elec. Devices*, ED-26, 1979, p. 1032
4. DELAGEBEAUDEUF, D., ETIEENE, P., LAVIRON, M., CHAPLART, J., and LINH, N.T.: 'Two-dimensional electron gas MESFET structure', *Electron. Lett.*, vol.16, no.17, 1980, pp. 667-668
5. LANG, D. V., and LOGAN, R. A.: 'Trapping characteristics and a donor-complex (DX) model for the persistent-photoconductivity trapping center in Te-doped $Al_xGa_{1-x}As$', Physical Review B, vol. 19, no. 2, pp. 1015-1030 (1989)
6. REZAZADEH, A. A., THOMASIAN, A., SMITH, R. S., and KIRBY, P. B.: 'Microwave device comparisons for MMIC applications', Proceedings of IEEE Int. Symp. on MMICs in Communications Systems, King's College London, September 1992
7. REZAZADEH, A. A., THOMASIAN, A., SMITH, R. S., and KIRBY, P. B.: 'A direct comparison of DC and microwave performance between HEMTs and GaAs MESFETs', 16th European Workshop on Compound Semiconductor Devices and Integrated Circuits, Madrid, May 1992
8. AHMAD, T., REZAZADEH, A. A., and GILL, S. S.: 'Current dependence of small signal base-collector capacitance in microwave AlGaAs/GaAs HBTs', Proceedings of 24th European Solid State Device Research Conference (ESSDERC), Edinburgh, 11-15 September 1994
9. KREN, D. E., REZAZADEH, A. A., and TOTHILL ,J. N.: 'Temperature dependence of current gains in high C-doped base HBTs', *Electron. Lett.*, vol. 30, no. 10, 1994, pp. 825-826
10. KREN, D. E, and REZAZADEH, A. A.: 'Defect and temperature characterisation of InGaP/GaAs and AlGaAs/GaAs HBTs', Proceedings of the 2nd IEEE Int. Workshop on High Performance Electron Devices for Microwave and Optoelectronic Applications (EDMO), King's College, London, 1994, pp. 20-25
11. MILLER, S. S., and KAMINOW, I. P.: *Optical Fiber Telecommunications II* Academic Press Inc., London, 1988
12. BASHAR, S. A., SHENG, H., and REZAZADEH, A. A.: 'Prospects of InP/InGaAs HBTs for optoelectronic telecommunications', Proceedings of the 2nd IEEE Int. Workshop on High Performance Electron Devices for Microwave and Optoelectronic Applications (EDMO), King's College, London, 1994, pp. 1-6

13. KING, C. A., HOYT, J. L., and GIBBONS, J. F.: 'Bandgap and transport properties of Si1-xGex by analysis of nearly ideal Si/Si1-xGex/Si heterojunction bipolar transistors', *IEEE Trans. Elec. Dev.*, ED-36, no. 10, 1989, pp. 2093-2104
14. WANG, G., PIERSON, R. L., ASBECK, P. M., WANG, K., WANG, N., NUBLING, R., CHANG, M. F., SALERNO, J., and SASTRY, S.: 'High performance MOCVD-grown AlGaAs/GaAs heterojunction bipolar transistors with carbon-doped base', *IEEE Electron Device Lett.*, vol. 12, no. 6, 1991, pp. 347-349
15. HO, W. J., CHANG, M. F., SAILER, A., ZAMPARDI, P., DEAKIN, D., MCDERMOTT B., PIERSON, R., HIGGINS J. A., and WALDROP, J.: 'GaInP/GaAs HBTs for high-speed integrated circuit applications', *IEEE Trans Elec. Dev.*, vol. 14, no. 12, 1993, pp. 572-574
16. LIU, W., FAN, S., HENDERSON, T., and DAVITO, D.: 'Microwave performance of GaInP/GaAs heterojunction bipolar transistor', *IEEE Electron Device Lett.*, vol. 14, no. 4, 1993, pp. 176-178
17. CRABBÉ, E. F., COMFORT, J. H., LEE, W., CRESSLER, J. D., MEYERSON, B. S., MEGDANIS, A. C., SUN, J. Y.-C., and STORK, J. M. C.: '73-GHz self-aligned SiGe-base bipolar transistor with phosphorus-doped polysilicon emitters', *IEEE Elec. Dev. Lett.*, vol. 13, no. 5, 1992, pp. 259-261
18. WILLIAMS, R. E.: *Gallium arsenide processing techniques*, Chapter 4, Artech House, Inc., Dedham MA, 1984, ISBN 0-89006-152-1
19. HIGASHISAKE, A., and HASEGAWA, F.: 'Estimation of fringing capacitance of electrodes on S.I. GaAs substrate', *Electron. Lett.*, vol. 16, 1980, pp. 411-412
20. CHEW, W. C., and KONG, J. A.: 'Effects of fringing fields on the capacitance of circular microstrip disk', *IEEE Trans. Microwave Theory Tech.*, MTT-28, 1980, p. 98
21. PROMIS System Coroporation, Toronto, Ontario, Canada M5X 1E3
22. WORKSTREAM is a product of Consilium Inc., Mountain View, California, USA
23. HOFSTEIN, S. R., and HEIMAN, F. P.: 'The silicon insulated-gate FET', *Proc. IEEE*, vol. 52, no. 12, 1963, pp. 1190-1202
24. MURPHY, B. T.: 'Cost-size optima of monolithic ICs', *Proc. IEEE*, vol. 52, no. 12, 1964, pp. 1537-1545
25. NORRIS, G. B., and BARRATT, C. A.: 'GaAs MMIC yield modeling', IEEE GaAs IC Symp. Dig., 1990, pp. 317-320
26. DEVINY, I.: Private communication
27. HILL, T.: *Manufacturing Strategy*, Macmillan Education Ltd., Hampshire, England, 1985
28. TURNEY, P. B. B.: *Common cents: The ABC performance breakthrough*, Cost Technology, Hillsboro, OR, USA, 1991, ISBN 0-9629576-0-7

Chapter 3

Passive components

M. Gillick and I. D. Robertson

3.1 Introduction

It is the extensive use of passive components that makes MMIC design so different to conventional integrated circuit design and layout. In MMIC design, passive components are used for impedance matching, DC biasing, phase-shifting, filtering, and many other functions. This chapter describes the key passive components that are used in MMIC design. These components include not only the basic lumped inductors, capacitors, and resistors but also a wide range of distributed transmission-line components. These transmission-line components include microstrip lines and elements such as bends and Tee-junctions, as well as standard building blocks like couplers and power splitters/combiners. Most often, these building blocks are realised in the microstrip transmission-line medium. However, other types of transmission line such as coplanar waveguide (CPW), slotline, and 'thin-film microstrip' are being used more and more. The principle advantages to be gained from using CPW are increased packing density of circuits and reduced dispersion for millimetre-wave operation. Transitions between the different transmission-line media can be used to realise components such as baluns. At the lower microwave frequencies transmission-line components are often too large to be practical. To overcome this, the standard microstrip components such as couplers and power splitters can be realised with lumped-element equivalent networks or by using the lumped-distributed miniaturisation technique. These approaches are an interesting example of how the precisely defined passive components available on MMICs can be used to develop new circuit techniques which overcome the fundamental size limitation not encountered with MIC circuits.

There are relatively few design equations in the description of the passive components since these are very process-dependant. For a particular MMIC fabrication process the manufacturer's design manual is the essential authority on the exact design rules for component layout as well as the appropriate design curves and equations. Nevertheless, in many cases the design manual is little more than a reference guide to the process data, and this chapter gives a great deal of background information which is essential for all but the most experienced designer.

3.2 Inductors

Depending on the inductance required, MMIC inductors can be realised either as straight narrow tracks (ribbon inductors), as single loop inductors, or as multi-turn spiral inductors. A microstrip ribbon inductor and its equivalent circuit are shown in Fig. 3.1. For short lengths ($<\lambda_g/4$) the inductance and shunt end capacitances can be calculated from the following well known equations [1]:-

$$L = \frac{Z_0}{2\pi f} \sin\left(\frac{2\pi l}{\lambda_g}\right) \quad (3.1)$$

and

$$C = \frac{1}{2\pi f Z_0} \tan\left(\frac{\pi l}{\lambda_g}\right) \quad (3.2)$$

A narrow track with high Z_0 is needed to achieve high inductance with low parasitic capacitance. However, in practice the choice of track width is determined by fabrication limits, by the DC current carrying capacity, and by the high resistance of very narrow tracks. The track length is limited simply by the need to ensure a realistic and economical chip size. The ribbon inductor is thus limited to values of less than 1 nH, but is a relatively 'pure' inductor with low parasitics. Hence it is often used in distributed amplifiers where very large bandwidths are required. Distributed amplifiers are described in detail in Chapter 5.

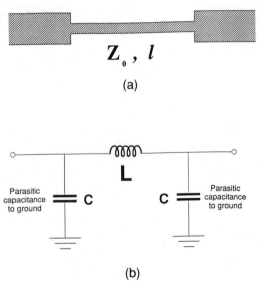

Fig. 3.1 The ribbon inductor: (a) Microstrip layout; (b) Equivalent circuit

3.2.1 Loop inductors

Single loop inductors like the one shown in Fig. 3.2 were used extensively in the pioneering days of MMICs. This is probably because the processing did not at first offer air-bridges for spiral inductors and because the limited experience of coupled lines on GaAs made designers reluctant to use meandered ribbon inductors. It is fair to say that in recent years loop inductors have been used very little because of their inefficient use of chip area. A number of texts give useful design equations for loop inductors [2-4].

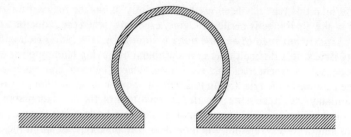

Fig. 3.2 Layout of a single loop inductor

3.2.2 Spiral inductors

Spiral inductors are essential for values above approximately 1nH. Comparing a spiral inductor with a meandered ribbon, one can see from Fig. 3.3 that for the meandered inductor (a) the current in adjacent tracks is flowing in opposite directions, whereas for the spiral inductor (b) the current in adjacent tracks is in the same direction. The resulting mutual inductance yields a significant increase in the spiral inductor's self-inductance. In the meandered case, however, the inductance is reduced and the overall inductance is significantly less than that of a straight track equal to the unwound length.

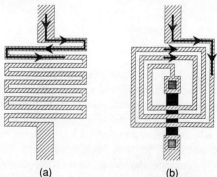

Fig. 3.3 Large value inductance using (a) meandered track and (b) spiral

70   *Passive components*

The drawback of the spiral is that the need to connect the centre turn back to the outside circuit dictates that either air-bridge crossovers or dielectric spaced underpasses must be used. There are a number of different solutions to this same fundamental connection problem and these are illustrated in Fig. 3.4. Note that square inductors are shown; rectangular ones could be used if layout constraints made them necessary, and round (approximately) ones would have slightly superior performance at the cost of greatly increased layout complexity and a less convenient shape for integration with other components. However, these alternative inductors would have to be characterised first.

Inductor type (a) has a single air-bridge span from the centre of the spiral to the outside. Whilst this type has been used successfully, it has the major drawback that the span of the air-bridge is perilously large, especially when the inductor has many turns. The maximum span of an air-bridge is limited by fabrication technology and reliability issues. It is difficult to ease the problem by having landing posts between the turns because adjacent tracks are very close. Hence, the more commonly accepted air-bridge solution is the type (b). In this type of spiral inductor the connection to the centre turn stays at the substrate level, and each turn of the inductor jumps over it using a short air-bridge. This type thus has more air-bridges but they are each of minimum length so that reliability is improved. The third type of spiral inductor (c) actually forms the entire spiral out of air-bridges, which jump from landing post to landing post. The advantage of this structure is that since the track spends much of the time off the substrate surface the parasitic capacitance between turns and to ground is reduced, giving a higher useful maximum operating frequency.

The alternative to using air-bridges is to employ a two-level metal process with a spacer dielectric. Polyimide is often chosen as it has a low ε_r and because thick films can be deposited very quickly. In this type of inductor (d) the turns are normally on the top metal layer above the dielectric so that the parasitic capacitance is reduced and so that the turns can be plated up for less series resistance. The underpass is necessarily of a much thinner metal, and may thus be made quite wide in order to maintain a usable current carrying capacity and minimise series resistance. The connection between the upper metal and the underpass is made with a via in the dielectric(s). These vias should not be confused with the through-substrate vias used for grounding components.

3.2.3 Stacked spirals

Stacked spiral inductors have been used exclusively by GEC-Marconi. They comprise a pair of inter-wound spirals placed on separate metal layers. The major advantage is that the turns are much more tightly packed than normal photolithography and metal patterning would allow for a single layer spiral, and since they are separated vertically to some extent there is less capacitance between adjacent turns than there would normally be with such small gaps. However, since the dielectric spacer can only planarise the circuit to a certain extent, the lower metal thickness is limited to 1µm or less and this means that the stacked spiral has a high series resistance. The inductance per unit area of stacked spiral inductors is very high, but their resonant frequency is much lower and so they are limited to frequencies below 5 GHz or so. They often find application as bias chokes, where high inductance is required, and in this application the high series resistance can actually be beneficial to the amplifier's stability, matching, and bandwidth.

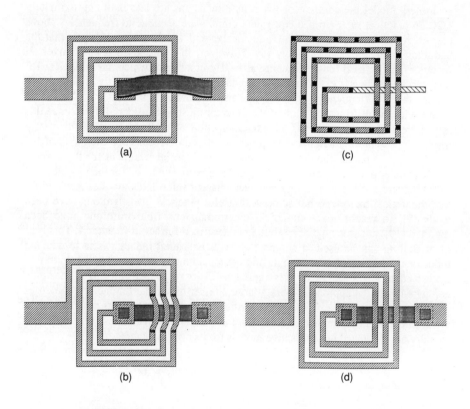

Fig. 3.4 Spiral inductors: (a) Single air-bridge; (b) Air-bridges over an underpass; (c) Formed entirely of air-bridges; (d) Using two metal levels for an underpass

3.2.4 Spiral inductor models

The basic inductor equivalent circuit consists of a prime inductance along with its associated series resistance, inter-turn and crossover feedback capacitance, and some capacitance to ground. This lumped element model is shown in Fig. 3.5. Direct calculations of prime inductance can be made using the classical formulas given by Grover [4], with appropriate correction factors [5]. It is more common, however, for the equivalent circuit data to be found for a number of different inductor geometries, and to use empirical curve-fitted expressions for the values of the equivalent circuit elements [6]. However, any equivalent circuit using lumped elements can only accurately model the component when its dimensions are less than approximately $\lambda/20$. In practice, such lumped equivalent circuits are accurate to frequencies above the self-resonant frequency of the inductor, beyond which one could argue that the 'inductor' is no longer useful anyway. However, in a circuit such as a mixer an inductor may be used for IF matching at 1 GHz in a circuit which has a 20 GHz RF input. In this case it is essential that the inductor is modelled properly at both the IF and at the RF frequency. In order to achieve an accurate model at such high frequencies the distributed nature of the inductor must be considered. In essence this means treating the inductor as an interconnection of multiple coupled microstrip lines. For a square inductor it is surprisingly effective to consider each side of the inductor as a separate coupled-line problem, with special treatment for the corners [7]. A number of quite rigorous approaches have been developed for the modelling of spiral inductors [8]. The coupling between adjacent spiral inductors has received far less theoretical treatment, but some very useful practical investigations have been made [9]. In recent years numerical electromagnetic field solutions have been reported [10], and this type of modelling is described further in Chapter 4. This type of modelling can be used in conjunction with measured inductor data to develop inductor models which are accurate and computationally efficient for incorporation into standard CAD packages. This approach allows more flexibility to the designer compared with the approach where a large number of standard inductor geometries are characterised by the foundry. Nevertheless, the foundry's own models should always be used in preference to those from any CAD package, unless that particular foundry has evaluated the alternative models for its own inductors.

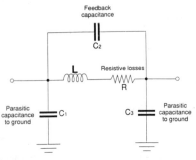

Fig. 3.5 Inductor model

3.2.5 Planar spiral transformers

The planar spiral transformer was first employed on MMICs by Honeywell [11, 12]. Fig. 3.6 shows a transformer employing underpasses, although air-bridge techniques can also be employed. The transformer offers DC blocking, matching, and DC bias injection functions in a very small size. As a result, transformer-based circuits can achieve exceptionally high packing density, and Pacific Monolithics made a successful business based on circuits using this technique. In particular, they sold the first commercially successful MMIC for C-band DBS TV receivers.

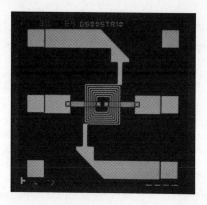

Fig. 3.6 Photograph of a spiral transformer (courtesy of GEC Marconi)

The basic equivalent circuit of a planar spiral transformer is shown in Fig. 3.7. The two coupled inductors have self-inductance and mutual inductance. There is series resistance in the tracks, inter-turn capacitance and shunt capacitance to ground. The most serious parasitic is usually the capacitance between the two spirals since this makes the transformer resonate as the capacitive coupling becomes dominant at higher frequencies. To minimise this parasitic capacitance whilst achieving a high mutual inductance, the turns need to be very narrow and close together. Typically a transformer would have 5μm tracks and 5μm gaps. Inevitably, the narrow tracks lead to high series resistance and so planar transformers are rather lossy. In a circuit such as an amplifier, matching can be achieved by resonating the self inductance with the transistor capacitance, rather than by employing a turns ratio for impedance transformation. Employing a different number of turns in the basic transformer does not necessarily give the desired effect because the magnetic field of the larger coil is not intimately coupled to the smaller coil. However, impedance transformation can be achieved by employing the distributed nature of the transformer. The lumped element equivalent circuit is only valid to frequencies up to and just over the resonant frequency of the transformer. At higher frequencies the tracks must be treated as multiple coupled microstrips. The transformer of Fig. 3.6 has a resonant frequency of approximately 6 GHz [13], and is only usable as a true lumped transformer up to 4

74 Passive components

GHz or so. However, as a distributed element the same component can be used as a transmission-line balun or coupler. Balun operation is very hard to achieve, largely due to shunt capacitance to ground which lowers the even-mode impedance. However, the spiral readily operates as a quadrature backward-wave coupler with a centre frequency roughly corresponding to an unwound length of $\lambda_g/4$. Further extensions of the planar transformer are possible: The 'tri-filar' transformer [14] offers broad-band balun operation, and multi-level techniques can realise tighter coupling, increased performance, and scope for many new ideas.

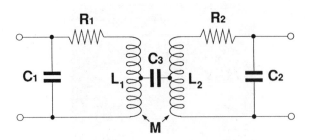

Fig. 3.7 Spiral transformer model

3.3 Capacitors

Both overlay capacitors (metal-insulator-metal (MIM) capacitors) and interdigital capacitors can be used on MMICs. Interdigital capacitors can be used for values up to approximately 1pF, above which their size and the resulting distributed effects prevent their use. Overlay capacitors are therefore essential for all but the simplest circuits, and are particularly important for DC blocking and decoupling, where large capacitor values are required.

3.3.1 Overlay capacitors

These consist of a metal-insulator-metal 'sandwich' with the most common insulators being silicon nitride, silicon dioxide, and polyimide. Silicon nitride is popular since it has a fairly high ε_r and can also be used for passivating the exposed GaAs in the active devices. In processes which use polyimide as the spacer dielectric for spiral inductors, the polyimide can also be used to realise small value capacitors for applications such as matching networks and filter inter-resonator coupling, and they are significantly smaller than interdigital capacitors. The type of connection used from the capacitor to the rest of the circuit depends on whether air-bridges or a polyimide-based two metal level process are used. In a two metal level process the main microstrip circuitry is normally on the upper metal, and so the lower plate has to be brought up to the top metal through a via-hole in the dielectric, as shown in Fig. 3.8 (a). Alternatively, with air-bridges, the upper plate is connected to the rest of the circuit with an air-bridge as shown in Fig. 3.8 (b). This is preferable to the method shown in Fig. 3.8 (c), because the sharp edge along the interconnection would make a short-circuit failure more likely and reduce the breakdown voltage rating of the capacitor. Since the width of air-bridges is limited by the need during fabrication to remove the temporary photoresist support, often several air-bridges would be put in parallel in order to achieve a low inductance connection, rather than using a single wide air-bridge.

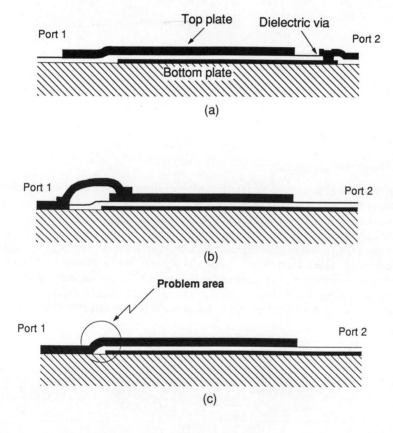

Fig. 3.8 Overlay capacitors: (a) Using a dielectric via; (b) With an air-bridge; (c) Without an air-bridge or spacer dielectric

76 Passive components

Fig. 3.9 shows the typical equivalent circuit model of a monolithic capacitor. The major parasitic is often the shunt capacitance to ground, especially for larger capacitances. This simple model is not truly distributed, however, and cannot represent real layout situations where asymmetrical or multiple connections are made to the capacitor. Process-oriented models have been developed which model the distributed nature accurately and model realistic layout scenarios [15].

Finally, it should be mentioned that the exact size of the fabricated capacitor may often be slightly different to the layout dimension, and this is a particularly important effect for small capacitors. The manufacturer's foundry design manual should be consulted to make sure that the correct allowances are made at the layout stage.

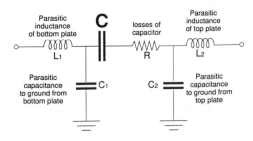

Fig. 3.9 Capacitor model

3.3.2 Interdigital capacitors

These consist of a number of interleaved microstrip fingers coupled together as shown in Fig. 3.10. The maximum value of an interdigital capacitor is limited by its physical size, and its maximum usable operating frequency is limited by the distributed nature of the fingers. They certainly cannot be used for values above 1 pF, and even a 0.5 pF interdigital capacitor will measure approximately 400×400 μm^2. Nevertheless, since interdigital capacitors do not use a dielectric film, their capacitance tolerance is very good and is limited only by the accuracy of the metal pattern definition. Hence they are ideal as tuning, coupling, and matching elements where small capacitor values are required and where precise values are necessary.

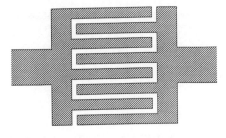

Fig. 3.10 Interdigital capacitor

3.4 Resistors

Resistors on MMICs can either use the doped semiconductor layer (mesa resistors or implanted planar resistors) or a deposited thin-film resistive layer. In either case, since the layer or film thickness is fixed it is very convenient to quote resistivity in terms of an ohms-per-square figure. Hence the value of the resistor is chosen by selecting a suitable aspect ratio, as shown in Fig. 3.11. The absolute size of the resistor is determined by the tolerance required and the expected power dissipation. The area of the ohmic contact pads is particularly important in determining the tolerance of the resistor as their contact resistance is difficult to control precisely and must be minimised by making the pads suitably large. A practical limit is imposed by the higher parasitic capacitance of large pads and the resistor's physical size.

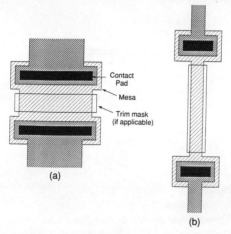

Fig. 3.11 Resistor examples: (a) Small value ($\approx 50\ \Omega$); (b) Large value ($\approx 3000\ \Omega$)

3.4.1 GaAs resistors

The term 'mesa' refers to the situation where the whole wafer is doped and subsequently selectively etched so that active regions remain only where required and are left standing proud of the substrate surface. This is the case when epitaxial growth or blanket ion-implantation are used. However, selective ion-implantation or proton isolation are often used, and in this case the active regions remain flush with the wafer surface. Hence resistors using the GaAs active layer can be either mesa resistors or implanted planar resistors. Fig. 3.12 shows the typical form of a mesa resistor. The GaAs resistor relies on the linearity of the semiconductor's current-field characteristic at low electric field values. Hence it is important to consider how much current is to be passed through the resistor. A design rule specifying the maximum electric field allowed in the resistor is normally given as a figure of Volts-per-unit-length for a certain permissible percentage deviation from perfect linearity. If a resistor is used for the biasing of a FET gate (or bipolar transistor base) this does not

78 *Passive components*

present a problem, but for drain or collector biasing it is very easy to exceed the maximum ratings if care is not taken. Resistor biasing in any case greatly increases the total DC power consumption, and the operating temperature of the resistor should be considered. For mesa resistors, the orientation of the resistor must be considered when calculating the exact layout dimensions of the resistor. The crystallographic orientation means that the mesa slope along the edges will be either positive or negative depending on whether the resistor is horizontal or vertical on the layout. This makes the effective width different and necessitates a small correction factor. For mesa resistors the active region may undergo an additional etching process (trimming) intended to increase the ohms-per-square figure and improve the tolerance on resistors. In this case the resistor will require an additional mask level to facilitate this etching without affecting the active devices.

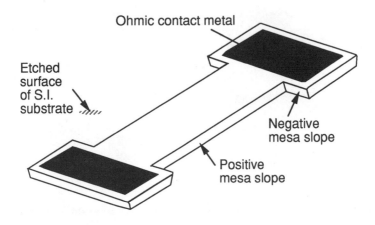

Fig. 3.12 Mesa resistor view showing positive and negative mesa edges

3.4.2 Thin film resistors

Sputtered thin film resistors offer improved linearity and lower temperature coefficients compared with the mesa types. In addition, the ohms-per-square figure can be optimised for the circuit designer without any limitations imposed by the requirements of the active devices. The most commonly used materials are Tantalum Nitride, Cermet, and Nickel Chrome. Their temperature coefficient of resistance is less than one tenth that of GaAs, and ohms-per-square figures of 50 Ω can be produced, which is convenient for the circuit designer.

3.4.3 Resistor models

Fortunately, the monolithic resistor is relatively free of parasitics. However, a frequency dependant resistance results from the frequency dependence of the skin-depth. In all but the most broadband circuits this can be accounted for at the design stage. The parasitic shunt capacitance to ground of the resistor and its contact pads, and the small end-to-end feedback capacitance of the resistor have to be included in any simulation. The typical equivalent circuit model of a resistor is shown in Fig. 3.13. The distributed nature of the resistor is taken into account with the series inductance. For greater accuracy it may be necessary to model the resistor as a proper distributed RLC ladder, or as a lossy transmission line.

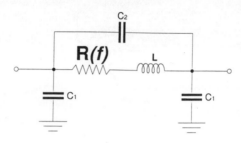

Fig. 3.13 Resistor equivalent circuit

3.5 Via-holes and grounding

The through-substrate via-hole, shown schematically in Fig. 3.14 (a), is an essential component for all but the most simple circuits as it enables a low inductance ground to be established within the circuit. Without via-holes, devices would have to be placed very near the edge of the chip and grounded with either wrap-around metallisation or bond-wires, as illustrated in Figs. 3.14 (b) and (c). These methods of grounding impose severe restrictions on the topology of the circuit. However, even when via-holes are used there are several pitfalls regarding grounding to be avoided at the design and layout stage. The most important principle to remember is that a mass of metal can never be assumed to be a 'ground bar'. The grounding connection must always be considered as a transmission-line element in its own right and modelled accordingly. Fig. 3.15 shows a classic example of a circuit layout based on the 'ground bar' principle which may not work in practice. In Fig. 3.15 (a) the FETs of a two-stage amplifier share a common grounding pad, which is grounded with a via-hole. As the equivalent circuit shows, this grounding method provides a direct feedback path which is almost certain to cause instability. The solution in this case is to provide each FET with its own ground pad, as shown in Fig. 3.15 (b). The source inductance is not removed, but there is no longer a feedback path.

80   *Passive components*

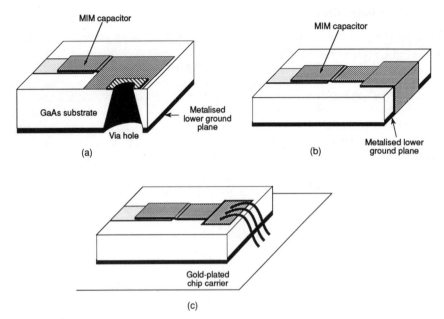

Fig. 3.14 Grounding methods: (a) Through-substrate via-holes; (b) Wrap-around grounding; (c) Bond-wires

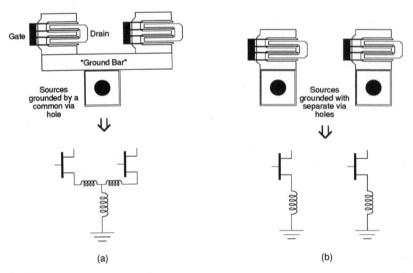

Fig. 3.15 An example of improper grounding: (a) FETs with a common source ground pad, leading to unwanted feedback; (b) FETs with separate source ground pads

A second example of the inappropriate use of a 'ground-bar' is shown in Fig 3.16. The 'ground bar' is used as a neat common ground for the DC bias decoupling capacitors in a two-stage amplifier. However, the distributed nature of the ground bar means that none of the decoupling capacitors are truly grounded, and so a dangerous feedback path is again created. In many cases, when two components share a single via-hole ground even the inductance of the via-hole itself is enough to allow the two components to interact undesirably. This can normally be detected by proper modelling.

As MMICs now operate in the millimetre-wave range even more subtle grounding effects have become apparent, such as the mutual coupling between via-holes placed in close proximity. In this situation, the inductance is much higher than expected as the current crowds together down one side of each via-hole. Hence, because grounding is so critical in millimetre-wave circuits, even the through-substrate via-hole ground may not offer low enough inductance. CPW offers a very attractive solution to this problem, and is becoming increasingly popular as the design tools improve. CPW circuits are described in detail in Chapter 12.

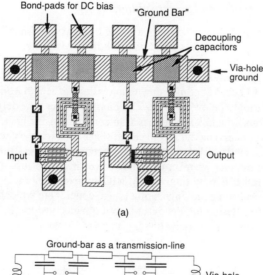

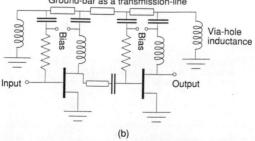

Fig. 3.16 A second example of improper grounding: Bias decoupling in a two-stage amplifier: (a) Layout of the bias networks; (b) Equivalent circuit

3.6 Microstrip components

For many years microwave CAD packages have offered a wide range of models for microstrip lines and discontinuities. However, most of these have been developed principally for MIC circuits on substrates such as alumina and Duroid™. MMICs have much smaller features than MICs and it is important to ensure that the standard models are not used outside their range of validity. The models are further limited because MMICs often have extra dielectric passivation and spacer layers. Many CAD packages have addressed these limitations and now provide models which are suitable for MMIC microstrip elements. However, the problem remains that on MMICs the components are packed more tightly and the coupling between elements must be accounted for. As a result, it may be necessary to employ field-based simulation for some parts of a circuit layout. All these aspects of CAD and modelling are described in more detail in Chapter 4, but it is necessary to highlight the problem here before describing the use of transmission-line components in MMIC design.

3.6.1 Microstrip couplers and power splitters

Microstrip couplers and power splitters are used extensively in MMIC balanced amplifiers, power amplifiers, mixers, and phase shifters. Many well known texts give extensive accounts of the design of Lange, branch-line and rat-race couplers, and Wilkinson power splitter/combiners [1]. For completeness, these components are illustrated in Figs. 3.17 to 3.20. Since they employ lines of length $\lambda_g/4$ or more, they tend to be used sparingly below 10 GHz. The Lange coupler is widely used since it readily provides a broad-band 90° phase split. In comparison, the branch-line coupler has a narrow bandwidth and has an inconvenient shape which is very wasteful of chip space. Nevertheless, the branch-line coupler has been used successfully, particularly in millimetre-wave circuits where it can offer lower insertion loss than the Lange coupler which has rather narrow fingers in MMIC implementations: As a guide, on a 200 μm thick GaAs substrate the fingers of a Lange coupler are typically 20 μm wide with 10 μm gaps. In determining the exact layout dimensions, frequency dispersion and the effect of any dielectric passivation layers must be taken into account [15].

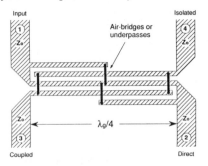

Fig. 3.17 Layout of the Lange coupler

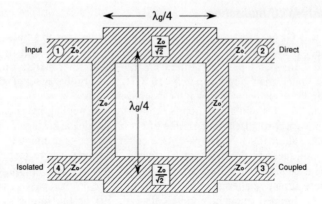

Fig. 3.18 The microstrip branch-line coupler

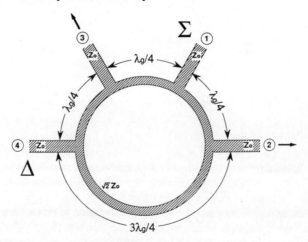

Fig. 3.19 The microstrip rat-race coupler

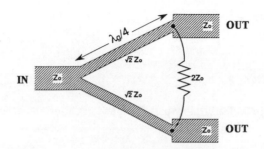

Fig. 3.20 Microstrip layout of the Wilkinson power splitter

3.6.2 Special MMIC realisations

On a 200 µm GaAs substrate, the branch-line coupler, for example, measures approximately 6 × 6 mm² at 5 GHz. It is apparent, therefore, that the standard microstrip couplers and power splitters are not ideal for MMICs operating at frequencies lower than approximately 10 GHz simply because of their size. Miniaturising these elements by using meandered microstrip lines can make a useful size reduction, but this miniaturisation is limited by the coupling between the lines, which will have a particularly undesirable effect on the phase balance of a coupler. There are other alternatives in this case: One could use active techniques, and many excellent techniques are described in Chapter 6. However, active techniques can impose severe penalties in circuit complexity, DC power consumption, and noise figure. A more attractive alternative is to miniaturise the original passive component by employing lumped element equivalents [16, 17] or the lumped-distributed technique.

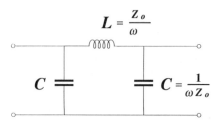

Fig. 3.21 The lumped element equivalent of a quarter-wave transmission line

The inductor-capacitor pi-network of Fig. 3.21 can be used to replace a transmission line of characteristic impedance Z_o which is a quarter wavelength (90°) long at a frequency f if their values are calculated as follows:-

$$L = \frac{Z_o}{2\pi f} \qquad (3.3)$$

and

$$C = \frac{1}{2\pi f Z_o} \qquad (3.4)$$

However, the structure is clearly a low-pass filter, and its cut-off frequency is given by:-

$$f_c = \frac{1}{2\pi \sqrt{LC}} \qquad (3.5)$$

As an example, the equivalent of a 50 Ω quarter wavelength line at 10 GHz would require $C=0.32$ pF and $L=0.8$ nH. Although the lumped-element equivalent network cuts off quite rapidly after 10 GHz, in practice components such as branch-line couplers and Wilkinson power splitters can achieve comparable performance in lumped element form to the transmission-line circuit. The major penalty with this technique is that the insertion loss will tend to be higher, especially at low frequency when multi-turn spiral inductors are required. Fig. 3.22 shows the lumped-element form of the Wilkinson power divider, and Fig. 3.23 shows a lumped-element branch-line coupler, and these have both been used extensively in many MMIC designs.

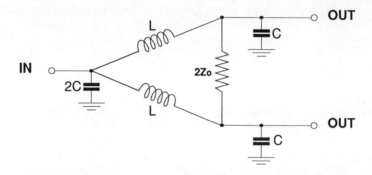

Fig. 3.22 Lumped element Wilkinson power divider

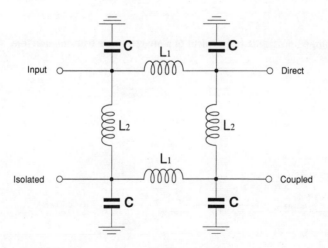

Fig. 3.23 Lumped element branch-line coupler

86 Passive components

In the lumped-distributed or 'reduced-size' [18] technique, each quarter-wavelength line required in the conventional coupler or power splitter is replaced with an equivalent pi-network consisting of a short high-impedance line with a shunt capacitor at each end, as shown in Fig. 3.24. If the characteristic impedance of the quarter-wavelength line to be substituted for is $Z_{\lambda/4}$, and the short line has electrical length θ, then at the design frequency f, the characteristic impedance of the short line is given by:-

$$Z_o = \frac{Z_{\lambda/4}}{\sin(\theta)} \qquad (3.6)$$

and the shunt capacitor value is given by:-

$$C = \frac{\cos(\theta)}{2\pi f Z_{\lambda/4}} \qquad (3.7)$$

From eqn. 3.6 it is evident that the minimum length of the short line is dictated by the maximum value of characteristic impedance that can be used, which is typically 90 to 100 Ω. Notice that when θ is 90° (a quarter wavelength), then $Z_o=Z_{\lambda/4}$ and $C=0$. The lumped-distributed equivalence is exact at the design frequency, and in practice components such as branch-line couplers will have performance very similar to the standard version. A lumped-distributed branch-line coupler is shown in Fig. 3.25.

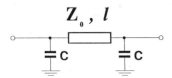

Fig. 3.24 Lumped-distributed equivalent of a quarter-wave transmission line

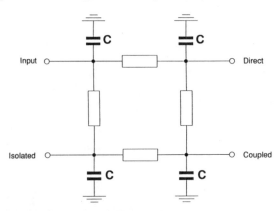

Fig. 3.25 Lumped-distributed branch-line coupler

In microstrip form the lumped-distributed technique is not so attractive since it requires extra via-hole grounds for the shunt capacitors. However, in uniplanar or CPW MMICs the lumped-distributed technique is very attractive since the capacitors can readily be grounded using the ground-planes adjacent to the signal track.

3.7 Other transmission-line media

Until recently, almost all MMIC designs were based on the microstrip transmission-line medium. This is largely because of its convenience, and because so much work has been done to characterise the electrical parameters of microstrip elements. Microwave CAD packages incorporate a large range of microstrip standard components and discontinuities. However, the major disadvantage of microstrip for MMICs is that the coupling (or crosstalk) between adjacent microstrip transmission lines and components requires that they be carefully spaced apart, increasing the chip size and leading to higher costs. Also, components can only be grounded effectively at microwave frequencies by employing through-substrate via-holes. As a result of the effort to reduce chip sizes and increase operating frequency into the millimetre-wave band, there is now considerable interest in the use of alternative transmission-line media. The main transmission-line types of interest are coplanar waveguide (CPW), slotline, thin-film microstrip (TFMS) [19], and coplanar strips. These can be mixed together on the same circuit and the generic term 'uniplanar' MMIC [20] is used to describe such circuits because the circuitry is confined to just the front-face of the chip, and the back-face ground-plane and through-substrate vias are no longer needed. The first 26 GHz-band uniplanar MMIC receiver and transmitter package modules were demonstrated in 1990 by NTT [21]. Each module, including a phased lock loop (PLL)-stabilised local oscillator, was integrated into a single package that measured only one cubic centimetre. Other notable CPW/uniplanar MMICs include a 5-100 GHz Indium Phosphide (InP) distributed amplifier with an average gain of more than 5.5 dB [22], and a Q-band monolithic AlGaAs/GaAs HEMT downconverter with a measured conversion gain between 8 and 11 dB [23]. CPW and multilayer techniques are described in detail in Chapter 12.

3.7.1 Transitions and baluns

Here we refer to the type of transition between two different transmission-line media that one might actually use as part of a circuit, and not to the transition from the circuit to the outside world. With the most common transmission-line media being microstrip, CPW, slotline, TFMS, and coplanar strips, there are many transitions which could be of interest to the circuit designer. The CPW-to-slotline, TFMS-to-slotline, and TFMS-to-coplanar strips transitions have the important feature of providing a transition from an unbalanced (CPW, TFMS) to a balanced transmission line (slotline, coplanar strips). These are often used to realise baluns for balanced mixers and push-pull amplifiers. However, this type of component cannot readily be

Passive components

modelled with the traditional microstrip-based CAD, and field-based simulation is virtually essential for them. The CPW-to-slotline transition is the most popular and can be realised with standard MMIC processing. Other components, such as the 'magic-T' can be designed based around this basic CPW-slotline structure. The TFMS medium is still rather esoteric, but is likely to find increasing popularity as the drive towards higher packing density continues. It can be shown that the TFMS-to-slotline and TFMS-to-coplanar strip transitions are equivalent to the classical Marchand balun [24, 25]. Baluns are described in more detail in the discussion of mixers in Chapter 6.

3.8 References

1. EDWARDS, T. C.: *Foundations for microstrip circuit design*, 2nd Ed., John Wiley & Sons, New York, 1992
2. PENGELLY, R. S.: *Microwave field effect transistors- theory, design and applications*, Research Studies Press Ltd., England, 1986
3. LADBROOKE, P. H.: *MMIC design: GaAs FETs and HEMTs*, Artech House, London, 1989
4. GROVER, F. W.: *Inductance calculations*, Van Nostrand, Princeton, N. J., 1946, Reprinted by Dover Publications, 1962
5. GREENHOUSE, H. M.: 'Design of planar rectangular microelectronic inductors', *IEEE Trans. on Parts, Hybrids, and Packaging*, PHP-10, no. 2, June 1974, pp. 101-109
6. PETTENPAUL, E., KAPUSTA, H., WEISGERBER, A., MAMPE, H., LUGINSLAND, J., and WOLFF, I.: 'CAD models of lumped elements on GaAs up to 18 GHz', *IEEE Trans. Microwave Theory Tech.*, MTT-36, no.2, February 1988. pp. 294-304
7. HILL, A., and TRIPATHI, V. K.: 'Analysis and modeling of coupled right angle microstrip bend discontinuities', IEEE Int. Microwave Symp. Dig., 1989, pp. 1143-1146
8. SCHMÜCKLE, F. J.: 'The method of lines for the analysis of rectangular spiral inductors', *IEEE Trans. Microwave Theory Tech.*, MTT-41, no.6/7, June/July 1993, pp. 1183-1186
9. HOWARD, G. E., DAI, J., CHOW, Y. L., and STUBBS, M.: 'The power transfer mechanism of MMIC spiral transformers and adjacent spiral inductors', IEEE Int. Microwave Symp. Dig., 1989, pp. 1251-1254
10. BECKS, T., and WOLFF, I.: 'Analysis of 3-D metallization structures by a full-wave spectral domain technique', *IEEE Trans. Microwave Theory Tech.*, MTT-40, no.12, Dec. 1992, pp. 2219-2227
11. JAMISON, S. A., PODELL, A., HELIX, M., NG, P., and CHAO, C.: 'Inductively coupled push-pull amplifiers for low cost monolithic microwave ICs', IEEE GaAs IC Symp. Dig., 1982, pp. 91-93

12. FERGUSON, D., BAUHAHN, P., KEUPER, J., LOKKEN, R., CULP, J., CHAO, C., and PODELL, A.: 'Transformer coupled high-density circuit technique for MMIC', IEEE Int. Microwave Symp. Dig., 1984, pp. 34-36
13. WIEMER, L., JANSEN, R. H., ROBERTSON, I. D., and SWIFT, J. B.: 'Computer simulation and experimental investigation of square spiral transformers for MMIC applications', IEE Colloquium on Computer Aided Design of Microwave Circuits, Digest no. 99, Nov. 11th, 1985, pp. V1-5
14. BOULOUARD, A., and LE ROUZIC, M.: 'Analysis of rectangular spiral transformers for MMIC applications', *IEEE Trans. Microwave Theory Tech.*, MTT-37, Aug. 1989, pp. 1257-1260
15. JANSEN, R. H., ARNOLD, R. G., and EDDISON, I. G.: 'A comprehensive CAD approach to the design of MMICs up to millimetre-wave frequencies', *IEEE Trans. Microwave Theory Tech.*, MTT-36, Feb. 1988, pp. 208-219
16. GUPTA, R. K., and GETSINGER, W. J.: 'Quasi-lumped-element 3- and 4-port networks for MIC and MMIC applications', IEEE Int. Microwave Symp. Dig., 1984, pp. 409-411
17. VOGEL, R. W.: 'Analysis and design of lumped- and lumped-distributed-element directional couplers for MIC and MMIC applications', *IEEE Trans. Microwave Theory Tech.*, MTT-40, Feb. 1992, pp. 253-262
18. HIROTA, T., MINAKAWA, A., and MURAGUCHI, M.: 'Reduced-size branch-line and rat-race hybrids for uniplanar MMIC's', *IEEE Trans. Microwave Theory Tech.*, MTT-38, Mar. 1990, pp. 270-275
19. HIRAOKA, T., TOKUMITSU, T., and AKAIKE, M.: 'Very small wide-band MMIC magic-T's using microstrip lines on a thin dielectric film', *IEEE Trans. Microwave Theory Tech.*, MTT-10, Oct. 1989, pp. 1569-1575
20. HIROTA, T., TARUSAWA, Y., and OGAWA, H.: 'Uniplanar MMIC hybrids- A proposed new MMIC structure', *IEEE Trans. Microwave Theory Tech.*, MTT-35, June 1987, pp. 576-581
21. MURAGUCHI, M., HIROTA, T., MINAKAWA, A., IMAI, Y., ISHITSUKA, F., and OGAWA, H.: '26 GHz-band full MMIC transmitters and receivers using a uniplanar technique', IEEE Int. Microwave Symp. Dig., 1990, pp. 873-876
22. MAJIDI-AHY, R., NISHIMOTO, C. K., RIAZIAT, M., GLEEN, M., SILVERMAN, S., WENG, S. L., PAO, Y. C., ZDASIUK, G. A., BANDY, S. G., and TAN, Z. C. H.: '5-100 GHz InP coplanar waveguide MMIC distributed amplifier', *IEEE Trans. Microwave Theory Tech.*, MTT-38, Dec. 1990, pp. 1986-1993
23. TON, T. N., CHEN, T.H, DOW, G. S, NAKANO, K., LIU, L. C. T., and BERENZ, J. : 'A Q-band monolithic AlGaAs/GaAs HEMT CPW downconverter', IEEE GaAs IC Symp. Dig., 1990, pp. 185-188
24. MARCHAND, N.: 'Transmission-line conversion transformers', *Electronics*, Vol.17, no.12, Dec. 1944, pp. 142-145
25. PAVIO, A. M., and KIKEL, A.: 'A monolithic or hybrid broadband compensated balun', IEEE Int. Microwave Symp. Dig., May 1990, pp. 483-486.

Chapter 4

Advanced CAD techniques

R. G. Arnold and D. M. Brookbanks

4.1 Introduction

The penetration of commercial radio frequency systems into the microwave and millimetre-wave bands has put increased pressure on reducing MMIC costs at all stages of production and design. In the field of MMIC design, this translates into successful first-pass design as well as reduced time to production. These needs can only be met by more design stage integration into CAD software and more accurate circuit simulation both for chip yield and electrical performance.

With the advent of powerful 32-bit RISC workstation computers and the recent initiatives in the USA and Europe on MMIC CAD, many new software techniques and systems have become available for interactive design. The main result of these initiatives has been the development of commercial software which integrates the various stages of MMIC design: schematic capture, simulation, and layout. A feature of this software is the availability of user-defined component models. This leads to the possibilities of MMIC manufacturers being able to customise models and integrate the topologies and electrical characteristics of MMIC components specific to their own technology into the software.

Increased computing power has also resulted in the development of electromagnetic field based simulators. The objective of this class of CAD software is to analyse component microwave behaviour from the solution of Maxwell's equations, based on the electrical and physical properties of the metal and dielectric layers defining GaAs, or similar, processes.

In this chapter the status of MMIC design process integration and its availability within major commercial CAD tools is discussed. Emphasis will be placed on the integration of the design process through the use of schematic entry and layout driven techniques, and the advantages of a hierarchical approach to MMIC design will be discussed within the context of making the design process easier to understand and manage as circuit complexity increases.

A discussion on the limitations of foundry-based MMIC design and the rise in importance of electromagnetic field simulation is included. A brief description of the background of electromagnetic field analysis is given with examples of simulator types and their use in MMIC simulation.

4.2 Design integration

All of the new software packages contain sophisticated database management schemes that offer a security of design in a complex system and incorporate report generating procedures that make control and traceability of the design simpler than in the past. We shall return to this point later when we see that the boundaries between the individual parts of the CAD design packages are becoming quite indistinct. This new software represents quite a significant change to the way in which design is undertaken, but also requires new approaches to component characterisation for efficient inclusion in the simulation environment.

The major development that has taken place in MMIC, and microwave, CAD is the concept of design integration where all aspects of a design are considered as part of a larger entity. Some aspects of this can be seen by reference to Fig. 4.1 which shows the relationship between many of the different aspects of the design process. From a software point of view, one of the features that can be exploited here is the use of flexible licenses where we do not have to purchase all of the design suite but only those parts that we need, the software checking the available licenses in and out as required. We may thus purchase a large number of schematic entry frameworks but fewer, or more varied, simulation tools. This concept of flexibility in software licensing can generate significant cost savings for an organisation with a large design team without significantly reducing an individual's access to the design tool that he/she needs at any point in time.

We shall concentrate our discussion in particular on the three aspects of the design process which are most commonly encountered in a MMIC design. System simulation, where we consider the effect of interconnecting individual circuit blocks (which may be individual MMICs) is a significant field in its own right. Our chosen aspects are:

(1) circuit entry
(2) simulation (including optimisation, statistical analysis and yield analysis)
(3) layout.

The question that we shall answer in the following sections is how do we link these three aspects in an efficient manner in order to maximise the benefits to the MMIC design process? So, what features are to be included in this new integrated environment? Whereas previously we would have had linear circuit simulators, non-linear simulators (both frequency- and time-domain) and system simulators, the boundaries between them are beginning to fade as we see single simulation environments becoming available where one can move easily between the different methods of analysis.

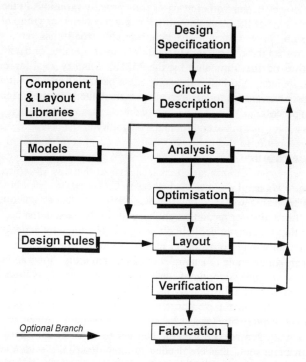

Fig. 4.1 The MMIC design process

4.3 Simulator customisation

One of the features of the major CAD packages that can be used for efficient MMIC design is customisation of the simulator. By this we mean the addition of elements to the simulator that are specific to a MMIC process, or a range of processes. In this way the designer has access to component (both active and passive) descriptions that are known to be valid for the process being used; thus ensuring certainty of design and the ability to remain within the acceptable limits of the process. For maximum efficiency of design these process-specific components will be in a compiled form and be an integral part of the simulator; a method which not only ensures the fastest simulation but also enables the MMIC foundry to maintain control of the component simulation. Many foundries will also supply tolerance or statistical data for these components for use in yield or sensitivity analysis. The designer should be aware of the fact that in many cases the variations within a wafer, or within a wafer batch, may be correlated and less than the published figures for the process. For example, the total measured spread of capacitance may be quoted by a foundry as ± 12% of the nominal value, as a result of measurements made on the process over an extended period of time. However, this variation arises from process variations in the thickness and dielectric constant of the insulating film. For any one wafer, or group of wafers in a deposition batch, this variation will be considerably smaller and so a simulation must be able to reflect these correlated variations across the wafer, or batch.

Customisation of the simulator by the MMIC foundry enables component variations to be included and builds the component parasitics into the simulator in a way which is both transparent to the user and preserves the inherent correlations that exist within the process.

4.4 Schematic capture

Schematic capture is one of the major advances in CAD that has been brought about by the new fast graphic workstations. In fact, some simulators no longer accept design file input in netlist format, although this will still be used in the background to control the simulation. The required circuit design is generated by placing schematic symbols for the individual components on the working page and then wiring the symbols together. An example of a simple schematic circuit is shown in Fig. 4.2. For this method to be of any value, there must be schematic symbols associated with every component that can be placed in the design file. It is in this area that customisation of the simulator is important so that all the foundry-specific MMIC components become available. One of the major advantages of schematic capture is that when a schematic symbol is placed on the working page the designer is prompted for the relevant parameters that are needed to fully describe the component. This immediately overcomes the disadvantage of netlist entry whereby the designer had to remember, or constantly look up, the syntax for the component being included in the circuit. As all the simulators used different syntax for the same function, and some simulators would only accept inputs in a specific order, this became confusing and

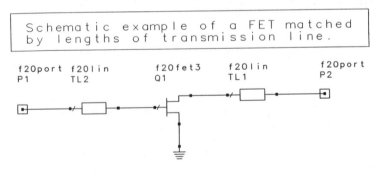

Fig. 4.2 Schematic entry of circuit elements

frustrating for the designer who had to use more than one simulation tool. Entry of the component parameters in a schematic entry system is usually through a series of fill forms (or drop down boxes) that prompt for each parameter required, and can present pre-defined default values for some of the parameters. Often, a descriptive string is associated with each highlighted component which can give the designer an indication of the parameter required (which could be a numeric value or a YES/NO string to include/exclude an option) and the range of acceptable values. Where default values apply, these can already be entered so reducing the number of entries that the designer has to make. When all the appropriate entries have been made the designer closes the fill form and the component will be placed on the working page and the schematic symbol shown with the associated component values. In order to avoid unnecessary clutter on the working page, the display of these component parameters can often be hidden, although the information is of course still available in the simulator's design database. Choosing the MMIC component for placement in the design is also done using drop down boxes or descriptive icons which can be arranged in sets appropriate to a process or subset of a process. The schematic circuit will be completed by the addition of ports to the design so that the simulator can apply the necessary external circuit connections for analysis and presentation of results. In these schematic diagrams the wires that connect the schematic symbols do not represent real wires but represent the nodal connections that will be needed by the simulator to in order to generate the netlist for simulation. If true interconnect wires are needed in the design these must be included as a wire element with an associated electrical model.

Amongst the many benefits of schematic entry of the circuit is the usual representation of the circuit in the form of a conventional circuit diagram which removes the need for paper sketches as a record of the circuit construction. In addition, the nodes are connected together where intended, ensuring that the electrical connectivity is exactly as the designer required. Incorrectly numbered nodes in a netlist are not uncommon and can be difficult to find in a complex circuit. The majority of the schematic capture frameworks will highlight unconnected or floating nodes which can be of considerable help in avoiding unintentional design errors. It is for this reason that there are schematic symbols shown in Fig. 4.2 for the input and output ports so that these are not treated as floating nodes. Similarly, bias inputs have to be treated as special cases.

4.4.1 Hierarchical design

The concept of hierarchical design in a schematic entry framework is in many ways similar to the use of sub-circuits in the netlist approach to circuit design. However, the implementations that are available now extend this approach and make hierarchical design a powerful tool for efficient circuit design. The complete sub-circuit can be given a unique name and it's own schematic symbol which can enhance the readability of the circuit and again help to avoid errors. In a properly constructed schematic framework these sub-circuits can be accessed by a large number of

96 Advanced CAD techniques

independent circuits. In many ways we have the start of a cell based approach to the design, where we can use existing designs as parts of a more complex overall design. Fig. 4.3 shows a simple example of a hierarchical design where the feedback network has been treated as sub-circuits, enabling changes to be made to the amplifier passband response by simply substituting alternative feedback networks from an available list. Hierarchical concepts may be applied to a design simply as a way of making the documentation easier to control by limiting the amount of detail that appears on each schematic page. As we shall see later, the concept of hierarchical design can also be applied to the layout and analysis outputs from the simulator.

4.4.2 Control parameters

The circuit schematic, or one of its associated pages, depending on the implementation, will normally contain any control parameters that are needed in order to simulate the circuit. These must of course include the frequency range of the simulation and the input/output port impedances if these are not the default value, which is normally 50+j0 Ω. In non-linear analysis the required DC bias values will also be entered on the schematic page, and the frequency information will be extended to include harmonic information (for the harmonic balance simulators). Fig. 4.4 shows a stimulus control block for a linear *S*-parameter analysis. Documentation can also be applied to the schematic pages not only to enhance the appearance but also as an aid to revision, or status, control of the design and, if necessary, to explain the concept behind the design as part of an overall quality control plan for the design project.

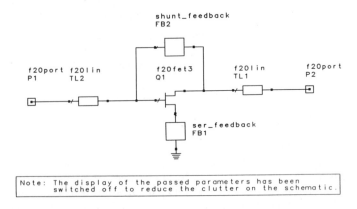

Fig. 4.3 Schematic entry of a circuit with hierarchical entities

In some simulators, control parameters may be placed on the schematic pages to determine the outputs that the simulator will display. In this case we have a predetermined list of outputs from which graphical or tabular displays may be chosen. Some other simulators allow the outputs to be set up after analysis, giving a greater flexibility in the choice of display. It is now usual for the simulators to support post-processing of the results of the simulation and to be able to treat these mathematical manipulations as valid results for presentation. A simple example is the ability to display the difference in phase between two circuits rather than the absolute phase; this feature is obviously of considerable help to those designers who are engaged in phased array developments. As this 'fictitious' circuit can be treated by the simulator as a 'real' circuit we then have the possibility of optimising for phase difference directly, as well as constraining the absolute phase variations if this is required. Other examples of alternative output requirements are circuits used in optical receivers where S-parameters are not the natural choice for the calculation, but instead parameters such as the equivalent input capacitance (which can be determined by normalising the input admittance to the admittance of a 1pF capacitor) or the equivalent input noise current spectral density rather than the noise figure are desired. In these latter circuits the noise figure is often high as the input can be a high capacitive impedance.

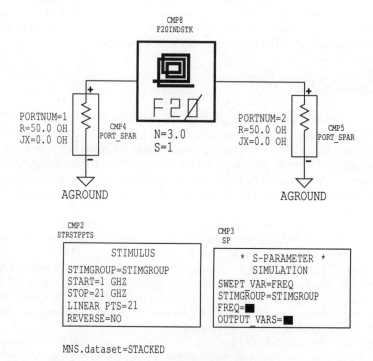

Fig. 4.4 The inclusion of simulation control in a schematic circuit

4.4.3 Back annotation

One of the important features of schematic capture frameworks is the availability of back annotation, which is the ability to update the schematic with the results of an optimisation or yield analysis in the simulator. The designer can then ensure that the schematic correctly reflects the status of the simulation by updating the design database with the new component values that may have resulted from the optimisation or sensitivity analysis. In design suites such as Libra™ and HP-MDS™ which do not use an external netlist, back annotation is good housekeeping, whereas in more open design suites such as Serenade™ and Harmonica™ where a real netlist does exist it is essential that the schematic (Serenade™) is back annotated from the simulator (Harmonica™).

Not all schematic capture frameworks are linked directly to the simulators from a single vendor and some can be used to provide netlists for a range of simulators. The Cadence Analog Artist™ framework, which has bi-directional links to Libra™, HP-MDS™ and Harmonica™, is an example of the high performance realisation of this approach. At a simpler, but very effective, level is the schematic capture capability of the WaveMaker™ suite from Barnard Microsystems which can generate netlists for Libra™, Harmonica™, Linmic™ and Octopus™, and any other simulator which has a Libra™ compatible netlist input. The Serenade™ framework, from Compact Software, is based upon a framework developed for Si ICs and can be used to generate netlists for many of the popular Spice programs that are available, in addition to its links to the Compact Software simulation suite.

4.5 Layout-based CAD

So far we have only discussed the generation of a circuit schematic for the simulation of an MMIC. However, this would be a fruitless exercise if we did not generate a layout from which a set of masks can be made. There are two ways in which we can generate a layout from the electrical circuit which use very different design routes and philosophies. The first method, which has been the method employed for many years, is to analyse the electrical circuit with a circuit simulator (e.g. Harmonica™ or Libra™) and then use an independent layout tool to produce the mask design. A variant on this method is to use an auto-processing tool that generates the layout from the netlist. Suitable layout packages are available from many sources which can be used to satisfactorily generate mask information for MMIC production. WaveMaker™ is an example of a package which can auto-process a range of netlist formats for the generation of layouts, and which has the capability of being customised for application to a particular foundry or process. The completed MMIC design output is usually in the form of a GDSII file, which is an industry standard format that mask manufacturers will accept.

More recently, integrated design suites have become available in which there is a direct link between the electrical circuit and the layout within the one CAD software suite. Examples of this are Academy™ (HP-EEsof), Serenade™ (Compact Software) and the Microwave Design System™ (HP). In addition, the Cadence Microwave Musician™ package has hooks into the EEsof, Compact and HP circuit simulators, bringing the complete activity together within a common framework independent of the provider of the simulation engine. EEsof coined the phrase SMART Libraries (Simulateable Microwave Artworks) for the environment that links all parts of the design process together for a single MMIC process.

The structure of a SMART Library is shown in Fig. 4.5 where there are links between all three parts of the design process, with the final result being a GSDII file that can be sent for mask manufacture. The links between the schematic entry and the electrical simulation have already been discussed, but we also have the additional ability to generate a layout representation directly from the schematic symbols. Once generated, the SMART Library gives the designer the following advantages:

(1) foundry elements are immediately available
(2) simulation can be from schematic or layout
(3) electrical connectivity is implicit
(4) artworks scale directly with component parameters
(5) design is correct by construction
(6) hierarchical design methodology is supported
(7) new components can easily be included
(8) data is back-annotated from layout to schematic.

The inclusion of the foundry elements directly in the SMART Library shields the designer from having to generate models (or include models from equivalent circuits) for the simulation and gives the foundry greater flexibility in the preparation of these models as mathematical methods can be used that may not be available in the simulator. Also, the electrical and the layout representations are generated from the same set of input parameters and thus are known to have the correct relationship. As

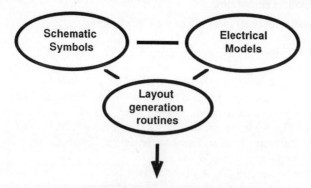

Fig. 4.5 Relationship between the component parts of a SMART Library

the framework is generating the artwork automatically, all the required layers will be present and all the layer spacings within the individual component will be correct. Simulation can be from either the schematic or the layout representation, which gives the designer the freedom to choose the most appropriate method at any particular point in the design cycle.

We have already seen that the schematic capture approach contains an implicit electrical connectivity check and we now have the alternative option of generating the design from the layout representation, which not only ensures the correct connectivity but also that the circuit is topologically valid. This direct link to the layout removes the need for sketches and permits the viewing of potential topological problems at an early stage. In the previous method of MMIC design it would not be uncommon for a design to be fully optimised at the electrical stage and then to find out that this could not be laid out as components would overlap. Of course, the final optimisation still has to take place at the electrical level, but at least the topological constraints can be ascertained at an early stage and the circuit modified accordingly. Back annotation from the layout to the schematic is essential if we are to have this seamless integration. The two representations are normally kept separate and the changes that have been made to either one synchronised to the other when necessary. The benefits of the SMART Library to the MMIC designer can be summarised as:-

(1) Schematic entry:
 -avoids netlist errors
 -no need to remember component syntax
 -default values can be set
(2) Direct link to layout:
 -no paper sketches to lose
 -topological problems can be seen at an early stage
(3) Reduces errors
(4) Reduces design time
(5) Makes more efficient use of the CAD facility.

Figs. 4.6 and 4.7 show screen dumps from a workstation running the HP-EEsof Libra™ framework with both the circuit schematic and layout representations of a a single-stage feedback amplifier shown simultaneously. In this implementation the customised elements are appropriate to the GEC-Marconi Materials Technology F20 foundry and are present as icons on the left hand side of both the schematic and layout windows. Fig. 4.8 is the circuit schematic page for this design and shows all of the components that have been included in the design. These can then be related to the equivalent components on the layout page.

The circuit schematic in Fig. 4.8 shows that a large number of elements have to be included even for a fairly simple layout. Some of these components are present so that the layout can be generated but may not be required in a purely electrical simulation. Other elements are present to ensure the accuracy of the simulation by including the effects of the appropriate junction discontinuities. The component types that are required fall into three broad categories:-

(1) Basic elements:
-capacitors, resistors, inductors, FETs etc.
(2) Geometrical components:
-metal layer interconnects, discontinuities etc.
(3) Input/output components:
-I/O ports, grounds, I/O pads for RF and DC.

The I/O ports and grounds are necessary for simulation but have no associated artworks as they are not part of the MMIC mask layout.

The icons in Fig. 4.8 show that we have a large number of elements available in a SMART Library which are not present in a circuit only simulation. As mentioned above, we require the definition of elements which do not have associated artworks, such as input and output ports, and conversely we require artwork elements that correspond to layout components that are not necessary in a pure circuit simulator. A trivial example of this is the interconnect between the second and third metal layers (M2 and M3). In the circuit simulator these would be simply connected together at a common node, but in a layout the interlayer via must be defined if we are to have a true metallic connection. Many elements also require additional passed parameters for the layout representation that are not required for the electrical model. For example, a spiral inductor has a well defined electrical model but the layout model has to carry the additional information as to whether the spiral is to be wound clockwise or counter-clockwise; electrically these two elements are identical. In a real circuit, connections may be made to adjacent faces of a capacitor as well as multiple connections to a single face (the so called 'junction box') and the layout must be capable of reflecting these features rather than a simple co-linear arrangement of a single connection on each of the M2 and M3 layers.

We can also group elements together to form a layout representation of a commonly used function and to provide enhanced accuracy, and reduced uncertainty, by placing the group as a single entity with passed parameters. One example is the RF on wafer (RFOW) pad group, shown in Fig. 4.7, where we ensure the correct alignment of the ground-signal-ground pads by creating a single entity. The other example in Fig. 4.7 is the decoupling element, for bias connection, of a high value MIM overlay capacitor, a through-GaAs via, and a standard bond pad. This element can be placed with the capacitor dimensions as the passed parameter and the electrical and layout representations will scale accordingly.

The final element that we shall discuss in this section is the four-port FET, which is shown in Fig. 4.9. It is a common circuit design approach to place inductance in the source of a FET to move the optimum source impedance for gain and noise figure to the same point on the Smith Chart. In the independent approach to circuit design and layout this can be achieved by connecting the source of the FET to an inductor in the electrical design and then manually placing the source inductances

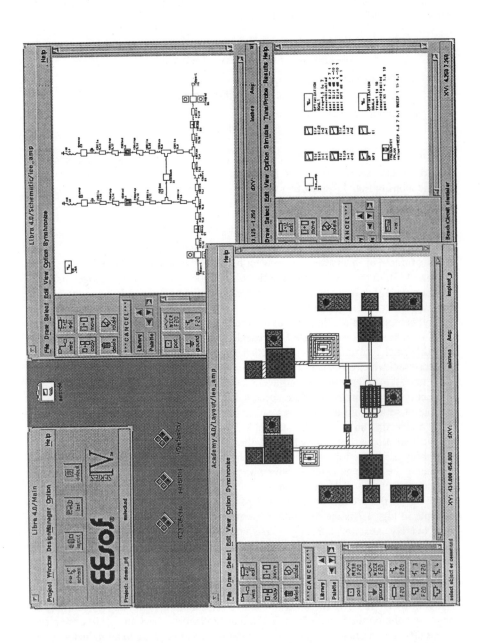

Fig. 4.6 Screendump showing circuit, test and layout windows for a feedback amplifier

Advanced CAD techniques 103

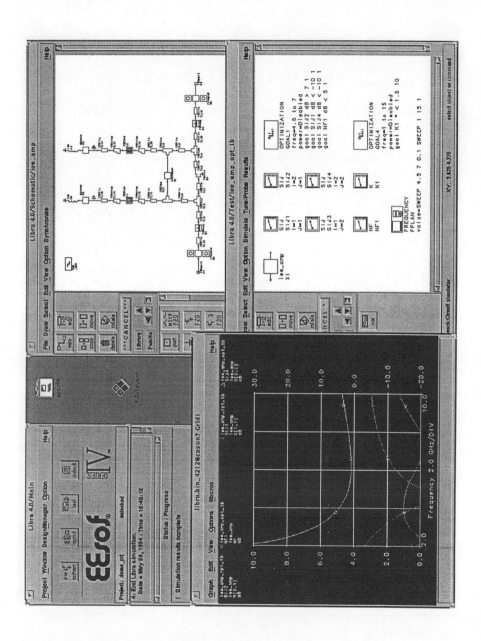

Fig. 4.7 An alternative view of Fig. 4.6 showing a simulation results window

104 Advanced CAD techniques

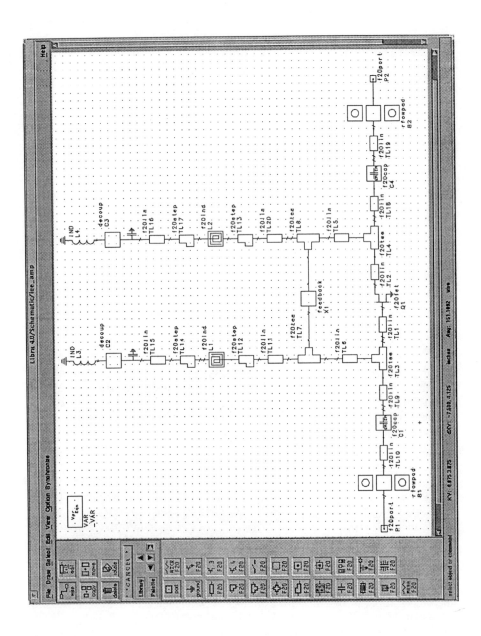

Fig. 4.8 The circuit schematic window of Figs. 4.6 and 4.7 shown in detail

Advanced CAD techniques 105

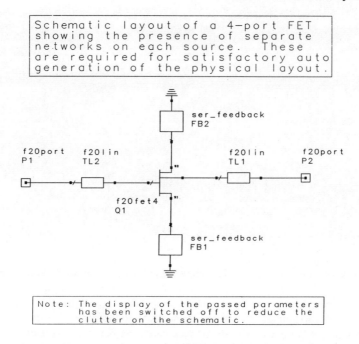

Fig. 4.9 The schematic representation of a dual source FET with feedback

necessary in the layout. However, in a SMART Library the electrical and layout descriptions must be consistent, and so the FET must have a four-port description of gate, drain and two sources. Electrically this requires the generation of the four-port Y and noise correlation matrices for the FET. The model is usually only strictly correct for equal impedances on both sources but it is necessary if we are to be able to include this element in the SMART Library.

4.5.1 Hierarchical design

SMART Libraries support the concept of hierarchical design in the layout representation in a similar manner to the schematic representation that was discussed earlier. In the electrical description this hierarchy is seen as individual circuits, which compare with the sub-networks present in the netlist design approach. When using a SMART Library each of these circuits also carries a full layout description. This is evident in Fig. 4.7, where the feedback network is shown as a sub-circuit in the schematic window and as the appropriate MMIC elements in the layout window. In the completed design database there is a one to one mapping of the hierarchy between the schematic and layout representations. This representation also appears in the GDSII file, which supports full hierarchical descriptions.

106   Advanced CAD techniques

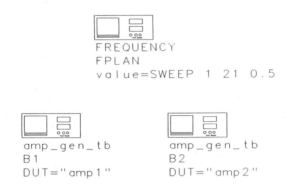

Fig. 4.10 A schematic test lab for the comparison of two *S*-parameter measurements

HP-EEsof also applies the concept of hierarchy to the generation of results from the simulator by the use of a series of schematic Test Labs and Test Benches. The simulation is in fact carried out from the Test Lab, which may contain a series of Test Benches each of which contains the tests (outputs) that are required for that particular circuit. A Test Bench can be seen in the screen dump shown in Fig. 4.7. Fig. 4.10 shows a Test Lab which can be used to compare the outputs of two circuits and may be used in device characterisation where one Bench contains measured *S*-parameters and the other Bench an equivalent circuit model. The Test Lab will then contain the control parameters that are required. When only one circuit is being considered the hierarchical approach is no longer necessary and all the detail can be included on a single Test Bench. Fig. 4.11 shows the detail of the Test Bench from Fig. 4.7 which contains display parameters, frequency plans and optimisation goals.

4.6 Completion of the design

The MMIC design will be completed, prior to sending to the mask manufacturer, by the placing together of all the necessary parts of the mask. These will include any circuits that have been designed (if the mask is to carry more than one design) and foundry-specific entities such as company logos, process control monitors (PCMs) and boxes for the X and Y numbers that give each cell on the mask a unique designator. Whereas all of this can be carried out using the layout capabilities of Libra™, Harmonica™ or MDS™, it is more common to use specialist layout tools such as Analog Artist™ or WaveMaker™. If the designs are being submitted to a foundry it is usual only to supply the hierarchical designs for the circuits so that the foundry can generate the array with the correct placement of the PCMs, numbers boxes, scribe channels etc. The foundry will also wish to carry out DRCs (design rule checks) to ensure that layer spacings are valid and that components have not been placed too close together. Obviously one of the major benefits of the SMART Library is that the chance of errors being found at this stage is minimised as each component should be inherently correct.

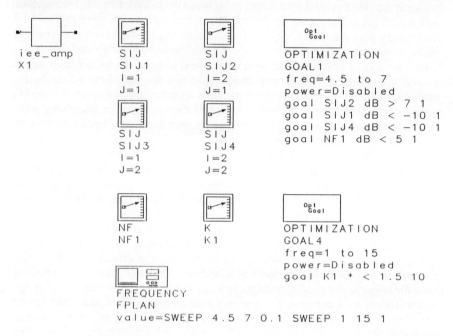

Fig. 4.11 A schematic test bench for the simulation and optimisation of the feedback amplifier of Fig. 4.6

4.7 Limitations in foundry-based MMIC design

CAD of MMICs as carried out on the main commercial simulation and optimisation tools requires user- or foundry-defined electrical models for components. These models cover a range of values for the common passive components; inductors, capacitors, resistors and transmission lines, together with models for active devices such as MESFETs, HBTs and HEMTs. The component values are sufficient to cover the major portion of MMIC designs in the low and medium microwave frequencies. However, the models are limited to a given process dielectric medium, taking the form of the component's main electrical characteristic embedded in a network of parasitic elements. These elements represent the modified electrical characteristics due to the components physical structure when manufactured within the medium, over a finite frequency range. As a result, an MMIC design is limited to a particular process and its response can only be predicted at frequencies below the maximum frequency of validity of the models.

As MMIC operational frequencies approach the high microwave and the millimetre-wave regions, the need to predict component behaviour without resorting to a costly characterisation and modelling programme increases. The inductor-capacitor matching networks employed at the lower microwave frequencies transform into transmission-line matching. Other components such as couplers and splitters increasingly enter into MMIC design. These components tend to have a finite band of usable frequency characteristics and for a particular circuit design are required to have a specific set of dimensions. As a result, these components are not amenable to generalised modelling. Furthermore, transmission-line discontinuities such as bends, junctions and steps in width, start to have a major influence on a circuit's performance. They have to be treated as part of the circuit design, especially when their dimensions become a significant part of the operational wavelength. As a result, their characteristics need to be known.

Implicit in the design of MMICs using a foundry's component library is the assumption that the circuit components will be laid out under the same conditions as when they were characterised. That is, components are spaced out in the layout to ensure no intercomponent electrical interaction. In practice, however, in order to reduce costs per MMIC, there is pressure on the designer to decrease the area taken up by a circuit on a GaAs wafer to a minimum. The result is that component density is pushed as high as possible without compromising circuit performance. Consequently, there will be instances when the electrical effects of two or more components in close proximity needs to be assessed, either for component separation rules for layout or for possible changes in a circuit's response. Another situation that commonly occurs is the meandering of transmission lines to save area. Bends in the line routing and line separation in the meander will have an effect on electrical behaviour.

Another concern is the effect of a packaging environment on a MMIC's performance. Although a circuit may be required to operate within a package, its designed performance is based upon that of the library components' measurement environment which is usually an open one. The type and position of package lid relative to the chip surface may have a significant effect on a MMIC's response. For example, a metal lid will reduce the prime value of an inductor, because of the introduction of a new mirror plane above it, and the use of a plastic lid or covering will increase the parasitic capacitances around a component.

It is these limitations in foundry-based MMIC design, and the requirement to be able to predict the electromagnetic behaviour of circuits and components, that has driven the development of more advanced computer aided design. As computational speed has increased, and with the introduction of more efficient algorithms, computer intensive methods of electromagnetic field analysis have allowed the characterisation of arbitrary transmission-line patterns to be calculated with moderate computation times. These methods have led to the establishment of two- and three-dimensional electromagnetic field simulators and are the foundation upon which process and environmentally independent layout oriented circuit simulation has developed.

4.8 Electromagnetic field simulators

Electromagnetic field analysis simulators are available commercially and are capable of solving Maxwell's equations for two-dimensional and three-dimensional transmission-line problems. The two-dimensional field simulators are used to calculate the modal characteristics and electromagnetic fields for a cross-section of transmission lines embedded between layers of different dielectrics. The frequency dependent modal characteristics generated can be impedances, voltages, currents, powers, propagation velocities and effective dielectric constants. The three-dimensional simulators can calculate the field patterns surrounding a metal structure embedded in various layers of dielectrics and can generate a set of S-parameters over a range of frequencies.

Electromagnetic field simulators fall into two distinct types. The first can solve the general metal-dielectric structure problem. The second type restricts its solution to a class of problems commonly found in MIC and MMIC design, that of infinitely thin metal patterns situated on the interfaces between continuous layers of dielectric. The restriction of the simulator to this category of problem means that faster, more specialised methods of analysis can be used in the software. The result is that the time taken to solution is many orders of magnitude less than for the more general problem solvers. The former type of simulator relies on solution methods which divide the problem space into a point mesh and solve for the fields at the vertices. The second simulator type, in the majority of cases, uses analytical techniques such as the spectral domain approach [1,2] and finite element method to derive the electromagnetic field.

4.9 The spectral domain approach

In the spectral domain approach, the electromagnetic field is formulated in terms of scalar LSM and LSE wave potentials in each of the dielectric layers in the transmission-line medium. This formulation is equivalent to the use of vector wave potentials having only one component in the direction perpendicular to the dielectric medium. These wave potentials are subject to the homogeneous Helmholtz equation in each of the layers. However, instead of considering the analysis in the space domain, a spectral domain equivalent of the Helmholtz equation is considered. This means that for a problem with conducting boundaries, the scalar potentials can be written in the form of an inverse Fourier transform, of dimension one less than that of the problem under analysis. The inverse transform description is dependent only upon the co-ordinate axes parallel with the dielectric surface and takes the form in the three-dimensional case, for each dielectric layer:

$$\phi_e(k_X, k_Y, z) = a_i(k_X, k_Y).\cos(k_{ze}(z - d_i)) + b_i(k_X, k_Y).\sin(k_{zi}(z - d_i)) \qquad (4.1)$$

where z represents the co-ordinate perpendicular to the layer surface, k_X and k_Y are spectral wave numbers and d_i depends upon the satisfaction of the ground conducting and shielding cover boundary conditions. In this formulation a_i and b_i are the spectral weighting functions.

The analysis is performed by obtaining the spectral domain equivalent of the z-dependent part of the electric and magnetic fields in terms of the scalar potentials. The total electromagnetic field, therefore, can be expressed in terms of the spectral LSM and LSE distribution functions $a_i(k_x, k_y)$ and $b_i(k_x, k_y)$. By enforcing the spectral equivalent of the continuity of the tangential electric and magnetic fields at those dielectric layer interfaces not containing metallisation, and representing the discontinuity in magnetic fields at these by tangential spectral surface current densities, the distributions a_i and b_i are either eliminated or expressed in terms of the current densities.

For a mixed stripline and slotline transmission structure, the spectral domain vector equations can be rearranged so that the tangential electric fields are related to the tangential current densities for the strip lines and vice versa for the slot lines by an immittance matrix [3]. In the case of an exclusively strip type structure, this system of equations simplified to:

$$\tilde{E}_T = \tilde{Z}(\lambda)\tilde{J}_T \tag{4.2}$$

where $\tilde{E}_T$ and $\tilde{J}_T$ denote the spectral tangential electric fields and current densities, respectively, and $\overline{Z}(\lambda)$ the immittance matrix. Similarly, for a slotline structure the relationship results in:

$$\tilde{J}_T = \tilde{Y}(\lambda)\tilde{E}_T \tag{4.3}$$

In these equations, $\tilde{Z}$ and $\tilde{Y}$ take a known analytical form dependent upon parameters defining the transmission-line structure, whose matrix constituents can be obtained by a straightforward method known as the spectral domain immittance approach. The two equations can be defined in the space domain by:

$$\overline{E}_T = L(\lambda).\overline{J}_T \quad \text{and} \quad \overline{J}_T = L^{-1}(\lambda).\overline{E}_T \tag{4.4}$$

where $\overline{E}_T$ and $\overline{J}_T$ are the tangential electric field and current density vectors in the plane of the metallisation. L and L^{-1} are linear integral operators with respect to the vectors and operate on the dyadic kernel defined by the spectral domain emittance matrices. As can be seen, there is a duality between the strip and slot line formulations.

The only boundary conditions that need to be satisfied are that the tangential electric field has to vanish on the metallisation, and that the surface current density does not exist in the regions between strips. In the strip line case the surface current density can be separated into an excited term, $\overline{J}_{ex}$, and an impressed source term, $\overline{J}_{imp}$, giving the boundary condition:

$$\overline{E}_T = 0 = L.(\overline{J}_{ex} + \overline{J}_{imp}) \tag{4.5}$$

for the metallised regions. For the slotline structures, a similar excited and impressed term for the tangential electric field can be defined, resulting in:

$$\bar{J}_T = 0 = L^{-1}(\bar{E}_{ex} + \bar{E}_{imp}) \qquad (4.6)$$

These integral equations also define, at the same time, the electric field and current density in the structure's complementary region; the region outside the strip metallisation region in the stripline case and outside the slot region in the slotline case. In transmission-line cross-section and resonator problems, these equations simplify in that the impressed terms are not present, thus defining an eigenvalue problem.

By replacing $\bar{J}_{ex}$ by a series expansion:

$$\bar{J}_{ex} = \sum_k \alpha_k \bar{j}_k \qquad (4.7)$$

in the integral equation, multiplying in turn by each of the expansion functions $\bar{j}_k$ and integrating over the metallisation region, a linear system of equations is obtained:

$$0 = \sum_k \alpha_k <\bar{j}_i, L(\lambda).\bar{j}_k> + <\bar{j}_i, L(\lambda).\bar{J}_{imp}> \qquad (4.8)$$

where $<,>$ defines a scalar product. This relationship translates in the spectral domain to:

$$0 = \sum_k \alpha_k <\tilde{j}_i, \tilde{Z}(\lambda).\tilde{j}_k> + <\tilde{j}_i, \tilde{Z}(\lambda)\tilde{J}_{imp}> \qquad (4.9)$$

The solution of these equations depends upon the type of problem involved.

4.9.1 Transmission-line solution

In transmission-line cross-section type problems, the objective is to obtain fundamental propagation velocities and transmission characteristics for component design [4]. Impressed currents do not occur and the right hand side of the above equations reduce to those terms in the excited current density expansion. The excited current density only exists if the determinant of the linear system of equations is zero. This results in a non-linear eigenvalue problem in λ. For transmission-line problems, this parameter is related to a propagation velocity. On obtaining an eigenvalue, λ, the current distribution can be calculated as well as the tangential electric field outside the metallisation region. The number of fundamental eigenvalues, and thus fundamental mode propagation velocities, to be found depends upon the number of transmission lines in the problem's cross-section.

112 Advanced CAD techniques

In the case of a multi-strip problem, the current distribution obtained for each eigenvalue can be integrated over each strip in turn to obtain a current eigenvector or modal current. The transported power associated with each mode can be calculated from taking the real part of the Poynting vector integrated over the cross-sectional area:

$$P_m = \frac{1}{2}\mathrm{Re}\int_A \overline{E}_m \times \overline{H}_m^* \, dA \tag{4.10}$$

where P_m, $\overline{E}_m$ and $\overline{H}_m^*$ are the modal transported power, electric field and conjugate magnetic field, with A being the cross-sectional area.

4.9.2 Three-dimensional field solution

Early analyses of three-dimensional line problems were performed using the eigenvalue formulation approach [5, 6]. However, it was found that apart from simple discontinuities, such as bends and open ends, the eigenvalue search required in the numerical computation was too time consuming for the analysis of more complex structures. As a result, a source type approach was developed [7, 8], in which for microstrip problems impressed current source regions, and for slotline problems impressed voltage generators, were introduced.

Following on from the above equations, in the case of a microstrip problem:

$$L(\lambda)\,\overline{J}_{ex} = -L(\lambda)\,\overline{J}_{imp} \tag{4.11}$$

in the plane of the metallisation, where there are no source regions. The introduction of source regions translates the formulation into a scattering problem, where the incident impressed and excited tangential electric fields have to cancel on the conducting metal pattern.

The problem is constructed in terms of impressed current regions feeding lengths of transmission lines attached to, and of widths equal to, each of the ports of the structure under analysis. For the characterisation of an N-port structure, N independent excitation states have to be considered. The excited current density is divided into two parts, the first representing the excitation of the feedlines, the second that of the structure to be analysed:

$$\overline{J}_{ex} = \overline{J}_{TR} + \overline{J}_{ST} \tag{4.12}$$

where $\overline{J}_{TR}$ and $\overline{J}_{ST}$ are the feed and structure current densities.

The current densities are represented by a linear set of two-dimensional functions which are not necessarily the same in the current source, feedstrip, and structure regions. The expansion functions chosen are in separable form and consist of one-dimensional functions in the plane of metallisation.

Advanced CAD techniques 113

To illustrate how S-parameters are derived from the expansion function coefficients, the early work of Jansen [9] is considered. Jansen, in order to reduce the number of unknowns in the problem, proposed the use of two expansion function classes in his three-dimensional simulator, which analyses general microstrip structures in an enclosed metal package. The first consists of a current density expansion of the source, feedline and any strip line regions internal to the structure under analysis, in which the transverse current distribution across the strip widths consists of pre-computed expansion functions generated by a two-dimensional microstrip cross-section field solver. This generator computes, for the three-dimensional analysis, model current density distributions in spectral form, strip characteristic impedances and propagation velocities. By choosing the impressed source distribution in modal form, the continuity of the current density from the sources into the feed regions is ensured with a minimum of disturbance. In the longitudinal direction, the current distribution takes the form of current standing waves. The second class of expansion functions is used to describe the current density distribution in regions containing and bordering microstrip discontinuities, junctions, steps, open ends, and bends. In such regions the currents are represented by overlapping one-dimensional piece-wise rooftop functions in the transverse and longitudinal directions. The continuity of the current distribution in the longitudinal direction between different expansion function regions is maintained by overlapping taper functions (semi-rooftop). The topology of a rooftop function used in the current

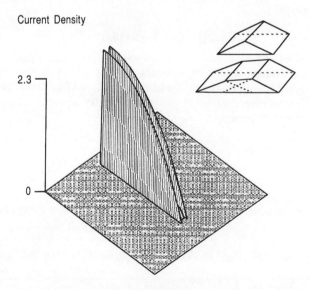

Fig. 4.12 Current density magnitude of a microstrip feed and open end

114 Advanced CAD techniques

density expansion is shown in the inset of Fig. 4.12. As can be seen, it is the product of a triangular function in one direction with a rectangular function in an orthogonal direction. The inset also shows two expansion functions as placed in two rectangular areas in the metallisation pattern, giving a linear piece-wise approximation to the current density in the direction of flow.

After assigning expansion functions to the various sub-regions of the metallisation pattern, the Galerkin method is applied to solve the electromagnetic fields and current distribution. This is done by calculating the scalar products associated with a spectral domain solution. The resulting linear system of equations takes the form:

$$[T][\overline{V}] = [\overline{C}] \tag{4.13}$$

where $[\overline{C}]$ is a matrix of excitation vectors associated with the different excitation states, $[\overline{V}]$ is a matrix of unknown weighting factors representing the current density expansion for each excitation state, and $[T]$ is a matrix of elements which consist of a double summation of spectral terms resulting from the two-dimensional scalar product of the expansion functions with the integral operator lid. The system of equations is solved by inverting the matrix $[T]$ to obtain the spectral domain expansion coefficients.

The extraction of the structure's characteristics is carried out by evaluating the current amplitudes at the user-defined reference planes on the feedstrips. Consider the modal current distribution, J_{nm}, part of the feedstrip expansion; it is proportional to the standing wave:

$$J_{y_{nm}} \sim a_{nm}\cos[\beta_n (y - d)] + b_{nm}\sin[\beta_n (y - d)] \tag{4.14}$$

where m is the excitation state, n the mode, β_n the propagation constant for mode n, and d the reference plane position on the feedstrip, a_{nm} and b_{nm} being the expansion function coefficients obtained from the solution of the linear system of equations. By reformulating this distribution,

$$J_{y_{nm}} = I_{nm}^+ e^{-j\beta ny} + I_{nm}^- e^{j\beta ny} \tag{4.15}$$

is obtained, where I_{nm}^+ and I_{nm}^- are current amplitudes belonging to forward and backward travelling waves. These current amplitudes can be transformed into complex power related amplitudes by multiplying by the modal characteristic impedance Z_n:

$$A_{nm} = \sqrt{Z_n}\, I_{nm}^+ = \sqrt{Z_n}/2\, (a_{nm} + jb_{nm})\, e^{j\beta nd} \tag{4.16}$$

and

$$B_{nm} = \sqrt{Z_n}\, I_{nm}^- = \sqrt{Z_n}/2\, (a_{nm} + jb_{nm})\, e^{-j\beta nd} \tag{4.17}$$

From these modal power amplitudes, column vectors A and B of complex wave amplitudes associated with each excitation can be set up. These vectors can then be combined into two matrices [A] and [B] from which the scattering matrix of the N-port structure can be determined by:

$$[S] = [B].[A]^{-1} \tag{4.18}$$

In current commercial simulators, the current density expansion on the feedline as well as the metallised structure is carried out in terms of rooftop functions [7, 10]. It has been found that this helps with the convergence of the solution. Fig. 4.12 illustrates the current density magnitude results from a three-dimensional analysis of a microstrip feed and open end at 42 GHz. The microstrip is 60 µm wide and 690 µm long on a 100 µm thick GaAs substrate. The figure shows the current density overlap envelope for functions defined on a 15 µm square grid across and along the structure. As can be seen, the current density magnitude increases from the strip centre to its edge and from the open end towards the current source region.

4.9.3 Spectral operator expansion approximation

As the complexity of the structure under analysis increases, the number of unknown expansion function coefficients, and therefore the matrix order to be solved for in the method of solution, increases. In order to decrease computation times, methods of simplifying the calculation have been developed. One of these, the spectral operator expansion technique [11, 12], gives a reduction of between one and two orders of magnitude in the time taken to solve the electromagnetic field problem.

The technique originates from the observation that the higher order spectral terms of the immittance matrix can be approximated by a simpler form without compromising the accuracy in the calculation of the electromagnetic fields in a problem. For strip structures the spectral impedance matrices take the form:

$$Z(k_x, k_y, k_o) = \begin{bmatrix} k_x^2 & k_x k_y \\ k_x k_y & k_y^2 \end{bmatrix} . Z_e + k_o^2 \begin{bmatrix} k_y^2 & -k_x k_y \\ -k_x k_y & k_x^2 \end{bmatrix} . Z_m \tag{4.19}$$

where Z_e and Z_m are the LSM and LSE modal input wave as seen in the medium above and below the metallisation plane. The wave numbers k_x and k_y are the wave numbers associated with the transverse directions in the problem and k_y is the free space wave number. The quantities Z_e and Z_m depend upon the wave numbers k_o and k_{zi}, the latter being associated with the direction perpendicular to the ith dielectric layer in the problem. These two wave numbers are defined in terms of k_x and k_y by:

$$k_p = \left(k_x^2 + k_y^2\right)^{\frac{1}{2}} \tag{4.20}$$

and

116 Advanced CAD techniques

$$k_{zi} = j(k_x^2 + k_y^2 - k_o \, \varepsilon_{ri} \, \mu_{ri})^{\frac{1}{2}} \tag{4.21}$$

where ε_{ri} and μ_{ri} are the ith dielectric layer's electric and magnetic permittivities.

For the two-dimensional transmission-line cross-section structure problem, k_y is the propagation constant, k_y/k_o being the unknown effective dielectric constant. The wave number, k_x, forms an infinite set of discrete values related to the lateral boundaries of the package. This wave number is integer dependent increasing in magnitude with value. For values of k_x much greater than k_y and k_o, it can be seen that Z_e and Z_m have only a weak second-order dependency on the latter variables in the layer. This is the case for many microwave and millimetre-wave problems for spectral contributions $n = N = 10...20$, Z_e and Z_m being replaced for $n > N$ over a wide range of k_y and k_o by their respective second order Taylor expansion:

$$Z(p,q) = Z_o + Z_p \cdot (p - p_o) + Z_q \cdot (q - q_o) + \tfrac{1}{2} Z_{pp} \cdot (p - p_o)^2$$
$$+ Z_{pq} \cdot (p - p_o)(q - q_o) + \tfrac{1}{2} Z_{qq} \cdot (q - q_o)^2 \ldots \tag{4.22}$$

where p and q are the respective variables k_y and k_o, and subscripts p and q imply differentiation with respect to these variables. Z_e, Z_p, Z_q, Z_{pp}, Z_{pq} and Z_{qq} are impedances dependant on the stationary values p_o and q_o. The fixed parameter p_o is chosen as the static propagation constant and k_o that value determined by the highest frequency of interest.

Substituting into the spectral impedance equation results in:

$$Z(k_y, k_o) = \tilde{j}_0(k_y, k_o) \cdot Z_{oe} + \tilde{j}_1(k_y, k_o) \cdot Z_{pe} + \tilde{j}_2(k_y, k_o) \cdot Z_{qe} \ldots$$
$$+ \tilde{g}_0(k_y, k_o) \cdot Z_{om} + \tilde{g}_1(k_y, k_o) \cdot Z_{pm} + \tilde{g}_2(k_y, k_o) \cdot Z_{qm} \ldots \tag{4.23}$$

where the terms Z_{oe}, Z_{pc}... and Z_{om}, Z_{pm}.... represent scalar spectral operators and $\tilde{j}_0$ and $\tilde{j}_1$...., and $\tilde{g}_0$ and $\tilde{g}_1$.... are algebraic dyadics. The space-domain linear integral operator L can therefore be represented by:

$$L \sim L_N + \sum_S \tilde{h}_S(k_y, k_o) L_S \tag{4.24}$$

where L_N is the original operator truncated after the Nth term of its kernel followed by a sequence of operators L_S containing the expanded remainder of the kernel.

The advantage of this spectral operator expansion is that, apart from the truncated operator L, it decouples the space dependence (x, k_x) in the problem from the model propagation constant and free space wave number (k_y, k_o) for the majority of the spectral contributions (typically 98%). As a result, the operators only need to be numerically calculated once in a problem. The resultant near analytical operator representation avoids the repeated full generation of L during the search for the unknown eigenvalue, the propagation constant. The possibility of increasing the number of spectral contributions with little computational time penalty results in an accurate localisation of modal propagation constants in cases of near degeneracy in transmission-line structures.

The spectral operator expansion technique can be applied to three-dimensional problems. In such problems the wave number k_y as well as k_x forms an infinite set of discrete values. These values may be related to boundaries or subsectional expansions. For large integer values of k_x and k_y, the k_{zi} wave number relationship can be written as

$$k_{zi}^2 = -k_p^2(1 - k_o\,\varepsilon_{ri}\,\mu_{ri}\,/\,k_p^2) \qquad (4.25)$$

which approximates to

$$k_{zi}^2 = -k_p^2 \qquad (4.26)$$

showing that the higher order contributions to the solution for a given frequency have a static behaviour, the Helmholtz operator having reduced to that of Laplace. As in the two-dimensional case, the full-wave operator can be expanded into a small leading contribution and a few frequency independent partial operators. The spectral impedance expansion, in the three-dimensional case above certain k_x and k_y value thresholds, is in terms of k_o, the free space wave number about a value appropriate to the centre frequency of analysis. A three term Taylor series expansion is sufficient to obtain accurate results over a multi-octave frequency range.

4.10 Transmission-line simulators

In order to simulate the electrical responses of transmission-line based components, such as inductors and couplers, the characteristics of coupled multi-strip line lengths are needed. To obtain these a two-dimensional field analysis of the multi-strip cross-section has to be performed.

The most sophisticated of the commercial programs that can perform this type of analysis are those included in the LINMIC+™ suite. This set of integrated microwave circuit and structure analysis programmes has two two-dimensional strip cross-section field solvers. The first of these is a single level transmission line analysis programme; the second can analyse a strip structure on two levels. These programmes can analyse a general structure of up to ten strips per metallisation level, or twenty strips for a symmetric structure, the structure being enclosed by top, ground and lateral metal boundaries. The multi-dielectric medium, in which a set of strips can be embedded, can have three dielectric layers above and below for single level strip analysis and two layers above, two between, and two below for a two level strip analysis. The two programmes generate modal frequency dependent strip characteristics by employing the spectral domain approach with the spectral operator expansion approximation to increase the speed and accuracy of computations.

The inputs to the programs are the dielectric constant, tanδ and layer thicknesses of the transmission medium, the metal thicknesses and conductivities of the strips and

118 Advanced CAD techniques

metal package boundaries, together with the strip positions and widths of the strip pattern under analysis. The accuracy of the computation is governed by defining the number of current expansion functions on each strip and the spectral density in the Fourier analysis.

These programmes generate for each mode, the number of fundamental modes and strips being equal, total transported power, conductor and dielectric power loss per unit length, and the normalised strip currents. Although modal powers are calculated, characteristic impedances can be derived for each mode for use in circuit simulators by employing the following definition:

$$Z_i^M = V_i^M / I_i^M \quad \text{with} \quad [V] = [P].[I]^{-1} \qquad (4.27)$$

where Z_i^M is the characteristic impedance for strip i and mode M. V_i^M and I_i^M are the strip voltage and current, [I] and [V] are current and voltage vector matrices and [P] is a diagonal matrix of modal transported powers. This definition of characteristic impedance gives the reciprocity condition required in the admittance, impedance and scattering matrices of lengths of multiconductor line. This definition is better than the power-current characteristic impedance definition which can lead to asymmetry in the above matrices.

As an example of the usefulness of two-dimensional field analysis, Figs. 4.13(a) and 4.13(b) show the physical dimensions of a microstrip for two different types of MMIC technology. The microstrip characteristic impedance and effective dielectric constant for a range of linewidths are shown in Figs. 4.14(a) and 4.14(b), respectively. The electrical characteristic curves A are for a microstrip line on the surface of a 150 µm thick GaAs substrate (ε_r = 12.9), a substrate typical of airbridge microstrip crossover technology. The curves B in the figures, relate to a microstrip line on the surface of a 2 µm polyimide dielectric layer (ε_r = 3.0), covering a 150 µm thick substrate. Such a complex substrate is employed in technologies in which airbridging is not used, but metal vias are produced to connect a metal strip on the GaAs-polyimide interface to underpass the top surface transmission line. As can be seen, the electrical characteristics are very different. The values of the effective dielectric constant of the microstrip on the complex substrate are lower relative to those of the simple GaAs substrate due to the reduction in capacitance per unit length of the line caused by the low dielectric constant polyimide. This can be deduced by calculating the capacitance per unit length from the effective dielectric constant, ε_{eff}, and characteristic impedance, Z_c, values on substitution into the equations:

$$\varepsilon_{eff} = c_o^2 LC \qquad (4.28)$$

and

$$Z_c = \sqrt{\frac{L}{C}} \qquad (4.29)$$

where L and C are the capacitance and inductance per unit length and c_o the velocity of light in vacuo. As the inductance per unit length in the two cases is similar, because the substrate materials are non-magnetic and the substrate thicknesses differ by 2 μm in 150 μm, the capacitance reduction translates into a higher characteristic impedance value for the complex substrate line. An immediate conclusion on comparing these results is that components, such as inductors, manufactured on the surface of the complex substrate medium will have a smaller parasitic capacitance to ground and a higher usable frequency than in the case of the simple air-bridge technology substrate.

In addition to giving characteristic impedance and effective dielectric constant values, the programmes calculate the transverse and vertical electric fields and transverse and longitudinal current distributions on each strip for each mode of the transmission-line structure. The calculations are performed for fields and currents across a horizontal plane at a user-defined vertical position in the package.

Fig. 4.15 shows the physical geometry for an asymmetric two microstrip structure, positioned upon a 150 μm thick GaAs substrate with 1.5 mm layer of air above. The structure has strips of width 20 and 80 μm, respectively, separated by a

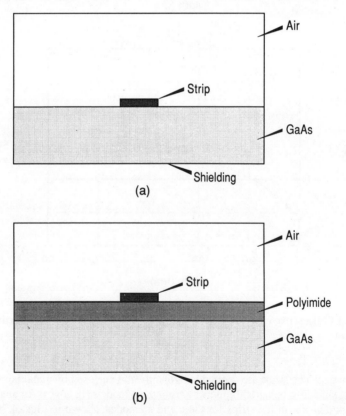

Fig. 4.13 Typical MMIC substrate cross-sections: (a) An air-bridge process; (b) A multilayer dielectric process

120 Advanced CAD techniques

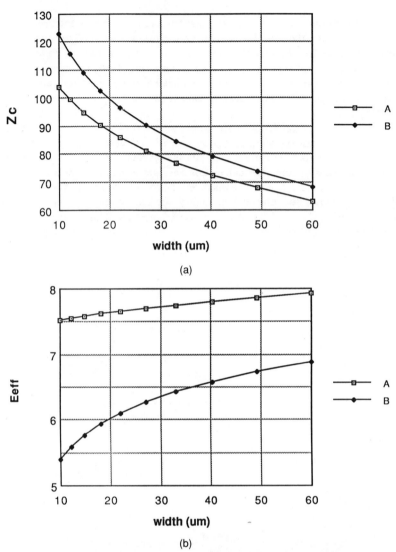

Fig. 4.14 Microstrip characteristics: (a) Characteristic impedance; (b) Effective dielectric constant

40 μm gap. The strips are centred horizontally and enclosed by metal boundaries, the lateral shielding being 600 μm away from the outside strip edges. An analysis of the current density on the strips for the c and π modes, shown in Figs. 4.16 and 4.17, reflects the geometry of the strip structure. Fig. 4.17 shows the current density for the π mode, and it can be seen that this mode shows the inflection in longitudinal current

density magnitude. The figures show that the current density distribution on the strips increases rapidly towards the strip edges, falling to a minimum in the gaps. No current should be expected in the regions between the strips, the residue levels computed being due to the finite expansion that occurs in the analysis.

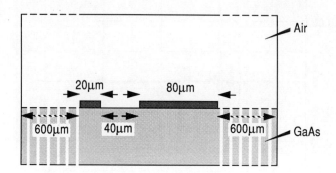

Fig. 4.15 Asymmetric coupled microstrip cross-section

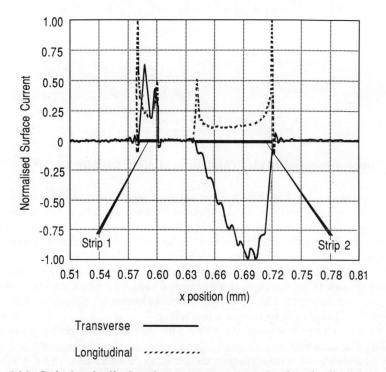

Fig. 4.16 Strip longitudinal and transverse current density distributions on the asymmetric coupled microstrip structure for the π mode

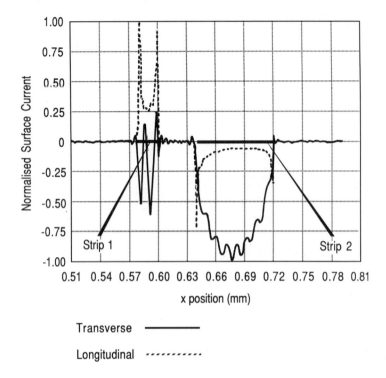

Fig. 4.17 Current density distributions for the c mode

4.11 Three-dimensional electromagnetic field simulators

All of the major commercial microwave software suites have at least one three-dimensional electromagnetic field simulator. These simulators can analyse a microstrip conducting pattern of arbitrary shape in an enclosed metal shielded environment. The conducting structure can be located at the interface between two dielectric layers within a multilayered medium.

The metallisation pattern's geometry is entered into the simulator either by reducing the pattern into rectangular elements and entering their dimensions and co-ordinates into an input file or, in the more sophisticated simulators such as Explorer™ and Sonnet™, by using a layout editor. The metal pattern usually has to conform to a regular rectangular grid defining the position of the rooftop expansion functions. For some problems a regular grid can be a constraint, because the grid size will be governed by the smallest dimension defined in the metal area. This results in a large number of expansion functions, leading to long analysis computation time. Simulators do exist in which the pattern can be defined within an irregular grid, which reduces the number of expansion functions. In order to obtain an accurate characterisation of the metal pattern, the lateral shielding dimensions must be sufficiently large to ensure there is no significant perturbation of the fields in the vicinity of the metal pattern, but small enough to exclude resonant cavity mode propagation.

These simulators can display field intensity and current density plots. Their main application, however, is the calculation of S-parameter data, between user defined reference planes on the metallisation area, for use in circuit design. To illustrate three-dimensional field analysis, the simulator UNISIM™ was used to characterise an open ended stub which has been folded, an arrangement that may be used in order to save GaAs chip area. Fig. 4.18 shows the conductor pattern involved in the analysis.

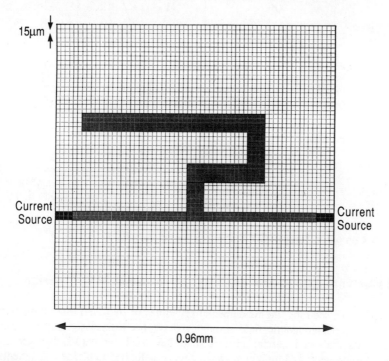

Fig. 4.18 Folded open ended stub conductor pattern showing feed and source areas and analysis grid

The folding of the stub length has resulted in the inclusion of three bends into the conductor pattern. As shown in the figure, the stub can be defined by four lengths of 60 μm wide microstrip. The stub is attached to two equal length feedlines of 30 μm stripwidth. This simulator uses the impressed current method of analysis and so a source region is present at one end of each feedline. The conductor pattern conforms to a 15 μm square grid, which encloses an area 0.96 mm square. The pattern is defined on a 100 μm thick GaAs substrate with 1mm layer of air above. Fig. 4.19 shows the S-parameter characteristics of the stub over the frequency range 2 to 42 GHz. As can be seen, the stub is resonant at approximately 29 GHz, the S_{21} transmission curve showing a classic null at this frequency.

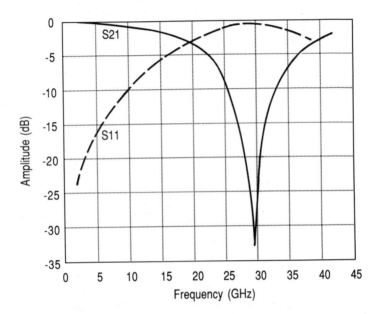

Fig. 4.19 Computed S-parameter characteristics for the folded stub

4.12 Electromagnetic field analysis and its use in MMIC design

In the previous sections, the fundamental principles of two- and three-dimensional microwave electromagnetic field simulators have been discussed and illustrated. Although the main commercial software suites have such simulators, their use in the design of microwave and millimetre-wave circuits is limited to planar components of fixed geometry. Interactive design involving geometric parameter optimisation, especially in the case of three-dimensional structures, has not been contemplated. This is because of the limited computational speed and lack of algorithms able to handle the large number of unknown expansion function coefficients, and thus the large number of linear equations, required to be solved, even in structures of moderate complexity.

Geometric optimisable component elements can be obtained for simple microstrip structures, such as a bend, by using an electromagnetic simulator to generate a data set of electrical characteristics over a range of frequencies and structural geometries based on a particular GaAs process. By fitting the data set to a generic model, a user defined electrical model can be developed for MMIC design and integrated into a simulator, such as Libra™.

The limitation of static geometric simulation is overcome in the software suite LINMIC™. This suite of CAD programmes consists of two-dimensional and three-dimensional electromagnetic field simulators, active and passive component modellers and a circuit simulator. The two-dimensional field simulators generate electrical data for microstrip elements, which are used in the circuit simulator. The circuit simulator's passive component library features microstrip elements of two classes of complexity. The first class of components includes the basic elements: coupled microstrip length sections, coupled corners, multi-strip crossovers and underpasses, bends, gaps, junctions, and microstrip stubs of various types. The second class is of components which contain elements constructed from the basic element set; couplers, planar capacitor, resistor and spiral inductor elements. All these elements are dimensionally dependent and therefore optimisable. Except the for the first three elements, each element's electrical description is based on three-dimensional electromagnetic field analysis.

In the case of the coupled microstrip section elements, two geometric variables are definable across a strip structure as well as the section length. The electrical properties of these elements are obtained for a particular set of geometric values and frequency by interpolating between sets of modal characteristic impedances, effective dielectric constants, powers and eigencurrent vectors found in a look-up table. The look-up table is calculated for a user-defined transmission-line structure cross-section and dielectric medium by a two-dimensional field simulator for a given set of fixed values of the variable parameters and frequency.

The coupled bend and overpass/underpass elements have variable parameters in strip width and gap. Electrical characteristics are calculated by interpolating between sets of capacitances in a look-up table. The look-up table is generated by a quasi-static two-dimensional field solver for each of the above structures, with the capacitances being computed for fixed value sets of strip width, gap, and dielectric medium parameters.

The main philosophy behind this circuit simulator is that it is not process medium specific and circuit design can be undertaken both for single dielectric as well as multi-dielectric substrates. This process independence derives from the circuit description used in the simulator. The simulator relies on a circuit description in which the design and thus layout can be represented by the basic and complex elements in the library. At frequencies at which this segmentation approach is valid, the major influences on circuit performance are the single and coupled microstrip sections, which rely on look-up tables of electrical characteristics, derived from a transmission medium's physical parameters.

To illustrate the segmentation approach, take for example a common microwave and millimetre-wave component, the Lange coupler. This component realises quadrature coupling over an octave bandwidth with a coupling factor of around -3dB. Its layout, shown in Fig. 4.20, can be reduced into three major sections: an input feed, an output feed and a coupled microstrip section. This coupled section can be segmented further into basic elements in which each individual line is terminated by an open end, with alternate lines connected by an underpass or overpass depending upon the metallisation connect layers in the MMIC process. The major basic element is a section of four coupled lines. The feed sections in the structure can be represented by a series of T-junctions connected by short lengths of single microstrip.

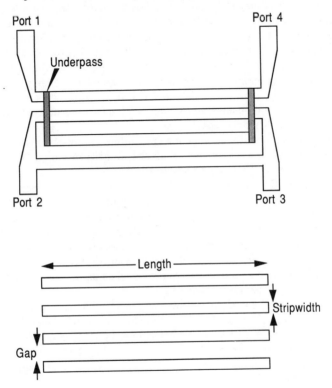

Fig. 4.20 Lange coupler layout

The component's performance depends upon the coupling section, which compensates for even and odd mode phase velocity dispersion. This is determined by the electrical modal behaviour of the four strip coupled line section. This is determined by the section length and the cross-section modal characteristics, which are related to the transmission medium and the geometric parameters, strip width and strip separation.

For a MMIC process described by a 200 µm thick GaAs substrate and 2 µm of polyimide covered by 20 µm of air, the required modal characteristics for the correct coupling lie within the strip width and gap width range 6 to 20 µm assuming metallisation is on the polyimide/GaAs interface. After generating a table of characteristics for a four-strip structure embedded in this transmission medium, optimisation can then be carried out on the geometric parameters, strip width, strip separation and length for the desired coupler specification: 19 GHz centre frequency, 14 to 24 GHz bandwidth and better than 20 dB return loss. The results show that this performance can be achieved with a width of 9 μm, a gap of 7 μm and a length of 1.4 mm. The Lange coupler's characteristics are shown in Fig. 4.21. As is shown, the S_{12} and S_{13} show an insertion loss of approximately 3.75 dB, the extra 0.75dB is caused by losses in the metallisation.

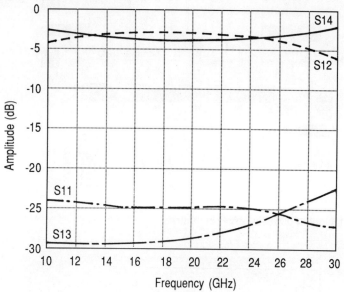

Fig. 4.21 *S*-parameter response for optimised Lange coupler

Fig. 4.22 shows the layout of a travelling wave amplifier designed using the GEC-Marconi Materials Technology Ltd foundry process. This process is multi-dielectric layered with a 200 μm thick GaAs substrate. It has both a silicon nitride and a polyimide layer above the GaAs substrate to support planar capacitors and two layers of metallisation. The design of this amplifier consists of four FETs connected by meandered microstrip transmission lines between the drains and gates of the devices. A simple reactive inductor-capacitor network has been used for the gate and drain biasing. This circuit has been designed using a coupled microstrip section element from the simulator's library to describe the meandered microstrip. Single bend and coupled bend characteristics have been included in the design, their electrical behaviour being ascertained from three-dimensional simulation. The response of this circuit is shown in Fig. 4.23. As can be seen, the bandwidth is approximately 2 to 18 GHz, the gain being around 6 dB. The neglect of bend compensation in the design has shown that the above bandwidth would be 1 GHz lower at the high frequency band edge of the response.

128 *Advanced CAD techniques*

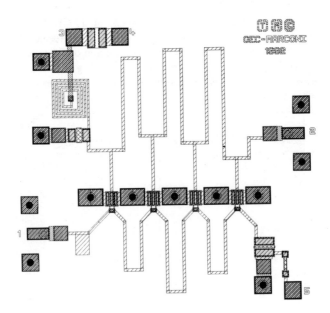

Fig. 4.22 Layout of travelling-wave amplifier

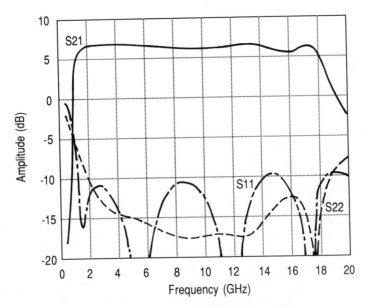

Fig. 4.23 S-parameter response of the travelling-wave amplifier

4.13 Latest commercial software packages

There are many commercially available software packages which can be used for MMIC design, but we shall concentrate our discussion on those which offer the full range of functionality that is necessary for MMIC design. In general terms this means that we shall only consider those packages which offer both linear and non-linear simulation and links to layout. There is little doubt that the most commonly used simulators are those from the USA with the honours being shared between Compact Software, EEsof and Hewlett Packard. The acquisition of EEsof by Hewlett Packard leads to a merging of their respective products with a single user interface in 1996. These companies offer products covering the full range of capabilities that we have been discussing throughout this chapter. For specialist applications we shall discuss two products from Germany: LINMIC™ from Jansen Microwave and Octopus™ from ArguMens. All of these products are available for high performance UNIX workstations. Out of the preceding companies, Compact Software is probably the most active in the PC area, with the suite of programs being available under the Windows environment. For the PC, the linear simulator MMICAD™ from Optotek is also popular. We shall only discuss the UNIX workstation variants of these packages in detail, although the PC version of Harmonica™ only lacks access to user-defined models and the results post processor. To summarise, the available options are:

- **Compact Software range**
    - Harmonica™ — non-linear analysis
    - Super Compact™ — linear analysis
    - Super Spice™ — time-domain analysis
    - Scope™ — optical extensions to Harmonica™
    - Explorer™ — pseudo 3D electromagnetic simulation
    - Serenade™ — schematic capture and layout editor
    - Success™ — system simulation

- **Hewlett Packard range**
    - MDS™ — linear (MLS) and non-linear (MNS) analysis
    - Impulse™ — time-domain analysis
    - Momentum™ — pseudo 3D electromagnetic simulation
    - HFSS™ — full 3D electromagnetic simulation

Note that the Hewlett Packard (HP) simulators operate under the Design Capture System framework which provides schematic entry and links to layout.

- **HP-EEsof range**
    - Libra™ — non-linear analysis
    - Touchstone™ — linear analysis
    - Academy™ — layout editor
    - EMSIM™ — pseudo 3D electromagnetic simulation
    - OmniSys™ — system simulation

130 *Advanced CAD techniques*

Note that the HP-EEsof simulators all operate under a schematic entry framework and time-domain analysis is carried out by the time-domain Test Bench extension (formerly Microwave Spice™).

- **Jansen Microwave**
  - LINMIC+/N™   non-linear analysis
  - LINMIC+™   linear analysis
  - UNISIM™   pseudo 3D electromagnetic simulation
  - SFPMIC+™   pseudo 3D electromagnetic simulation

The LINMIC™ suite has links to WaveMaker™ for schematic entry and layout and can be used in conjunction with other drawing packages such as AutoCAD™.

- **ArguMens**
  - Octopus™   linear analysis
  - Stingray™   pseudo 3D electromagnetic simulation
  - Shark™   layout editor

4.13.1 User (custom) model capability

The three main simulators all offer a user model capability, which we have seen is essential if a SMART Library is to be generated. In all three cases, linear models can be compiled into the simulator for security and speed using either FORTRAN or C code. HP also offer an extensive equation capability on the circuit schematic page which can be used to generate models but can be cumbersome for complex models and is slower as the models have to interpreted at run-time. For non-linear simulation, Compact and HP-EEsof require compiled models for customisation whereas HP have an interpretative model called the Symbolic Device Description (SDD) which is excellent for prototyping. At present this cannot be converted into a compiled model. This customisation of non-linear models is important as the built-in descriptive non-linear models in the simulators are often not sufficiently accurate. Each simulator has its own set of models and a foundry will often have to provide model descriptions for each of the simulators. The optimum model choice (Curtice, Materka, TOM etc.) is often dependent upon the active device being considered.

4.13.2 Time-domain analysis

The major non-linear analysis tool that is offered by the principal vendors is the harmonic balance method (frequency-domain) which integrates well with the linear circuit simulators. However, for strongly non-linear elements or when the time development of a circuit is required, time-domain (SPICE like) analysis is needed. Again, the principal vendors now offer convolution (or impulse) methods which combine the traditional SPICE style analysis with convolution methods to convert frequency dependent components (for example measured S-parameters or dispersive transmission lines etc.) into the time-domain.

4.13.3 System simulation

System simulation is a building block approach to the analysis of complex systems that uses the outputs from linear and non-linear simulators as inputs. These may be power dependent S-parameter files or tables of gain against input power. Inputs can also be from manufacturer's data sheets. The Success™ and OmniSys™ tools are the most comprehensive, with the System model library in HP-MDS™ giving a level of system capability to this simulator. System simulators can use complex waveforms and perform distortion analysis in a way which is not possible for harmonic balance programs and which is only possible for time-domain programs if a full non-linear model is available for each device in the chain.

4.13.4 Electromagnetic simulation and field theoretical methods

The pseudo 3D electromagnetic simulators use the fact that for MMIC design we are interested in the solution of problems that are essentially planar, and this can reduce calculation time dramatically compared to a completely general three-dimensional solution to Maxwell's equations. Explorer™, Momentum™ and Sonnet™ use the method of moments approach to the solution of this problem, which is fast and efficient. Jansen Microwave and ArguMens use field theoretical methods for the solution of the electromagnetic wave propagation and are able to obtain fast and accurate analysis of quite complex transmission-line structures and MMIC components. Examples of the use of LINMIC™ were given earlier in this chapter. By combining a per unit length description of a FET equivalent circuit into the LINMIC™ field theoretical descriptions, Jansen Microwave have been able to generate linear and non-linear FET models which implicitly include the effects of voltage and current distributions within the active devices [13] thus extending the frequency range over which accurate simulation can be achieved.

4.14 Summary

In this chapter we have shown that analysis and working methods are now in place for MMIC design that enable complex structures to be analysed, and it has been shown how the simulators can be customised for a given MMIC process. As a result, the designer is able to use efficient schematic entry routines with direct links to layout and mask manufacture for the accurate and rapid design of MMICs.

4.15 References

1. JANSEN, R. H.: 'The spectral domain approach for microwave integrated circuits', *IEEE Trans. Microwave Theory Tech.*, MTT-33, 1985, pp. 1043 - 1056
2. HARRINGTON, R. F.: *Field Computation by Moment Methods*, MacMillan, 1968.
3. ITOH, T.: 'Spectral domain immitance approach for dispersion characteristics of generalised printed transmission lines', *IEEE Trans. Microwave Theory Tech.*, MTT-28, 1980, pp. 733 - 736
4. JANSEN, R. H.: 'Unified user-oriented computation of shielded, covered and open planar microwave and millimetre wave transmission line characteristics', *Microwave, Optics and Acoustics*, Vol. 3, 1981, pp. 14 - 22
5. JANSEN, R. H.: 'Hybrid mode analysis of end effects of planar microwave and millimetre wave transmission lines', *IEE Proc.*, Vol. 128, pp. 77 - 86
6. KOSTER, N. H. L., and JANSEN, R. H.: 'The microstrip step discontinuity : a revised description', *IEEE Trans. Microwave Theory Tech.*, MTT-34, 1986, pp. 213 - 223
7. RAUTIO, J. C., and HARRINGTON, R. F.: 'An electromagnetic time harmonic analysis of shielded microstrip circuits', *IEEE Trans. Microwave Theory Tech.*, MTT-35, 1987, pp. 726 - 729
8. JACKSON, R. W.: 'Full-wave, finite element analysis of irregular microstrip discontinuities', *IEEE Trans. Microwave Theory Tech.*, MTT-37, 1989, pp. 81 - 89
9. JANSEN, R. H., and WERTGEN, W.: 'A 3D field-theoretical simulation tool for the CAD of mm-wave MMICs', *Alta. Frequenza*, 1988, pp. 203 - 216
10. HILL, A. and TRIPATHI, V. K.: 'An efficient algorithm for the three-dimensional analysis of passive microstrip components and discontinuities for microwave and millimetre-wave integrated circuits', *IEEE Trans. Microwave Theory Tech.*, MTT-39, 1991, pp. 83 - 91
11. JANSEN, R. H.: 'Recent advance in the full-wave analysis of transmission lines for application in MIC and MMIC design', Proc. SMBO International Microwave Symp., 1987, pp. 4667 - 4675
12. JANSEN, R. H., and SAUER, J.: 'High speed 3-D electromagnetic simulation for MIC/MMIC CAD using the spectral operator expansion (SOE) techniques', *IEEE Trans. Microwave Theory Tech.*, MTT-39, 1991
13. JANSEN, R. H., and POGATZKI, P.: 'Non linear distributed modelling ....', Proc. 1991 European Microwave Conference.

Chapter 5

Amplifiers

I. D. Robertson and M. W. Geen

5.1 Introduction

This chapter describes the most important techniques used for MMIC amplifier design. The classical theory and analysis of S-parameters, two-port stability, and unilateral transducer gain are briefly reviewed. The key impedance matching techniques and DC biasing circuits are described before introducing the five major MMIC amplifier topologies, these being the reactively matched amplifier, the lossy match amplifier, the feedback amplifier, the distributed amplifier, and various forms of actively matched amplifier. Finally, the design of low-noise and high-power amplifiers is considered. For further information on the mathematical treatment of microwave amplifier design, the reader is referred to one of the many excellent texts on this subject [1-6]. The justification for not covering this topic in great detail is that MMIC amplifier design techniques are rather different from conventional microstrip MIC amplifier design. This is partly because the size limitations of MMICs have forced designers to develop new topologies, and partly because the low component parasitics on MMICs have enabled more complicated circuit topologies to be used. In addition, the extensive use of CAD has removed the need for many of the original graphical and analytical methods of matching network synthesis and amplifier design [e.g. References. 7 and 8], although a good understanding of the principles of amplifier design is still vital.

Compared with hybrid-MIC amplifiers, MMIC amplifiers have the major advantages of low cost and excellent reproducibility when manufactured in large quantities. However, MMIC amplifier performance can suffer from their small size because this increases the matching network losses and restricts the choice of topology. Furthermore, because of the need for high yield, the active devices on MMICs often have inferior performance compared with their discrete counterparts. This is particularly true of high power devices, where the monolithic device performance suffers from a number of constraints. Generally, therefore, MMIC amplifiers do not achieve the state-of-the-art noise figure and output power performance. However, the removal of bond-pad and bond-wire parasitics in monolithic amplifiers means that very large bandwidths can be achieved, especially with the distributed amplifier. Also, millimetre-wave amplifiers can be designed very easily compared with hybrid MIC techniques, with which the required precision of mechanical assembly is almost impossible to achieve.

The performance of MMIC amplifiers has improved dramatically as high performance MESFET, HEMT and HBT devices have been developed. MMIC amplifiers using HEMTs have been demonstrated at over 100 GHz, and, at the time of writing, HBT amplifiers have been demonstrated at 30 GHz. A general comparison of the different MMIC technologies and devices can be found in Chapters 1 and 2. For amplifier designs, the MESFET finds applications at frequencies up to approximately 30 GHz, and has quite good noise figure performance up to 18 GHz. The HEMT has the best noise figure performance and is currently the best device for low-noise and millimetre-wave applications. The GaAs HBT has demonstrated excellent output power capability, with over 12 W of CW power having been achieved at 8.5 GHz from a monolithic amplifier [9]. The silicon bipolar has continued to improve, and amplifiers operating at over 10 GHz have been reported. Although the silicon bipolar device does have a relatively poor noise figure performance, it is extremely economical to produce in large volumes and can be integrated with CMOS circuitry in BiCMOS technology. SiGe HBT technology is developing very rapidly, and at the time of writing an f_t of 117 GHz has been reported and a device noise figure of less than 1 dB has been achieved at 10 GHz [10]. This high f_t and low noise were not from the same device, and very few SiGe HBT monolithic amplifiers have been demonstrated.

It is certain that device performance will improve further, but the amplifier design techniques presented in this chapter will continue to apply. However, as the gain available from the devices increases, and as different circuit functions become more closely integrated (rather than being individual 50 Ω units), the need for the classical matching techniques is reduced. As a result, designs at low microwave frequencies will use active matching and DC-coupled techniques more and more, because of their high packing density. There will be some exceptions to this trend: in the cases where low noise figure, high output power, or low DC power consumption is the overriding concern. In millimetre-wave applications, where transistor gain is limited, traditional matching techniques, such as stub matching and impedance transformers, will continue to be used for many years.

5.2 Classical stability and gain analysis

The gain provided by a transistor is very dependant on the source and load impedances presented to its input and output terminals. It is the job of the matching networks to ensure that maximum gain is achieved over the required frequency range, and the design of the matching networks is the major part of the amplifier design task. In addition, real devices have significant reverse transmission coefficients which means that the problem of stability must also be considered in amplifier design. In order to analyse the stability and gain of a small-signal amplifier, the transistor can be represented by its S-parameters, and Fig. 5.1 shows the flowgraph representation of a transistor with the impedance presented to its input and output being represented as source and load reflection coefficients Γ_S and Γ_L, respectively.

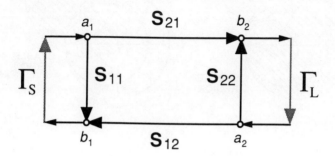

Fig. 5.1 Flowgraph representation of a two-port network

The first point to note is that the presence of the reverse transmission coefficient, S_{12}, means that the load reflection coefficient, Γ_L, will affect the input impedance of the transistor, and that the source reflection coefficient, Γ_S, will affect the transistor's output impedance. Using Mason's non-touching loop rule it can easily be shown that

$$\Gamma_{in} = S_{11} + \frac{S_{21}S_{12}\Gamma_L}{1-S_{22}\Gamma_L} \tag{5.1}$$

and
$$\Gamma_{out} = S_{22} + \frac{S_{21}S_{12}\Gamma_S}{1-S_{11}\Gamma_S} \tag{5.2}$$

There are two important consequences of eqns. 5.1 and 5.2: first, in general the input and output impedances to which one must match are only given by S_{11} and S_{22} when the transistor has $S_{12}=0$, which is not the case in practice. Secondly, there are certain values of Γ_S and Γ_L which can result in Γ_{in} and Γ_{out} becoming greater than unity, and the transistor can have negative input/output resistance. This means that the transistor can become unstable for certain source and load impedances. If we consider the input side, then by setting $\Gamma_{in}=1$ we can establish the boundary condition for stability:

$$\left| S_{11} + \frac{S_{21}S_{12}\Gamma_L}{1-S_{22}\Gamma_L} \right| = 1 \tag{5.3}$$

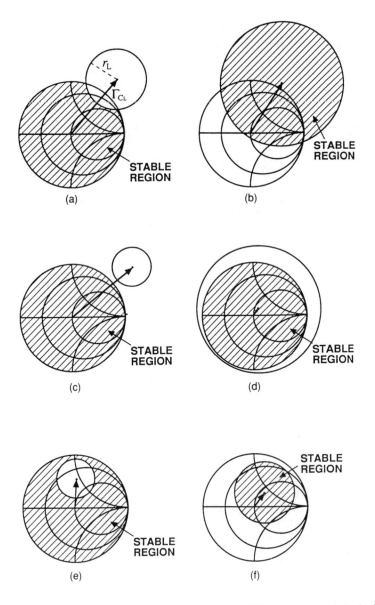

Fig. 5.2 Stability circles on the Smith chart: (a) Stability circle partially inside the Smith chart; (b) Partially inside and encompassing the 50 Ω point; (c) Completely outside; (d) Completely encompassing the Smith chart; (e) Completely inside but not encompassing the 50 Ω point; (f) Completely inside and encompassing the 50 Ω point

It can be shown that this equation can be rearranged to give the equation of a circle whose radius is given by:

$$r_L = \frac{|S_{12}S_{21}|}{|S_{22}|^2 - |\Delta|^2} \qquad (5.4)$$

and whose centre is given by:

$$\Gamma_{C_L} = \frac{(S_{22} - \Delta S_{11}^*)^*}{|S_{22}|^2 - |\Delta|^2} \qquad (5.5)$$

where $\Delta = S_{11}S_{22} - S_{12}S_{21}$.

This circle represents a region in the Γ_L plane (on the Smith chart) inside or outside of which all load impedances will make the transistor input reflection coefficient greater than unity, leading to instability. The circle is known as the load stability circle, and it can be partially inside the Smith chart, completely inside the Smith chart, or completely outside the Smith chart. Fig. 5.2 shows the possible cases graphically. Whether the stable region is inside or outside the circle can be determined by inspecting where the origin (the 50 Ω point) lies, since this point is within the stable region. Similar equations can be found to define the source stability circle: source impedances lying within the unstable region will lead to an output reflection coefficient greater than unity. Since the S-parameters vary with frequency, the centre and radius of the stability circles are different at different frequencies.

It is evident from Figs. 5.2 (c) and (d) that there are cases when there are no source or load reflection coefficients within the Smith chart which lead to instability. This is known as unconditional stability. Unconditional stability means that no combination of *passive* source and load terminations can lead to instability. However, it should be stressed that in multi-stage amplifiers one transistor might present negative resistance to another, and so the stability problem is more severe.

It is very important for the designer to know straight away whether the transistor is unconditionally stable or not: Rollet's stability factor [11] gives an immediate indication of stability and is given by:

$$K = \frac{1 - |S_{11}|^2 - |S_{22}|^2 + |\Delta|^2}{2|S_{12}S_{21}|} \qquad (5.6)$$

If $K>1$ then the transistor is unconditionally stable, and if $K<1$ the stability depends on the position of the source and load impedances relative to the stability circles.

5.2.1 Constant gain circles

The transducer power gain of the transistor, G_T, is defined as the ratio of the power delivered to the load to the power available from the source, and is given by:

$$G_T = \frac{|S_{21}|^2 (1-|\Gamma_S|^2)(1-|\Gamma_L|^2)}{|1 - S_{11}\Gamma_S - S_{22}\Gamma_L + \Delta\Gamma_S\Gamma_L|^2} \tag{5.7}$$

This equation determines how the source and load impedances affect the gain. However, in order to visualise this clearly it is necessary to simplify the equation by making the assumption that the transistor has $S_{12}=0$; i.e. that the transistor is unilateral and has no reverse transmission. This leads to an expression for the *unilateral transducer gain*, G_{TU}, which is as follows:

$$G_{TU} = \frac{|S_{21}|^2 (1-|\Gamma_S|^2)(1-|\Gamma_L|^2)}{|(1 - S_{11}\Gamma_S)(1 - S_{22}\Gamma_L)|^2} \tag{5.8}$$

This can be separated into three factors:

$$G_{TU} = G_S . G_0 . G_L \tag{5.9}$$

where

$$G_S = \frac{1-|\Gamma_S|^2}{|1 - S_{11}\Gamma_S|^2}$$

$$G_0 = |S_{21}|^2$$

and

$$G_L = \frac{1-|\Gamma_L|^2}{|1 - S_{22}\Gamma_L|^2}$$

With the unilateral assumption, and with this separation of the three gain factors, we now find that circles of constant unilateral transducer gain exist in both the Γ_S and Γ_L planes. Fig. 5.3 shows the constant gain circles (in the Γ_L plane) for a typical 300 μm MESFET at 8 GHz. These gain circles can be used to develop a matching strategy for broadband amplifier design. They must, however, be used in conjunction with the stability circles because the higher gain circles will lie either partially or wholly inside the unstable region when the K-factor is less than unity.

When the K-factor is less than unity, the point of maximum gain will be inside the unstable region. In this case we have to accept that the maximum transducer gain is not achievable due to instability. When K is less than unity the maximum gain that can be safely achieved is called the *maximum stable gain* (MSG) and is given by:

$$\text{MSG} = \left|\frac{S_{21}}{S_{12}}\right| \tag{5.10}$$

When K is greater than unity (and the magnitudes of S_{11} and S_{22} are both less than unity [5]) the device is unconditionally stable and the maximum gain that can be achieved is called the *maximum available gain* (MAG), given by:

$$\text{MAG} = \left|\frac{S_{21}}{S_{12}}\right|\left(K - \sqrt{K^2 - 1}\right) \tag{5.11}$$

Typically, a transistor is conditionally stable ($K<1$) at low frequencies, and the maximum stable gain rolls off at 3 dB per octave. At a certain frequency $K=1$, and beyond that the device is unconditionally stable ($K>1$) and the maximum available gain rolls off at 6 dB per octave.

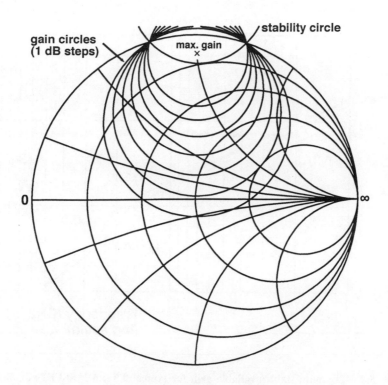

Fig. 5.3 Gain circles (in the Γ_L plane) at 8 GHz for a typical 300μm MESFET

The maximum gain response for typical 0.5 μm MESFETs of different gate-widths is shown in Fig. 5.4. Note that the number of fingers is fixed, and the unit gate-width is varied: apart from some differences due to distributed effects, a 4×100 FET has the same MSG/MAG response as a 2×100 or 6×100 FET, because identical device subsections are being placed directly in parallel. However, at low frequencies a small device has very high impedance and so a large gate-width device is easier to match to 50 Ω. Conversely, at high frequencies a large device has very low impedance and so a small gate-width device is easier to match to 50 Ω. The unit gate-width has a marked effect on the MSG/MAG transition frequency: a small unit gate-width leads to a lower gate resistance and a lower source inductance (in most cases). These changes in the FET parasitic elements are the reason why the 4×150, 4×100 and 4×50 have different MSG/MAG responses in Fig. 5.4. These important subtleties of maximum gain, stability, and device impedance should be closely studied early in the design process for the design frequency and devices of interest in order to find the optimum choice of device geometry.

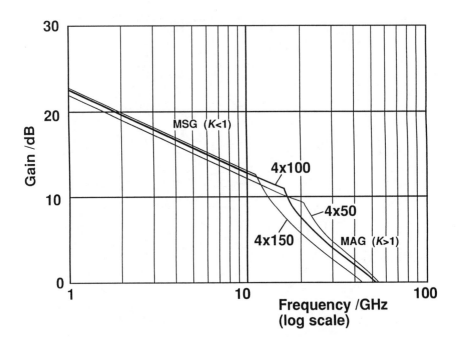

Fig. 5.4 Maximum stable/available gain for typical 0.5 μm MESFETs of different geometries

5.2.2 The practical implications of the theory

The classical analysis has been summarised here in order to illustrate that it is erroneous to think that amplifier design is simply a matter of matching to the transistor's S_{11} and S_{22}. In fact, although it might provide a useful starting point in the design, strictly speaking the designer should *never* match to S_{11} and S_{22}, and the cases of conditional and unconditional stability must be handled differently.

When the transistor is unconditionally stable, the designer is able to conjugately match the input and output in order to achieve maximum gain and good input and output matches. However, because of the transistor S_{12} the reflection coefficients that must be matched to are those given by eqns. 5.1 and 5.2 (and *not* S_{11} or S_{22}). These reflection coefficients, which are only defined when $K>1$, are known as the simultaneous conjugate match reflection coefficients. Hence, eqn. 5.1 gives the transistor's input reflection coefficient when its output is perfectly matched. It is this value which must be plotted on the Smith chart and then manoeuvred into the 50 Ω point with suitable matching elements. The values are most readily obtained from CAD programmes. However, care must be taken to establish whether the CAD programme is displaying the transistor's input reflection coefficient under conjugate matching conditions *or* the required matching network reflection coefficient that must be presented to the transistor's input to achieve conjugate matching. In Touchstone™ the parameters are called the *simultaneous match reflection coefficients* (*GM1* and *GM2*) and the values Touchstone™ gives are those that the matching networks must present to the transistor. Hence, for the matching methods described later in this chapter, the conjugate of the Touchstone™ values must be plotted on the Smith chart. Touchstone™ and other CAD programmes can also display stability and gain circles.

When the device is only conditionally stable it cannot be conjugately matched at the input and output because this would lead to oscillation. It is possible to design a working amplifier by deliberately mis-matching the input and output and staying a safe distance from the instability regions. However, such an amplifier would not be a very useful component as the mismatch would lead to gain ripples, and the amplifier might still oscillate if presented with non-50 Ω source and load impedances. Hence, it is necessary to take steps to stabilise the device.

5.2.3 Amplifier design under conditional stability conditions

5.2.3.1 Resistive loading

A transistor can be stabilised by adding small series resistors or large shunt resistors to its input and/or output as shown in Figs. 5.5 (a) and (b). These lossy elements ensure that the transistor cannot be presented with impedances inside the instability regions, irrespective of what source and load impedances are connected. However, this technique cannot be used for low-noise amplifiers because the resistors will degrade the noise figure. The resistor values can be tuned on the computer until the K-factor is just greater than unity. It is important to note that purely reactive elements cannot change the K-factor.

5.2.3.2 Parallel feedback
The introduction of negative feedback by adding a resistor network from the output to the input as shown in Fig. 5.5 (c) has a very beneficial effect on the transistor stability. In addition, the effect of feedback is to make the input and output impedances more convenient for matching. Feedback amplifiers using FETs are described in Section 5.7. Resistive feedback will, however, degrade the noise figure.

5.2.3.3 Series feedback
Series feedback entails inserting a resistor or inductor into the common-lead of the device. Most commonly an inductor is inserted into the source of an FET, as shown in Fig. 5.5 (d), in order to make the device stable at a lower frequency. With an inductor the noise figure may actually improve, and the noise matching impedance may be brought closer to the power matching point. The special case of the series feedback FET LNA is discussed in Section 5.11.

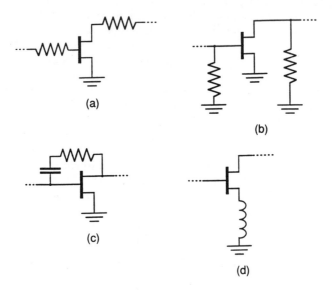

Fig. 5.5 Stabilisation methods: (a) Series resistors; (b) Shunt resistors; (c) Parallel feedback; (d) Series feedback

5.2.3.4 The balanced amplifier
The balanced amplifier is used extensively in MIC low-noise and power amplifiers. In low-noise amplifiers the perennial problem is that matching for low noise figure does not give a good 50 Ω input match. The balanced amplifier solves this problem by terminating the reflected signals in a matched load. In a similar way, if a conditionally stable device is deliberately mis-matched to avoid instability then the

balanced amplifier can be used as a means of terminating the unwanted reflected signals. Fig. 5.6 shows the schematic diagram of a balanced amplifier. Two identical amplifiers are placed between a pair of quadrature couplers (e.g. Lange couplers). The result of this arrangement is that the amplifiers are fed 90° out-of-phase. The reflected signals at the input become 180° out-of-phase after passing back through the coupler, and thus cancel out. The reflected signals at the coupler termination are absorbed. The result is that the balanced amplifier has excellent input and output matches, and the designer is free to optimise the amplifier for stability, flat gain, noise figure etc.

Compared with the single amplifier, the balanced amplifier has the following features:

Noise figure	Single amplifier's noise figure + the coupler loss
Gain	Single amplifier's gain – 2 × coupler loss
Output power	Single amplifier's + 3 dB – the coupler loss
DC power	Doubled
Reliability	Some redundancy: gain drops by 6 dB if one amplifier fails
Port matches	Excellent, easily cascaded without ripple
Stability	Excellent
Intermodulation	Third-order intercept is 9 dB higher

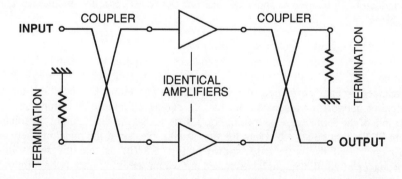

Fig. 5.6 Block diagram of a balanced amplifier

144 *Amplifiers*

5.3 Matching techniques

5.3.1 Lumped element matching

Lumped-element matching networks are attractive because of their small size and their smooth frequency characteristic. MMIC lumped elements are extensively described in Chapter 3. Here, we first show how an arbitrary impedance (such as a transistor input impedance, for example) can be matched to 50 Ω by using two lumped elements. Then we consider how two complex impedances can be matched, and finally how the Q-factor can be controlled by the use of three-element matching networks.

5.3.1.1 L-networks

In principle, any impedance can be matched to 50 Ω by using just two reactive lumped elements. On the Smith chart a series inductance will move the load clockwise along the constant resistance contour. A series capacitance will move the load anti-clockwise along the same constant resistance contour. A shunt inductance will move the load anti-clockwise along the constant conductance circle, and a shunt capacitance will move the load clockwise along the constant conductance circle. As the constant resistance and constant conductance contours are essentially orthogonal, a suitable choice of series and parallel components can move any load impedance into the centre of the chart.

In total there are eight series/shunt inductor/capacitor combinations, and at least one of these will be able to match any specific impedance to 50 Ω. Fig. 5.7 shows the eight combinations that can be used and summarises the required Smith chart operations for series and shunt inductors and capacitors. These networks are often called L-networks, and are the most basic type of lumped element matching network. Often, more than one of the L-networks can be used to match a given impedance and a choice must be made by considering factors such as the value of the components and the convenience of applying DC bias. For example, if it was suitable, network 4 would be convenient for a transistor matching network because the grounded end of the inductor can be used to apply DC bias and the series capacitor doubles up as the DC block.

5.3.1.2 Matching two complex impedances with L-networks

There are cases where two complex impedances are required to be matched, and this can be achieved with a simple extension of the L-network technique. The reactance/susceptance of one of the impedances can either be absorbed into the matching network or resonated out by introducing an additional component. An example of the former case, shown in Fig. 5.8 (a), is that the shunt output capacitance of an FET might be absorbed into the first shunt capacitor of the matching network. An example of the latter case, shown in Fig. 5.8 (b), is that the series input capacitance of an FET might be resonated out with a series inductor (which might in turn be absorbed into the matching network).

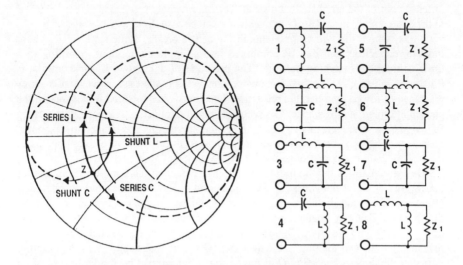

Fig. 5.7 Lumped element impedance matching with L-networks

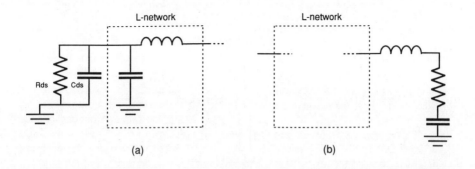

Fig. 5.8 Matching complex impedances with L-networks: (a) Absorb the reactance; (b) Resonate out the reactance

5.3.1.3 T- and π-networks

The major disadvantage of L-networks is that the Q-factor of the circuit is entirely fixed by the source and load resistances. In other words, the designer has no control over the bandwidth of the amplifier with this matching strategy. The use of T- or π-networks overcomes this limitation by using a two stage matching process. The load impedance is first transformed to an intermediate resistance, which can be chosen in order to control the amplifier bandwidth. The intermediate resistance is then matched to the source resistance. The T- or π-network can be visualised as a cascade of two L-networks, where the two components in the centre are the same type (series/shunt inductor/capacitor) and can be combined into a single component, resulting in a three-component matching network. The principle of a T-network, for example, is illustrated in Fig. 5.9. The first L-network, connected to the load, consists of a series reactance followed by a parallel reactance. This transforms the load impedance to the chosen intermediate resistance. The second L-network consists of a parallel reactance followed by a series reactance. This matches the intermediate resistance to the source resistance. The two shunt reactances are combined to form a single component, and the intermediate resistance stage doesn't exist at any real node.

Design equations can be derived in order to calculate the component values for the various T- and π-networks [5]. The source and load impedances and the operating Q are specified, and the equations yield the required component reactances. On the Smith chart, T- and π- matching networks can be designed by superimposing curves of constant Q onto the chart. The extremes of the Q-curves are the real axis ($Q=0$; no reactance at all) and the outside of the chart ($Q=\infty$; no resistance). It is important to note that the overall operating Q of the circuit is determined by the highest Q of all the nodes. The L-networks will give lower Q-factor and more bandwidth than these T- and π-networks. Any move towards the edge of the chart will increase the Q-factor. This is an important point to remember for matching network design: a topology which involves moving near the edge of the chart will have narrow bandwidth, require extreme component values, and be more sensitive to process parameter variations.

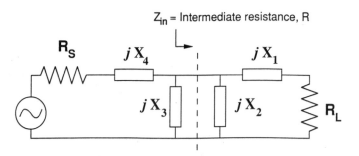

Fig. 5.9 The principle of matching with a T-network

5.3.2 Distributed matching networks

At high frequencies the stray capacitance, spurious resonances and distributed effects of spiral inductors mean that transmission-line matching elements are preferred to lumped elements. The three methods that will be described here are the single stub matching technique, the quarter-wave transformer and the short transformer. There are, of course, many other matching techniques, but the limited space available on MMICs makes many of these unattractive. When using distributed matching elements, it should be remembered that these have a cyclical frequency response, and the amplifier's stability should be checked at the harmonics and sub-harmonics of the design frequency.

5.3.2.1 Series line and parallel stub

In the microstrip medium, series stubs cannot be used because the ground plane cannot be accessed readily. Parallel (shunt) open circuit or short circuit stubs are used instead. This necessitates the use of the admittance chart for some of the operations, and some familiarity with the Smith chart is required to avoid mistakes. The open circuit stub has an input admittance given by:

$$Y_{in_{oc}} = jY_o \tan \beta l \qquad (5.12)$$

and the short circuit stub has an input admittance given by:

$$Y_{in_{sc}} = -jY_o \cot \beta l \qquad (5.13)$$

where β is the propagation constant of the line and is equal to $2\pi / \lambda_g$, λ_g being the guided wavelength. βl is the electrical length of the stub, normally expressed in terms of degrees or fractions of a wavelength, and Y_o is the characteristic admittance of the stub. The input admittance against electrical length for an open circuited stub is shown in Fig. 5.10.

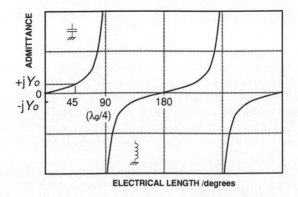

Fig. 5.10 Input admittance against electrical length for an open circuited stub

Note that in principle any value of susceptance can be achieved. This enables the stub to be used as a matching element for any impedance on the chart. As the shunt stub can only change the imaginary part of a load's admittance, an additional transmission-line element is required to adjust the real part. The most common method is to use a series line before the stub, as shown in Fig. 5.11 (a). The design procedure, referring to the Smith chart of Fig. 5.11 (b), can be summarised with an example load impedance of $25-j100\ \Omega$ as follows:

(1) Plot the load admittance on the admittance chart, normalised to the chosen characteristic admittance of the series line; point A (i.e. plot $Y_L \times Z_{0ser}$ using constant conductance and susceptance contours).

(2) Rotate the load admittance around the centre of the chart until it meets the $g=(Z_{0ser}\div 50)$ circle (point B). The electrical length of the series line, βl_1, is half this angle of rotation, giving a series line length of 41° for this example.

(3) The required stub susceptance can now be read by following the constant susceptance contour to the edge of the chart (point C). As the stub has to cancel the susceptance at point B, the opposite sign is used, remembering that in admittance the bottom half of the chart represents positive susceptance.

(4) The value should be de-normalised (divided by Z_{0ser}): in the example, the required stub susceptance is $b_{stub}=-3$, so $B_{stub}=-3\div 70\ \Omega^{-1}$.

(5) From eqns. 5.12 or 5.13 the required stub electrical length can be calculated for the desired stub characteristic impedance and type (short or open circuit). In the example, for an open circuit stub:

$$Y_{0_{stub}} \tan \beta l_2 = B_{stub}$$

$$\therefore \text{stub electrical length, } \beta l_2 = \tan^{-1}\left(B_{stub} \times Z_{0_{stub}}\right) = \tan^{-1}(-0.0428 \times 50)$$

This gives a stub length of 115°, after adding 180° to the negative result from the calculator. The design can then be converted to physical dimensions, and effects such as T-junctions and bends incorporated into the simulation. It is important to use the Smith chart first because the matching problem can be solved with many other combinations of stub length and position, which means that the CAD optimiser will have many local error function minima to contend with. Furthermore, some of the alternative solutions will involve longer lines, narrower bandwidth, and excessive sensitivity.

This stub matching problem can be solved with other graphical techniques, but if the above steps are followed exactly then mistakes will be avoided. The series line and stub can have arbitrary characteristic impedances with this method, which is important because 50 Ω lines are rarely used in MMICs due to their excessive width.

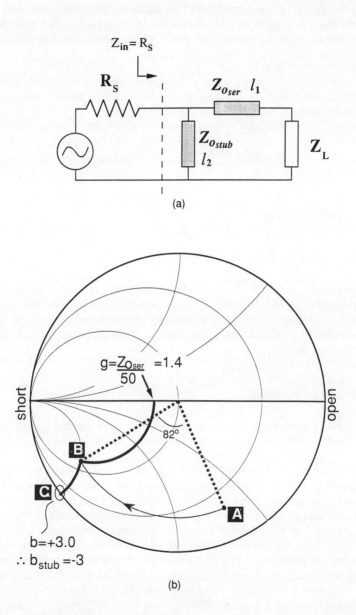

Fig. 5.11 Single stub impedance matching: (a) Network; (b) Smith chart operations

5.3.2.2 The quarter-wave transformer

This well known technique can match a load resistance, R, to Z_0 using a series transmission line one quarter-wavelength long (90°) with a characteristic impedance of $Z_T = \sqrt{Z_0 \times R}$. Since a transistor's impedance is rarely purely real, first a reactive element must be used to resonate out the imaginary part. This can be achieved with either a series or shunt inductor/capacitor or with a transmission-line stub.

5.3.2.3 The short transformer

The short impedance transformer technique can be used to directly match a load impedance $R+jX$ to Z_0 by selecting the transformer characteristic impedance, Z_T, and electrical length, θ, such that:

$$Z_T = \sqrt{Z_0 R - \frac{X^2 Z_0}{Z_0 - R}} \tag{5.14}$$

and

$$\theta = \tan^{-1}\left(\frac{Z_T(Z_0 - R)}{X - Z_0}\right) \tag{5.15}$$

provided that $X^2 < R(Z_0 - R)$ and $R \neq Z_0$.

This matching technique is more compact than the quarter-wave transformer, but often requires unrealistic values of Z_T. A useful solution to the length and width problems encountered with the quarter-wave and short transformer is to convert the transmission line into an equivalent lumped or lumped-distributed π-network, as described in Chapter 3.

5.4 DC bias injection

The transistor must have DC bias applied to set the operating point on its I-V characteristic about which the AC microwave signal swings. Fig. 5.12 shows schematically the I-V curves of a generic FET and indicates the four most common operating points used for amplifiers. V_{ds} is the drain-source voltage, V_{gs} the gate-source voltage, and I_d is the drain current. I_{dss}, the saturated drain current, is used as a reference current and is the drain current obtained when the FET is biased at its specified drain voltage and with V_{gs}=0 V. Point I represents the low noise operating point; with the FET biased at 3 V (or lower for many HEMTs) and 10 to 15 % of I_{dss} the best noise figure performance is obtained, along with low DC power consumption. However, because the gate is biased near to pinch-off, the gate voltage swing, and therefore the power handling, is limited. At point II the FET is operated with V_{gs} close to 0 V and I_d is close to I_{dss}. At this bias point the FET gives its maximum small-signal gain. However, the high DC power consumption and increased noise figure make this operating point unattractive. Point III is placed almost exactly in the middle of the I-V curves, and this operating point allows the maximum linear output power to be achieved for class A operation. For higher efficiency, but degraded linearity, operating point IV can be used for class AB or class B operation of the amplifier.

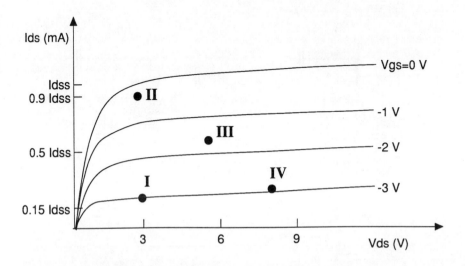

Fig. 5.12 Typical FET I-V curves and operating bias points

The gate and drain DC bias voltages can be applied to an FET in a number of ways, as illustrated in Fig. 5.13. In method (a) inductors act as bias chokes, with DC blocking capacitors used at the input and output to isolate the bias from other circuits, and with decoupling capacitors to prevent the leakage of RF signals into the power supplies. In this bias arrangement the drain voltage is required to be positive (e.g. +3 V), and the gate voltage is required to be negative (e.g. −1 V). If inductors are introduced as stand-alone bias chokes, then considerable chip area will be wasted. Hence, when designing the matching networks it is very convenient to find a topology which has shunt inductors, so that bias can be applied through them.

In method (b) high value resistors are used to apply the bias voltages. With the exception of low-noise and high-power amplifiers this is very suitable technique for the gate bias, since the gate draws no current and the resistor can be used to stabilise the device. However, on the drain side it is likely that the drain current will be too high to allow resistor biasing unless the device has a very small gate-width. For example, even a modest 20 mA drain current would lead to a 6 V drop and 120 mW

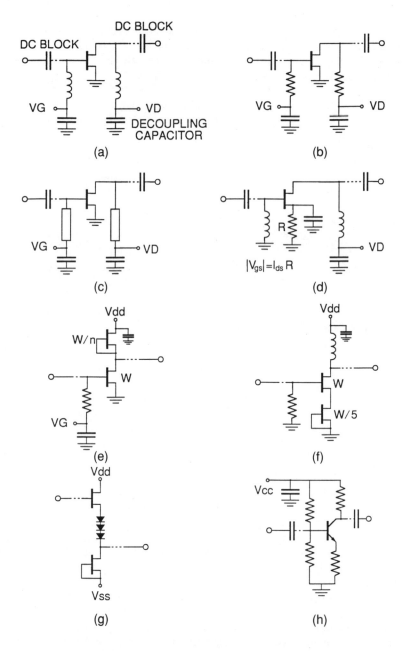

Fig. 5.13 DC bias networks: (a) Inductors as bias chokes; (b) High value resistors; (c) Microstrip stubs; (d) Self-biasing; (e) Active load; (f) Constant current source self-biasing; (g) DC coupling; (h) Bipolar transistor biasing

dissipated in a 300 Ω drain bias resistor. The use of smaller bias resistors will start to lower the amplifier gain considerably.

In method (c) short circuit microstrip stubs are used. The short circuit end is grounded with a decoupling capacitor so that DC can be applied. As stand-alone elements these bias lines would have to be a quarter wavelength long, and this would be too large in most cases. However, if the stubs are part of the input and output matching networks then the technique is quite attractive. The decoupling at the end of these matching stubs is very critical, and care must be taken to avoid oscillations.

Method (d) is known as the self-biasing technique and enables the FET to be biased from a single power supply rail. The gate is grounded at DC through either an inductor or high value resistor. The source is raised to a positive DC potential equal in magnitude to the desired gate-source voltage by inserting a small resistor in the source (whose value is equal to the required V_{gs} divided by the desired drain current). In order to prevent a loss of RF gain the source is grounded with a large decoupling capacitor. The advantage of the single power supply is considerable, particularly for battery operation, but there is a penalty because of the slightly increased DC power consumption. One drawback of the technique is that it removes one useful post-fabrication tuning mechanism from the amplifier, because the bias is fixed. Also, the amplifier is likely to be more sensitive to process variations such as changes in pinch-off voltage. Some foundries have facilities for laser trimming of resistors, and this would solve these problems (at a cost). Alternatively, a number of different resistors can be connected to the source and the optimum one selected during testing. Fig. 5.14 shows a photograph of a single-stage feedback amplifier using the self-bias technique: a very large decoupling capacitor can be seen on the FET source, this being necessary to maintain good low frequency performance.

Method (e) is the well known active load technique, in which the active load has $V_{gs}=0$ and hence runs at its saturated drain current. Therefore, if the active load FET has a gate-width of $W_g/2$, then the amplifying FET with gate-width W_g runs at a drain current of $I_{dss}/2$. The active load presents a large resistance in parallel with a small capacitance. It does cause some reduction in gain as a result, but is very compact and suitable for high packing density and DC coupled amplifiers. It is particularly useful for IF amplifiers, where bias choke inductors would be too large. However, with FETs the standard active load has been shown to be very sensitive to the gate-source voltage, and an improved active load has been proposed for MMIC applications [12]. Method (f) is the constant current source self-bias technique: the gate of the amplifying FET is at 0 V, but its source is raised to a positive DC voltage by the drain-source voltage of the current source FET. Thus the gate-source voltage is negative and the amplifying device runs at the I_{dss} of the current source FET.

Method (g) is to employ level shifting diodes to drop the DC output voltage of one stage down to the input voltage of the following stage. This technique is used for DC-coupled amplifiers and is discussed later. Finally, the bipolar transistor should not be forgotten as it is of ever increasing importance to MMICs. The bipolar transistor can be controlled via its base-emitter voltage with many of the same techniques described here for FETs. However, the base-emitter junction is forward biased, and so a positive supply is required (for an npn transistor). Also, the base

154 *Amplifiers*

current must be carefully controlled, with at the very least a current limiting resistor. Generally, the standard bipolar bias configuration shown in (h) is used. An emitter stabilisation resistor can be added if desired.

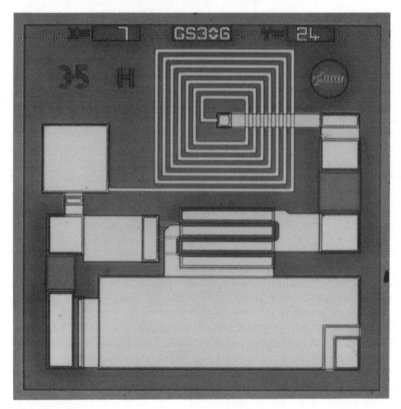

Fig. 5.14 Photograph of a single-stage amplifier employing the self-bias technique

5.4.1 Stacked bias

There are many applications where low DC current is more important than the ability of the MESFET or HEMT to operate at low drain voltages of 2 or 3 V. Such applications might include battery operated communications terminals in which the supply voltage is fixed but where ultra-low DC power consumption is required. For this type of application the stacked bias technique is very attractive. This is best illustrated by studying the three-stage amplifier circuit diagram shown in Fig. 5.15. This amplifier uses both self-biasing (evident on the source of the first stage's FET) and stacked biasing. The drain current from the third stage's FET passes from its source into the drain of the second stage FET, and then from the second FET's source into the drain of the first FET. This is achieved by having the sources grounded at ac. with decoupling capacitors and by using inductors to connect successive sources and drains at DC. Hence, if each FET is required to operate at 3 V drain voltage with 10 mA drain current ($V_{gs} \approx -1$ V) then the complete amplifier requires a single 10 V supply and will draw only 10 mA. The first stage FET has ~+1 V on its source, and 4 V on its drain. The second FET has 4 V on its source and 7 V on its drain. The third FET has 7 V on its source and 10 V on its drain. The amplifier only requires a single supply rail and draws a minimum of current. The operating points of the transistors can be adjusted during RFOW testing by control of the first FET's gate bias voltage and by the inclusion of select-on-test self-bias resistors. Fig. 5.16 shows a 1.5 to 2.3 GHz two-stage low-noise amplifier which employs this biasing technique.

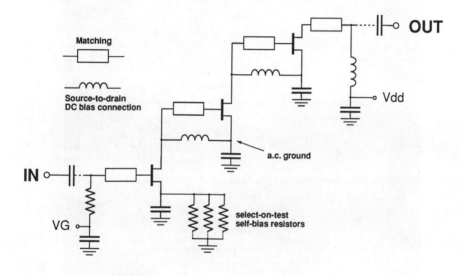

Fig. 5.15 The stacked bias technique used in a three-stage amplifier

5.4.2 Off-chip components

On-chip bias decoupling capacitors are limited in value to approximately 20 pF because of their large area. However, this is rarely enough to ensure that the amplifier is unconditionally stable at frequencies below 1 GHz and so off-chip decoupling capacitors and other bias circuitry usually have to be included. It is very important that this is considered early in the design procedure and included in the simulations.

156 *Amplifiers*

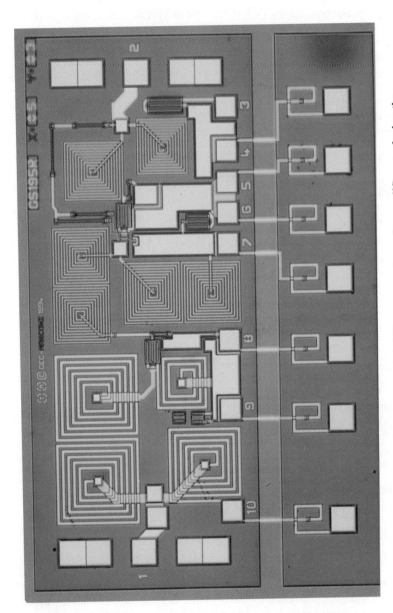

Fig. 5.16 Photograph of a 1.5 to 2.3 GHz two-stage low-noise amplifier employing the stacked bias technique (provides 22 dB gain with a 3 V, 4 mA supply)

For example, it is often found that an off-chip decoupling capacitor has to be placed directly next to the chip and connected with very little bond-wire inductance, and this may have an influence on the layout of the DC bias bond pads. Alternatively, it may be the case that off-chip components cannot solve the stability problem and that on-chip resistive loading is the only solution. As a result, it is very unwise to send a chip for fabrication thinking 'I can worry about the off-chip bias circuits later'.

When an off-chip decoupling capacitor is wire-bonded to the MMIC, you can be assured that its capacitance will resonate with the bond-wire inductance at some frequency. Such resonances must be kept out of band and closely scrutinised using CAD for signs of possible instability. Fig. 5.17 shows the typical arrangement of an MMIC amplifier mounted in a package with off-chip bias components.

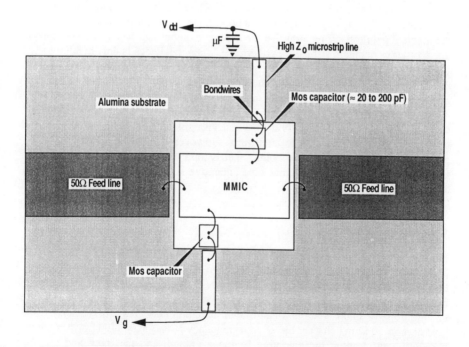

Fig. 5.17 Packaged MMIC amplifier with off-chip bias components

5.4.3 RFOW testing considerations

For RFOW testing of an amplifier it is not essential that it is unconditionally stable at all frequencies because the amplifier will be terminated with the 50 Ω impedances from the network analyser. However, it is of course essential that the amplifier does not go into oscillation during measurement! The inductance of DC probe needles is very high, however, and so it is very difficult to add effective extra decoupling capacitance to the supply lines. The amplifier must be modelled to take the RFOW scenario into account to make sure that low frequency oscillations will not result from this probe needle inductance.

5.5 Reactively matched amplifier design

The reactively matched amplifier uses lossless matching networks, with either lumped or distributed elements. Since the matching networks are lossless the reactively matched amplifier can be designed for optimum gain, noise figure, and output power. For a single stage narrowband amplifier using a transistor which is unconditionally stable, the design is a matter of matching the input and output simultaneous match reflection coefficients to 50 Ω, using the techniques described in Section 5.3, and designing suitable bias networks. However, for broadband amplifiers the inherent gain roll-off requires that the input and output matches have to be traded off in order to maintain flat gain. Furthermore, when the device is only conditionally stable, the input and output cannot be matched to 50 Ω because of instability. Hence, the disadvantages of the reactively matched amplifier are that it is difficult to achieve good input/output matches, flat gain, and good stability. This makes the reactively matched amplifier difficult to cascade, unless isolators or a balanced amplifier topology are employed. In hybrid-MIC design, isolators or balanced stages can readily be used for multi-stage amplifiers. However, for MMIC amplifiers it is preferable to use an amplifier topology that is more amenable to cascading: more suitable techniques are lossy matching, negative feedback, and the distributed amplifier.

5.5.1 Multi-stage design

In multi-stage reactively matched amplifiers the two main problems are stability and gain ripple. In a multi-stage design the interstage matching networks can be designed to give a positive gain slope which counteracts the gain roll-off of the transistors. Fig. 5.18 shows the block diagram of a generic two-stage amplifier. If the devices are unconditionally stable, the input and output matching networks are synthesised to give a good 50 Ω match over the desired frequency range: maximum power is thus transferred throughout the operating frequency range, and the input and output matching networks have a flat frequency response. However, the transistors themselves, shown as FETs, will have a 6 dB/octave gain roll-off. Hence, in order to achieve flat overall gain, the interstage matching network must introduce a

12 dB/octave positive gain slope. For a reactively matched amplifier the matching networks are lossless, and so this gain slope must be achieved by introducing frequency-dependent mismatch between the output of the first stage and the input of the second. If the transistor output and input impedance are represented as simple resistor-capacitor networks, the required interstage network can be synthesised from first principles by considering the poles and zeros needed to give the desired insertion loss response [2]. Nowadays, many CAD programmes exist which can perform filter and matching network synthesis, such as E-SYN™ by HP-EEsof. However, great care must be exercised in defining the desired frequency response parameters: these types of synthesis programmes will readily provide solutions which are overcomplicated and require extreme element values.

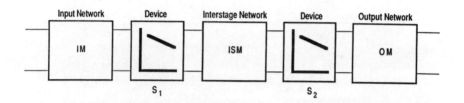

Fig. 5.18 A generalised two-stage amplifier

Stability is a major problem in reactively matched multi-stage amplifiers. The overall K-factor of an amplifier is not a sufficient indication of the amplifier's stability. For example, an intermediate stage may have $K<1$, and the stage before it (which is not necessarily matched to 50 Ω) may present it with an impedance which is in an unstable region. Interstage oscillations can thus exist which cannot be diagnosed from the overall two-port S-parameters. This is particularly serious at low frequencies, where the device has high gain and $K<1$. The stability of each amplifier stage should be investigated individually. Furthermore, as discussed in Chapter 3, unwanted feedback paths can result from improper grounding and poor decoupling of the DC power supplies. Fig. 5.19 shows a photograph of a two-stage X-band amplifier which employs reactive matching with a single inter-stage stabilising resistor. Note that this design features many of the good grounding practices described in Chapter 3.

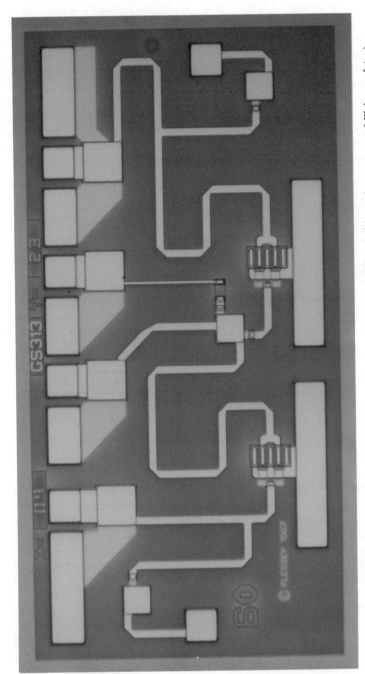

Fig. 5.19 Photograph of an X-band two-stage reactively matched amplifier (with an interstage stabilising resistor)

5.6 Lossy matching

The lossy match amplifier uses resistors within its matching networks to enable flat gain to be achieved over a broad bandwidth. The most typical topology is to employ resistors in series with high impedance stubs on both the input and output, as shown in Fig. 5.20 (a): At low frequencies the stubs have little reactance, and the resistors load the transistor and lower its gain. At high frequencies the stubs have high reactance (going to infinity when they are a quarter-wavelength long), and the resistors have little effect on the transistor. Hence, the matching networks can introduce a positive gain slope to compensate the transistor's gain roll-off without resorting to mismatching. The lossy match amplifier has quite high gain, which is very flat, and has good input and output matches. In addition, the resistors can greatly ease the stability problem at low frequencies. The disadvantages of this approach, compared to the reactively matched amplifier, are that it has lower gain, lower output power, and higher noise figure.

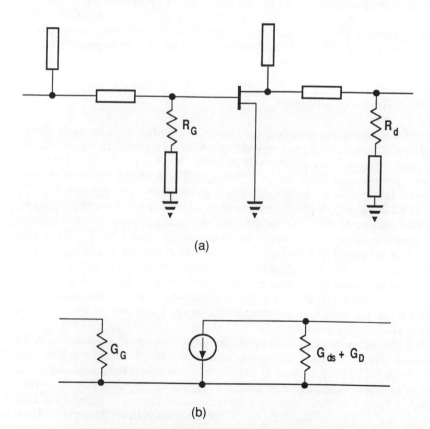

Fig. 5.20 Lossy matching: (a) Basic topology; (b) Low frequency model

From the low frequency model of a lossy match FET amplifier, shown in Fig. 5.20 (b), it can be shown that [13]:

$$S_{11} \approx \frac{1-G_G Z_0}{1+G_G Z_0}, \qquad (5.16)$$

$$S_{22} \approx \frac{1-(G_{ds}+G_D)Z_0}{1+(G_{ds}+G_D)Z_0} \qquad (5.17)$$

and

$$\text{Gain} \approx \left[\frac{g_m Z_0}{2}(1+S_{11})(1+S_{22})\right]^2 \qquad (5.18)$$

where G_{ds} is the drain-source conductance ($1/R_{ds}$) and G_D is the lossy match drain loading conductance. From eqn. 5.18 it can be seen that if $G_G=(G_{ds}+G_D)=1/Z_0$ (i.e. the gate loading resistor is 50 Ω and the drain loading resistor in parallel with R_{ds} gives 50 Ω), then $S_{11}=0$, $S_{22}=0$ and the gain (in dB) is equal to $20\log(g_m Z_0/2)$, which gives typically 8 dB for an 800 µm MESFET. This low frequency model clearly shows how the gate-width of the FET determines the low frequency gain of the lossy match amplifier in the ideally matched case.

5.7 FET feedback amplifier

The FET feedback amplifier is a very common solution for wide-band MMIC amplifiers. The technique gives very flat gain with good input and output matches and can achieve moderate power levels (hundreds of mW). It is highly suitable as a general purpose gain block and as a wideband IF amplifier for millimetre-wave systems. The basis of the technique is that negative feedback is applied to the FET (MESFET or HEMT) by connecting a resistance (of the order of hundreds of ohms) from the drain to the gate. This has the effect of stabilising the device and can make the input and output impedances much closer to the desired 50 Ω. This is very beneficial because FETs can have very low input resistance and high output resistance at low frequencies, and matching these to 50 Ω is rather difficult because of the limited range of component values provided by MMIC lumped elements. Through the use of feedback the matching networks can be kept simple [14]. Furthermore, the amplifier performance becomes less sensitive to process-related variations of the FET parameters.

For the best response, a number of additional components are required in the amplifier. Fig. 5.21 shows the schematic of a matched feedback amplifier [15] suitable for use up to approximately 10 GHz. Each component has a particular role to play in achieving maximum performance: R_{FB} is the key feedback element, and its value determines the basic gain and bandwidth, which have to be traded off against one another. L_{FB} introduces a degree of frequency dependence into the feedback loop; at the lowest frequencies it has no effect and R_{FB} controls the gain level, but at high frequencies the reactance of L_{FB} increases, which reduces the amount of negative feedback. Hence, the effect of L_{FB} is to maintain flat gain and give operation up to a higher frequency. L_D is chosen to compensate for the C_{ds} of the FET; the overall effect is to provide gain peaking at the upper edge of the frequency response, and this provides extended bandwidth. Although the feedback network achieves input and output impedances quite close to 50 Ω, there is still a need for additional matching elements to give good input and output return loss. These matching elements are L_{in}, C_{in}, and C_{out}. Finally, C_{FB} is a DC block, which is required to isolate the positive drain bias from the negative gate bias. This completes the basic matched feedback amplifier. In practice the matching networks may have to be slightly more complicated than this, and after the basic design has been established the whole circuit should be optimised using CAD. Bandwidths as high as 1 to 10 GHz and 6 to 18 GHz have been achieved [16, 17]. Fig. 5.22 shows a photograph of a basic 1 to 6 GHz feedback amplifier.

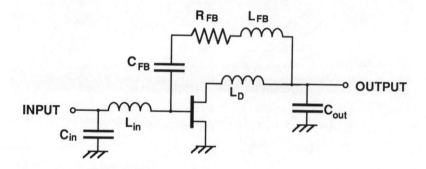

Fig. 5.21 The matched FET feedback amplifier

5.8 Distributed amplifiers

In the distributed (or travelling-wave) amplifier the problem of achieving a broadband match to the transistor input and output impedance is overcome by incorporating the input and output capacitances of a number of transistors into artificial transmission-line structures. The amplifier thus consists of an input line, incorporating the input capacitances of the transistors, and an output line incorporating the output capacitances. The signal on the input line is amplified and fed into the output line. The result of this arrangement is that the amplifier can operate over a very broad frequency range, from very low frequencies up to the cut-off frequency of the artificial transmission lines. The technique was originally patented by Percival [18] as a technique for realising broadband amplifiers with thermionic valve devices. In the

164 *Amplifiers*

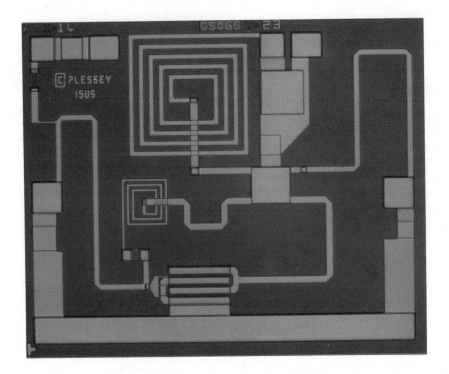

Fig. 5.22 Photograph of a basic 1 to 6 GHz feedback amplifier

early 1980s the technique suddenly became very popular with the realisation that GaAs MESFETs could be used in a monolithic distributed amplifier to achieve decade bandwidth microwave amplification [19, 20]. Its real strength is that with MMIC technology the distributed amplifier can achieve extraordinarily wide bandwidths with a simple circuit topology that is insensitive to process variations.

In the most common configuration the capacitances become the shunt capacitances in a constant-K low-pass filter ladder network. In the case of the FET distributed amplifier there is a gate-line and a drain-line, as shown in Fig. 5.23. The input signal travels down the gate-line, exciting each FET in turn, before being absorbed by a terminating resistor. The transconductance of the FETs amplifies the signal and feeds it into the drain-line. If the phase velocities in the gate-line and drain-line are roughly equal then the signals from each FET will add constructively at the output port. At high frequencies the signals will largely cancel out at the reverse end of the drain-line, but this is not the case at low frequencies. A drain-line termination absorbs any undesired signals present at the reverse end.

Amplifiers 165

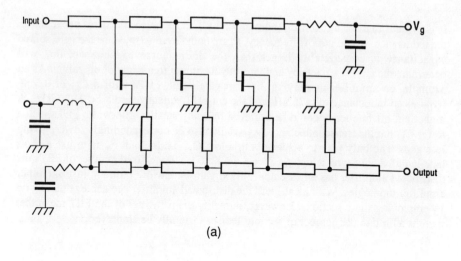

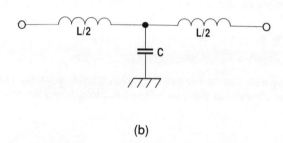

Fig. 5.23 The distributed amplifier: (a) Circuit diagram; (b) A single section of constant-K ladder

The basic design of a FET distributed amplifier can be carried out very easily by considering the characteristics of the constant-K ladder. The characteristic impedance of the artificial transmission line is given by:

$$Z_0 = \sqrt{\frac{L}{C}} \qquad (5.19)$$

and its cut-off frequency is given by:

$$f_c = \frac{1}{\pi\sqrt{LC}} \qquad (5.20)$$

It is evident that for a given Z_0 (normally 50 Ω) and a given transistor capacitance, the inductance, L, and cut-off frequency, f_c, are both fixed. Also, since the gate-source capacitance is considerably larger than the drain-source capacitance, this will determine the cut-off frequency and thus the maximum frequency of operation. As an example, an amplifier employing 300 μm gate-width FETs with a C_{gs} of 0.3 pF requires an inductance of 0.75 nH, and the cut-off frequency would be 21.2 GHz. A higher cut-off frequency could be achieved by using smaller gate-width FETs with a lower C_{gs} but the transconductance, g_m, would also be correspondingly lower, giving less gain. The only way to achieve both a low C_{gs} and a high g_m is to use a short gate-length device with a high f_t (since $f_t = g_m/2\pi C_{gs}$). For a fixed number of FETs the distributed amplifier has an essentially fixed gain-bandwidth product, just like other amplifier topologies. With ideal FETs the distributed amplifier can achieve more gain by employing more sections. However, in reality the parasitics of the FET mean that there is a limit to the number of sections than can usefully be employed.

5.8.1 Gate and drain-line losses

In the case of ideal lossless constant-K lines, it has been shown [20] that a distributed amplifier with n sections has a gain given by:

$$G = \frac{n^2 g_m^2 Z_0^2}{4} \qquad (5.21)$$

In the ideal case the gain can be increased by adding more sections. However, in practice the number of sections that can be used is limited by gate and drain-line losses.

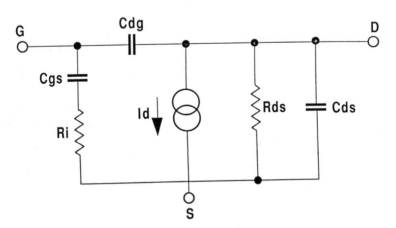

Fig. 5.24 A simplified FET equivalent circuit

Referring to the simplified MESFET equivalent circuit in Fig. 24, the resistance R_i of each FET will absorb some of the signal travelling down the gate-line, and the resistance R_{ds} will absorb some of the signal travelling up the drain-line. It is found that the drain-line loss is essentially constant with frequency, whereas the gate-line loss increases rapidly with frequency and is the more dominant limitation in most cases. The result of gate-line loss is that successive FETs along the gate-line receive a steadily decreasing level of input signal. The effect is so pronounced that after a certain number of sections the input signal becomes so weak that there is no benefit in adding more sections: there is found to be an optimum number of sections for given levels of gate and drain-line loss. Additional stages beyond this optimum number add virtually no signal to the drain-line because their input signal has been attenuated so much. Furthermore, these additional sections add more and more drain-line loss, so they are actually detrimental to the performance.

The effects of gate and drain-line losses have received considerable attention in the literature. Conventionally, the analysis is presented in terms of the gate and drain-line angular cut-off frequency, ω_c, and in terms of the FET's intrinsic cut-off frequencies ω_g and ω_d. The gate-line loss is given by [21]:

$$A_g = \frac{(\omega_c/\omega_g)(\omega/\omega_c)^2}{\sqrt{1-\left[1-(\omega_c/\omega_g)^2\right](\omega/\omega_c)^2}} \tag{5.22}$$

and the drain-line loss is given by:

$$A_d = \frac{\omega_d/\omega_c}{\sqrt{1-(\omega/\omega_c)^2}} \tag{5.23}$$

where $\omega_g = 1/R_i C_{gs}$, $\omega_d = 1/R_{ds} C_{ds}$,

and $\omega_c = 2/\sqrt{L_g C_{gs}} = 2/\sqrt{L_d C_{ds}}$.

Gate and drain-line losses against normalised frequency (from ref. [21]) are plotted in Fig. 5.25 for different relative values of ω_g and ω_d. The gate-line losses are generally found to limit the bandwidth and cause high frequency gain roll-off, whereas the drain line losses effect mainly the low frequency gain level. When these losses A_g and A_d are incorporated into the analysis, the gain of the distributed amplifier is found to be [20]:

$$G = \frac{g_m^2 Z_o^2}{4} \frac{\left[\exp(-A_g n) - \exp(-A_d n)\right]^2}{(A_g - A_d)^2} \tag{5.24}$$

168   Amplifiers

and the optimum number of sections has been shown to be [21]:

$$N_{opt} = \frac{\ln(A_g / A_d)}{A_g - A_d} \qquad (5.25)$$

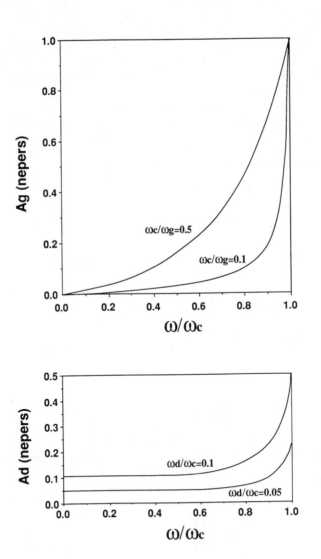

Fig. 5.25 Gate- and drain-line losses against normalised frequency [21]

In practice, with typical MESFETs and HEMTs the optimum number of sections has been found to be four or five, although as many as seven sections have been used to good effect. Fig. 5.26 compares the predicted frequency responses of 3, 4 and 5 section distributed amplifiers using typical 0.5µm MESFETs. Fig. 5.27 shows a photograph of a 2 to 20 GHz distributed amplifier with four sections. A frequency range of 0.5 to 20 GHz is typical of that which can be achieved with a 0.5µm MESFET MMIC process. The low frequency coverage is determined by the design of the bias networks, which would certainly require off-chip components to get down to 100 MHz, say. Realising 'simple' DC blocks over such multi-decade bandwidths is no small task, as the large capacitors required for low frequency operation will have very large parasitic elements which will destroy the high frequency performance. One of the most popular solutions to this problem is to mount the chip capacitors in a suspended-microstrip environment so that their shunt capacitance to ground is reduced.

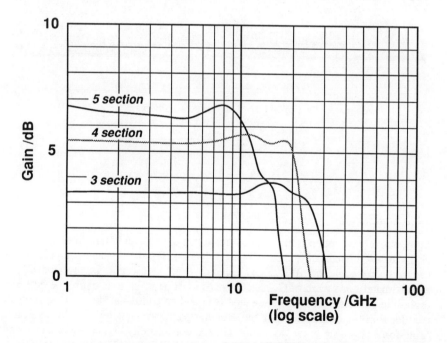

Fig. 5.26 Distributed amplifier simulated responses (0.5 µm MESFETs): 3, 4 and 5 sections

170 *Amplifiers*

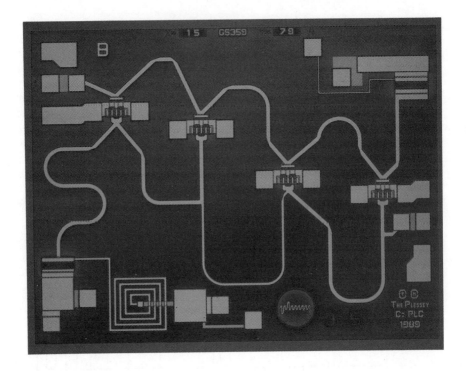

Fig. 5.27 Photograph of a 2 to 20 GHz distributed amplifier

5.8.2 Equalisation of gate and drain-line phase velocities

In the above description is has been assumed that the gate and drain-lines have equal characteristic impedance and equal phase velocities. For the constant-K topology this can only be achieved if C_{gs} and C_{ds} are equal. However, in practice C_{ds} is approximately one tenth of C_{gs}, and so steps must be taken to equalise the gate and drain-line phase velocities. One method is to add an additional shunt capacitance in parallel with C_{ds} in order to make the overall capacitance equal to C_{gs}. However, this brute-force approach is far from optimum and considerably higher performance can be achieved by employing m-derived sections on the drain line. In this topology, shown in Fig. 5.28, a series inductance αL_d is connected at the drain of the FET. This results in equal gate and drain-line phase velocities and improved performance.

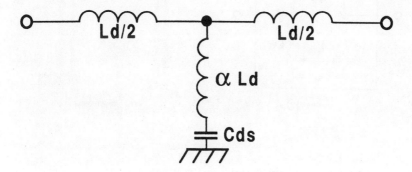

Fig. 5.28 Gate and drain-line velocity compensation using an *m*-derived section

5.8.3 Distributed amplifier with constant-R networks

The constant-*K* network is the most obvious choice for the artificial transmission lines of the distributed amplifier, but it does not give the best performance. Chase and Kennan [22] demonstrated that constant-*R* networks could be used instead. The constant-*R* network requires a bridged-T element using a mutual inductor, bridging capacitor, and the FET capacitance, as shown in Fig. 5.29. The cut-off frequency for the constant-*R* network is given by:

$$f_c = \frac{\sqrt{2}}{\pi R_L C} \qquad (5.26)$$

The advantage of this topology is that the cut-off frequency is increased by a factor of $\sqrt{2}$, which yields a significant improvement in bandwidth. Furthermore, the constant-*R* network presents a conveniently constant resistance to the transistors, whereas the constant-*K* presents a reactive impedance which changes rapidly near the cut-off frequency.

The real beauty of the circuit presented by Chase and Kennan is that the mutual inductor and bridging capacitor required were implemented with a single spiral inductor structure: the coupling between turns provided the mutual inductance, and the normally unwanted crossover capacitance provided the bridging capacitance. In addition, the four-section amplifier used cascode FET pairs in order to achieve higher performance and used the capacitive coupling technique (described later in this chapter) to achieve more power. Fig. 5.30 shows a photograph of a distributed amplifier employing constant-*R* networks.

172 *Amplifiers*

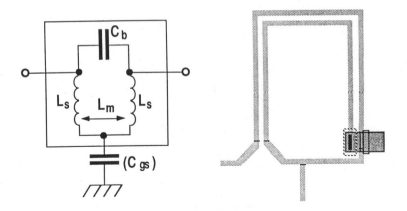

Fig. 5.29 Constant-R network using a bridged-T element

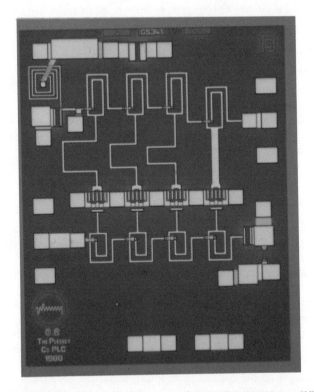

Fig. 5.30 Photograph of a four-section 2 to 18 GHz distributed amplifier employing constant-R networks

5.8.4 Cascode distributed amplifier

The cascode arrangement of two FETs provides high gain, high output resistance, and high reverse isolation. In addition, the maximum output voltage swing is higher than for a single FET because the common-gate FET's gate is grounded at AC. As a result of these advantages, the cascode FET is often used in distributed amplifiers, as shown in Fig. 5.31, and this yields a significant improvement in performance with very little additional circuit area. Furthermore, the common-gate FET's gate bias may be used as an effective means of achieving gain control: Fig. 5.32 compares the gain control range of a conventional distributed amplifier with that achievable from a cascode distributed amplifier. A small series inductance between the two FETs can provide a useful gain peaking mechanism to extend the bandwidth. Often, dual-gate FETs are used to implement the cascode FETs rather than two separate devices. This means that the increase in circuit area becomes negligible [23]. Dual-gate FETs in the cascode arrangement are routinely used as gain elements by many designers. However, care should be exercised regarding the stability of this device. In particular, grounding inductance in the common-gate FET can cause serious problems, and distributed amplifiers have been known to oscillate at frequencies above the cut-off frequency of the amplifier.

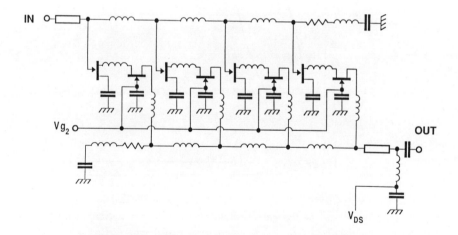

Fig. 5.31 A distributed amplifier using the cascode FET arrangement

5.8.5 Single-section distributed amplifier

The travelling-wave matching technique has been extended to the design of compact, broadband single-stage amplifiers by Minnis [24]. Whilst the distributed amplifier is attractive because of its excellent bandwidth and good port matches, a single stage conventional distributed amplifier would have very low gain because of the n^2 dependence of the gain on the number of sections (eqn. 5.21). The first step towards increasing the gain is to use drain-line tapering (see Section 5.10.4.3) and in fact completely remove the drain-line termination so that all the drain current passes into the load. In addition, by increasing the size of the transistor the larger g_m can compensate for the low gain. However, the increased capacitance will then lower the cut-off frequency of the conventional low-pass input and output networks. Minnis showed that this could be overcome by converting the low-pass networks to band-pass ones, using classical transformations for lumped element filters. The result is an amplifier topology which could be likened to the lossy matching technique. However, the reactive and lossy elements are treated as a single network, so that exact synthesis methods can be used. Typical results for 0.7 µm gate-length MESFETs show that a flat gain of 6 dB is achievable from 7 to 14 GHz.

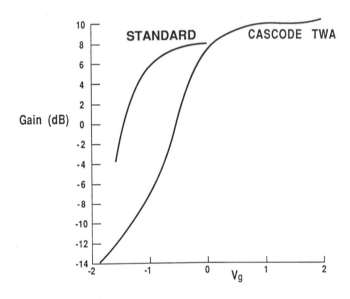

Fig. 5.32 Comparison of the gain control range of a conventional and cascode distributed amplifier

5.8.6 Matrix distributed amplifier

The matrix distributed amplifier [25, 26] consists of a number of 'tiers' of FETs connected into artificial transmission lines, as shown in Fig. 5.33. The distinctive feature of this topology is that the signal is amplified from one tier to the next before the final output signal leaves the uppermost drain-line output. This means that both multiplicative gain (one transistor's output feeding the next's input) and additive gain (the summing of transistor outputs on the same drain-line) are provided within the single module. A cascade of conventional distributed amplifiers also provides additive gain and multiplicative gain, but multiple amplifiers are required.

If their are *m* tiers, and *n* sections in each tier, then there is an *m*×*n* matrix of active devices, and hence the name matrix amplifier. In practice, a two tier matrix amplifier with four sections has very similar performance to a cascade of two conventional four-section distributed amplifiers. In fact, the cascade of conventional amplifiers can achieve slightly more bandwidth, possibly because the central artificial transmission lines of the matrix amplifier suffer from both gate and drain losses. However, the matrix topology is highly suitable for monolithic integration and a number of compact, high-performance realisations have been reported [27-30]. A further advantage of the matrix amplifier is that it can easily employ the stacked biasing technique to lower the current consumption.

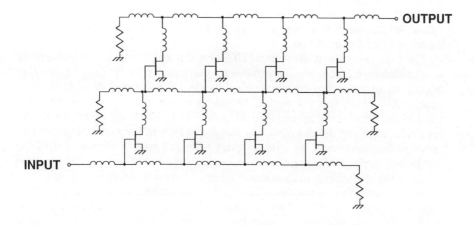

Fig. 5.33 The circuit diagram of a matrix distributed amplifier with four sections and two tiers (bias circuitry not shown)

5.8.7 Practical design guidelines

The most important ingredient to a successful distributed amplifier design is a high performance transistor. The MMIC designer has complete freedom over the transistor gate-widths, the number of sections, and the layout of the interconnecting lines, but the extent to which the basic transistor performance determines the maximum gain/bandwidth that can be achieved is remarkable. The transistors must be grounded effectively, and through-substrate via-holes are essential for a 20 GHz microstrip design. For operation into the millimetre-wave band, the inductance of these via-holes may become too high and so CPW is an attractive alternative.

The inductors required for the artificial transmission lines are normally realised as narrow microstrip tracks (ribbon inductors). Spiral inductors tend to have too high a parasitic capacitance to be used, except for designs which are only required to operate up to 10 GHz or so. Indeed, even ribbon inductors have some capacitance to ground and this will lead to a reduction in bandwidth compared with a simulation employing ideal inductors. This degradation can be reduced by making the tracks as narrow as possible. However, the minimum width is determined by the DC current handling requirement (which may be quite high for the drain-line), by the fabrication limits, and by the high series resistance encountered with very narrow tracks. The track lengths required usually necessitate the use of meandering. However, the effects of coupling between lines and the effects of bend discontinuities will both tend to add more capacitance into the circuit, further restricting the bandwidth. It is not uncommon for a design using straight tracks to start with a promising simulated performance of, say, 6 dB gain from 1 to 20 GHz, only for this to drop disappointingly to 5 dB gain up to 15 GHz once the tracks have been meandered and the resulting layout modelled properly. An important lesson is to not waste time optimising a circuit before these unavoidable discontinuities and coupling have been introduced into the simulation.

DC bias can be applied to the FET gates through the gate-line terminating resistor. However, the drain-line termination cannot normally be used for the drain biasing because the total current required by the FETs is rather high. Instead, an inductor bias choke network needs to be used. Because of the very wide frequency range of the distributed amplifier, it is difficult to find a single inductor that has both high inductance (for the low frequency range) and high self-resonant frequency (for the high frequency end). In order to extract the full bandwidth potential of the distributed amplifier, the drain bias network may have to be a fairly sophisticated low-pass filter structure, and a number of off-chip components may be required.

As an alternative, active loads have been demonstrated as a means of biasing distributed amplifiers [12]. This technique offers very wideband operation with negligible off-chip circuitry. However, it does increase the power consumption, lower the gain, and increase the noise figure. Whichever biasing method is used, it is important to realise that the DC current in the drain-line is highest at one end, but steadily drops as it flows into successive transistors. Hence, although the total DC current may necessitate the use of a inconveniently wide line at one end of the drain-line, this width does not need to be maintained throughout the drain-line, and the tracks can be made narrower later on down the drain-line in order to improve performance.

5.9 Active matching

This section describes a number of techniques in which the impedance matching problems are eased or overcome by using more than just a single common-source FET or common-emitter BJT.

5.9.1 Common-gate/common-source/common-drain amplifier

At low frequency the input admittance of a common-gate FET can be approximated to

$$Y_{in} = g_m + j\omega C_{gs} \tag{5.27}$$

For a transistor with g_m=20 mS and a low value of C_{gs}, the input impedance of the common-gate configuration is therefore very close to 50 Ω at low frequencies. In a similar manner, the common-drain (or source follower) configuration can have an output impedance very close to 50 Ω at low frequencies. As a result, it is possible to construct a broadband amplifier with good input and output matches using a minimal number of passive matching elements. This amplifier topology, illustrated in Fig. 5.34, uses a common-gate input stage followed by a common-source stage and a common-drain output stage. Resistors are employed both for DC biasing and for stabilisation: both the common-gate and common-drain stages are significantly more prone to oscillation than the common-source FET. The use of small inductances between stages provides a degree of gain-peaking to extend the high frequency response.

The advantage of this topology is its small size, but it does have quite high DC power consumption, and the noise figure of the common-gate configuration is significantly worse than the common-source FET. This type of amplifier can achieve operation from approximately 0.5 to 5 GHz with typical 0.5 µm MESFETs. Since the common-gate FET is non-inverting with an input impedance close to 50 Ω, whereas the common-source FET is inverting, with a high input impedance, common-gate and common-source FET pairs have found use in active baluns, and these are described in Chapter 6.

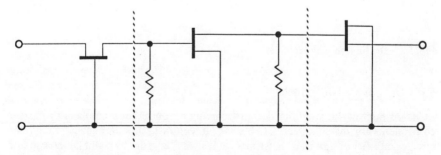

Fig. 5.34 Actively matched amplifier using common-gate and common-drain FETs

5.9.2 Darlington pair

The Darlington pair has become a very familiar amplifier configuration through its use in the ubiquitous silicon bipolar modular MMIC amplifier (e.g. the MODAMP™, which stands for *monolithic Darlington amplifier*). The Darlington pair has a current gain equal to the product of the individual transistor current gains and gives an increased input impedance and larger S_{21} bandwidth. The circuit diagram of a typical Darlington pair amplifier is shown in Fig. 5.35. The Darlington pair ensures a high gain-bandwidth product, and the resistor feedback ensures a smooth broadband response with good input and output matches. Emitter resistors are employed for stabilisation of the operating point and for series feedback. External decoupling capacitors and bias inductor or resistor are often used to enable DC coupling or operation down to low frequency. With either silicon- or GaAs-based heterojunction bipolar transistors, this same amplifier topology is capable of achieving a DC-to-20 GHz bandwidth [31]. The Darlington pair has also been used for broad-band transimpedance amplifier design for applications such as fibre-optic communications [32].

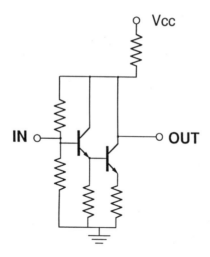

Fig. 5.35 Darlington pair amplifier

5.9.3 DC-coupled amplifiers

Standard microwave amplifier design techniques require large numbers of passive components for matching and DC biasing. In contrast, "low frequency" integrated circuit amplifier design uses transistors exclusively, in order to facilitate the use of a simple process technology and realise high packing density designs. Passive components such as capacitors, resistors and inductors are not used and are not part of the IC technology. Instead, DC-coupled amplifiers are designed using active loads for biasing, rather than inductors or resistors as bias chokes, and using level-shifting diode chains for interstage coupling, rather than using capacitors. The problem of impedance matching is avoided by the judicious use of small periphery devices and by employing common-gate or common-base input stages and source- or emitter-followers as buffer amplifiers. This type of amplifier design can be classed as precision analogue IC design. This is a huge field in its own right, and it is impossible to do this topic justice in the confines of this book. Nevertheless, the early work of Hornbuckle and Van Tuyl [33, 34] and Estreich [35] should be highlighted since it serves to demonstrate the great improvement in packing density that these techniques achieve compared to traditional microwave amplifier design approaches. Fig. 5.36 shows the circuit diagram of one of Hornbuckle's single-stage amplifiers (with feedback). These techniques are limited to fairly low microwave frequencies because of the matching limitations, but the high packing density means that this design approach can lead to very small (and therefore cheap) circuits. Hence, this type of design should be seriously considered for low-cost applications such as mobile radio communications in the low-GHz frequency range. However, for situations where performance is critical, such as low-noise and high-power amplifiers, the traditional lumped-element matching approach is superior.

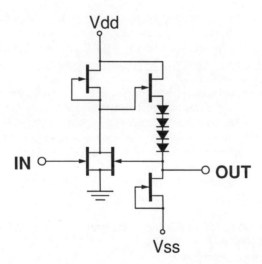

Fig. 5.36 DC-coupled feedback amplifier stage used by Hornbuckle

180 *Amplifiers*

5.10 Power amplifiers

In power amplifier design the over-riding concern is to ensure that the load impedance presented to the transistor gives the optimum load-line on the I-V characteristic in order to maximise the voltage/current swing. This load-line, for an FET operating in class A, is illustrated in Fig. 5.37. At the input the gate voltage swing is limited by the pinch-off voltage and the forward barrier voltage. If the transistor has low gain it is this maximum swing which determines the output power. Otherwise, the drain voltage swing becomes the limitation, since this is limited by the knee voltage and the gate-drain breakdown voltage. Similar limits occur in the bipolar transistor. If the transistor structure and material parameters are optimised for maximum power, then it is the breakdown voltages which ultimately limit the output power. In order to overcome the limited input and output voltage swing, a power transistor is a large periphery device which can handle a large current. For FETs this means using the largest gate-width possible, and for bipolar transistors it means using the largest device area possible. In a reactively matched power amplifier, the transistor operates in a low impedance environment (less than a few ohms) and matching transformers are required to bring this up to the desired 50 Ω level. However, there are four major limitations to this:

(1) the thermal problem of extracting heat
(2) the problem of matching to the resulting low impedances
(3) the yield problem
(4) distributed effects; non-uniform power distribution within the device

The thermal problem is compounded in any MMIC because a thick substrate is required for low loss passive elements, whereas a thin substrate is required for good heat dissipation. GaAs MMICs are even worse off because of the low thermal conductivity of GaAs. Perversely, because the HBT has achieved extraordinary power density, it suffers more from thermal limitations, and many quoted results are for pulsed operation. The matching problem is caused by the size limitation of MMICs: a hybrid MIC power amplifier can readily use, say, a 10 Ω microstrip transformer, even though it may be 8 mm wide. However, this is clearly not possible on a MMIC. The yield problem is encountered when a large power device is integrated onto a multi-function chip. There are a number of methods for easing the thermal and impedance-matching limitations:

(1) bath-tub vias (selective thinning of the wafer, with a plated heatsink)
(2) flip-chip mounting
(3) power dividing/combining
(4) cluster matching

The bath-tub via process is important for multi-function MMICs because it means that the substrate is thick, so low loss passive components are available throughout the chip, and the substrate is only thinned under the power devices. Flip-chip mounting is common with discrete power FETs, and the technique can be extended to MMICs although the passive elements would first need to be characterised when mounted in this way. Power combining and cluster matching are ways of distributing the signal among a number of smaller devices in order to overcome the problems encountered with a single large device. With power combining and cluster matching the signal is shared amongst a number of smaller devices, so that the heat is spread over a wider area and many via-holes can be used to improve heat dissipation.

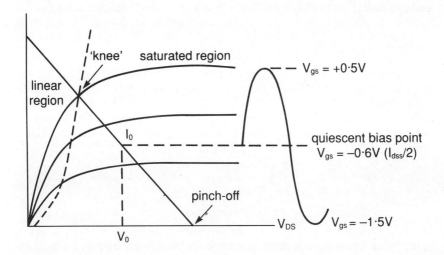

Fig. 5.37 Class-A load line for an FET

5.10.1 Device characterisation

The most important device parameter is the output load impedance required for maximum output power. This can be measured using the load-pull technique with either manual or automatic tuners. With manual tuners the device is tuned up for maximum output power and gain and the tuners are then removed and individually measured on a network analyser to determine the optimum source and load impedance. However, with manual tuners the optimum impedance may take some time to find, and there is no guarantee that the optimum point has been found. With automatic tuners, under computer control, the contours of constant output power can be plotted on a Smith chart, and this has the advantage that the optimum point can be found with accuracy and a high level of confidence. The source impedance is also important because it will be quite different to the small-signal value. Once these optimum source and load impedances have been determined, the matching networks can be designed and the amplifier gain and stability checked using the small-signal S-parameters. This is the basic design procedure for a narrowband power amplifier.

Load pull measurements are very time consuming and do not normally take into account the effects of the terminating impedances at the signal harmonics. Hence, it is better to use a large-signal device model if one is available. Using computer simulations the load pull 'measurements' can be conducted to determine the optimum source and load impedances. In addition, the harmonic frequency terminations can be optimised to improve efficiency and linearity. However, it has to be said that large-signal models often lack sufficient accuracy to make a purely CAD-based power amplifier design wise. CAD is essential for broadband power amplifier design, but load-pull data should still be used ideally in order to verify the model accuracy.

5.10.2 Power combining and cluster matching

There are many well-known techniques for power dividing/combining [36], but many of these are unsuitable for MMIC power amplifiers: N-way in-phase Wilkinson power dividers/combiners are not entirely suitable for planar constructions. Fork dividers/combiners are difficult to realise with equal path-lengths. The serial power combining method requires couplers with a logarithmically varying sequence of coupling levels. This leaves three main methods:

(1) corporate (or binary) power combining with Wilkinson dividers/combiners (Fig. 5.38 (a))
(2) travelling-wave power combining with Wilkinson dividers/combiners (Fig. 5.38 (b))
(3) dual balanced amplifiers with Lange or branch-line couplers (Fig. 5.38 (c))

Fig. 5.39 shows a photograph of an X-band transmitter chip consisting of a VCO feeding a two-stage power amplifier. The output stage consists of two 1.5 mm MESFETs, power combined using lumped-element Wilkinson splitters and combiners.

Two other methods are cluster matching and the distributed amplifier. However, with true power combining the individual amplifiers are isolated from one another due to the port isolations of the divider/combiners or couplers. This is not the case with cluster matching or with the distributed amplifier. As a result, the cluster matching technique does not have the stability advantage provided by the isolation between devices, and a distributed amplifier with N devices cannot achieve N times the power of a single device. The cluster matching technique does, however, greatly ease the thermal and impedance matching problems, and as a result this technique has become almost universally adopted in monolithic power amplifier design.

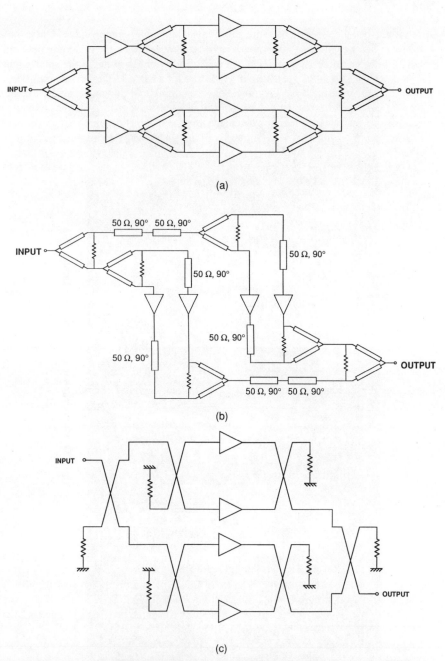

Fig. 5.38 Power combining techniques: (a) Corporate combining with Wilkinson dividers and combiners; (b) Travelling wave combining with Wilkinson dividers and combiners; (c) Dual balanced amplifiers

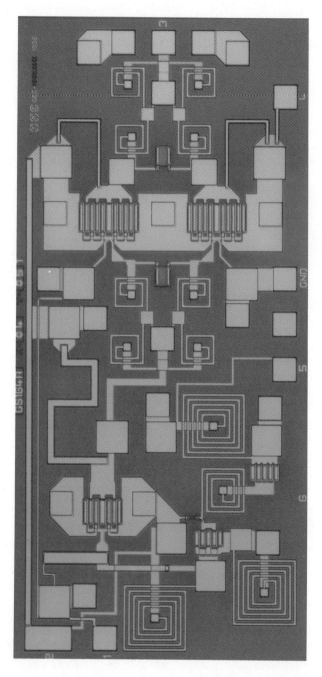

Fig. 5.39 Photograph of an X-band transmitter chip

In the cluster matching technique [37, 38] the amplifier is split up into cells, and the cells are connected with input and output manifolds which incorporate impedance matching. In this way the matching operation is carried out on clusters of small periphery devices, and extremes of impedance are avoided. In the enhanced method demonstrated by Pavlidis [39], the input manifold is replaced by an active splitting network: each cell has one input and two (or more outputs) feeding the next stage cells. Once the required number of paralleled devices has been reached, the outputs are combined together again using an output manifold. Impedance matching takes place within the individual cells, so that low impedances are avoided. A schematic of this topology is shown in Fig. 5.40. The advantage of this approach for MMIC power amplifiers is that the devices are spaced well apart for improved thermal dissipation, the power splitting/combining does not required large couplers, the impedance matching is simplified, and the problem of unequal phasing is overcome.

The relatively high loss of microstrip on MMICs means that the overall power efficiency of a power combined amplifier can be reduced considerably. For example, if four 2.5 W devices with individual efficiencies of 40% are combined using couplers with 0.5 dB loss each, then only 7.9 W is obtained. The other 2.1 W is lost as heat in the combining network and the efficiency drops to 32%. Not surprisingly, therefore, it is very common for the final output matching and combining to be performed off-chip with conventional microstrip lines. This has the additional benefit of overcoming the current handling limitation of monolithic microstrip tracks.

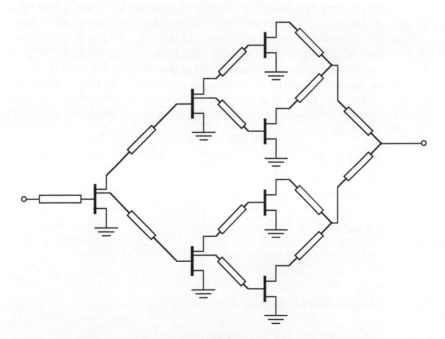

Fig. 5.40 The enhanced cluster matching technique

5.10.3 Class B operation

So far, class A operation has been assumed, and this gives the best linearity. However, class A operation has a maximum theoretical efficiency of 50%, and 30% power-added efficiency is typical of most MMIC GaAs FET amplifiers. With the increasing importance of high efficiency power amplifiers for mobile communications and phased-array antennas, there has been a corresponding increase in the study of high efficiency amplifiers. Class C amplifiers, while giving the best efficiency, are the most non-linear and are not well suited for most digital communication systems. Class B amplifiers can achieve good linearity and a theoretical efficiency of 78.5%. In class B, transistors are operated with almost zero quiescent current, which means that the efficiency is high, but a single transistor can only amplify one half-cycle of the signal. Two transistors are therefore employed in tandem; one amplifies the positive half cycles, and the other amplifies the negative half-cycles. At microwave frequencies it is not possible to employ complimentary transistor pairs (npn/pnp BJTs or n-channel/p-channel FETs) because of the large difference in the electron and hole mobilities. Hence, microwave class B amplifiers have to employ a push-pull approach, with two identical transistors fed by complimentary signals generated by transformers or baluns. Fig. 41 shows the schematic of a GaAs FET push-pull amplifier using planar spiral transformers. The major problem with the class B amplifier is that 6 dB of the transistor available gain is lost in the splitting and combining process.

A second problem encountered with GaAs FET push-pull amplifiers is that the g_m of a standard device drops rapidly towards pinch-off. Hence, the device can either be operated with a small quiescent current (10 or 20% of I_{dss}), which is actually class AB operation, or the channel doping profile has to be re-designed specifically for class B operation. The doping density has to increase with depth so that g_m is kept high towards pinch-off. Unfortunately, this is the opposite requirement to that of low-noise FETs, where high doping at the surface is required for low contact resistance. Some excellent results have been achieved with class B and push-pull power amplifiers [40, 41].

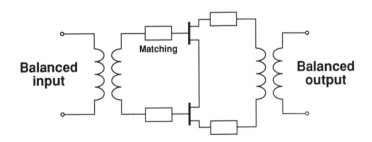

Fig. 5.41 A push-pull FET amplifier using spiral transformers

5.10.4 High power distributed amplifiers

The distributed amplifier suffers from a fundamental power limitation that is not encountered with reactively matched power amplifiers. In a reactively matched power amplifier the transistor can operate in a low impedance environment in order to limit the voltage swing: however, in the distributed amplifier the minimum impedance of the gate- and drain-lines is determined by other factors, such as the gate- and drain-line loss. In the standard distributed amplifier the first FET on the gate-line sees the 50 Ω source impedance almost directly, and the last FET on the drain-line sees the 50 Ω load impedance almost directly. As a result, with a typical FET breakdown voltage of 18 V one can calculate that the output power cannot be more than about 400 mW, unless some enhancements are made to the topology.

5.10.4.1 The capacitively-coupled distributed amplifier

A further limit to the achievable output power is the fact that large gate-width transistors, suitable for high power, have a high C_{gs} which leads to a reduction in bandwidth. In addition, gate-line loss means that the first FET on the gate-line will receive a considerably greater input voltage swing than the others further down the gate-line. Hence, this FET will go into gain compression too early. An extremely neat solution to these limitations is to capacitively couple the FETs to the gate-line [42], as shown in Fig. 5.42. This technique allows larger FETs to be employed without lowering the cut-off frequency of the gate-line. In addition, if the coupling capacitors are varied the effect of gate-line loss can be accommodated so that each FET has the same gate voltage swing applied. In other words, the first FET has a relatively small coupling capacitor because the input signal is large, whereas the last FET on the gate-line is almost directly coupled to the gate-line with a large coupling capacitor. The penalty, however, with the capacitive coupling is that the gain is reduced because of the potential divider action across C_{gs} and the coupling capacitor. The coupling capacitors are shunted with high value resistors in order to allow the DC gate bias to reach the gates. This can impose a restriction on the high power performance, however, because the gates need to be fed from a low impedance DC supply for maximum saturated power.

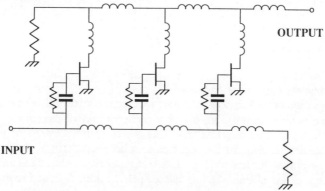

Fig. 5.42 The capacitively-coupled distributed amplifier

5.10.4.2 The split gate-line distributed amplifier

It has already been stated that C_{gs} is much larger than C_{ds} and that the gate-line cut-off frequency is hard to maintain with large gate-width FETs required for high power. A novel solution to this limitation is the split gate-line amplifier [43]. As shown in Fig. 5.43, the input line splits into two separate gate-lines, but the FETs are all connected to the same drain-line. The individual gate-line cut-off frequencies can be kept high, since each FET has half the C_{gs} of the standard amplifier. The cut-off frequency of the drain-line is still high because C_{ds} is much smaller than C_{gs}. Hence, the technique allows a greater total gate-width to be used. Ideally, a matched splitter should be used to split the signal; clearly if the two gate-lines are increased to a 100 Ω impedance level (in order to maintain an overall 50 Ω input) then some of the power advantage is lost because the voltage swing is fixed by the gate pinch-off voltage and the built-in barrier voltage.

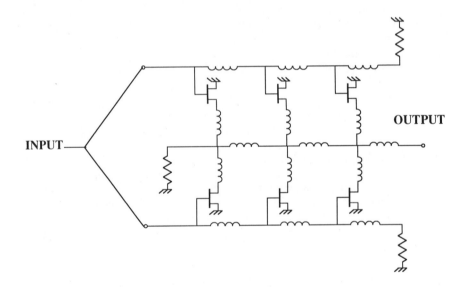

Fig. 5.43 The split gate-line distributed amplifier

5.10.4.3 The tapered drain-line technique

This topology, shown in Fig. 5.44, attempts to force the total FET output current at the drain-line reverse termination to zero and ensure that all the current from the FETs is fed in the forward direction only. This is accomplished by tapering the gate-widths of the FETs [44]. In the ideal case, with a properly designed tapered distributed amplifier the first FET feeds a section of line with impedance Z_0, the second feeds a section of line with impedance $Z_0/2$, the third $Z_0/3$, and so on [1]. This arrangement means that the proportion of current produced by section $n-1$ that is reflected from the impedance change at section n will cancel the reverse current produced by the FET in section n. Since the drain-line impedance is decreasing rapidly, there is a limit to the number of stages that can be employed. For a 50 Ω system the output impedance would need to be matched with an off-chip microstrip taper. However, an advantage of the low load impedance presented to the final FET in the drain-line is that it alleviates the problem of the limited voltage swing.

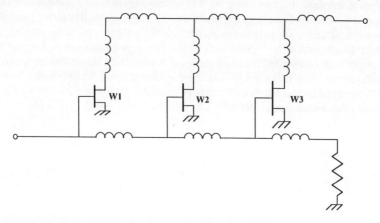

Fig. 5.44 A distributed amplifier with a tapered drain-line

5.11 Low noise amplifiers

The noise factor, F, of an amplifier chain or receiver can be calculated by Friis' well known formula:

$$F = F_1 + \frac{F_2 - 1}{G_1} + \frac{F_3 - 1}{G_1 G_2} + \frac{F_4 - 1}{G_1 G_2 G_3} + \dots \quad (5.28)$$

where F_n is the noise factor of component n in the chain, and G_n is the power gain (not in dB) of component n in the chain. This formula shows that the noise figure (which is given by $10 \log F$) of the first stage is most critical, because the following stages' noise contributions are reduced by the gain of the preceding stages. (An important exception to this is the millimetre-wave receiver where no LNA is used: in this case the overall noise figure is approximately equal to the mixer noise figure plus the IF amplifier noise figure). The noise measure, M, of an amplifier gives the noise factor that would result from an infinite cascade of identical amplifiers, and is calculated from:

$$M = \frac{F - 1}{1 - 1/G} \quad (5.29)$$

The noise factor is given by:

$$F = 1 + M \tag{5.30}$$

This is therefore the lower limit on the overall noise factor. Generally, the first amplifier stage should be designed for optimum noise *measure*, rather than optimum noise figure, because the gain must be kept sufficiently high to minimise the noise contributions from the following stages. In a multi-stage LNA, the second and third stages can be designed for higher gain, with their noise figure becoming steadily less important. If the gain of a single stage is quite low (significantly less than 10 dB), which might well be the case for millimetre-wave amplifiers, then all three (or more) stages may need to be designed for optimum noise measure. The first step in LNA design is to calculate the number of stages needed and the noise figure and gain that is required from each particular stage.

5.11.1 Noise matching

Conjugate matching between the source and the input impedance of a transistor achieves maximum power transfer, and hence maximum gain, but this does not generally give minimum noise figure. Instead, there is an impedance Z_{opt} (reflection coefficient Γ_{opt}) which needs to be presented to the transistor input to achieve the minimum noise figure, F_{min}. It can be shown that, as one moves away from this optimum noise matching point on the Smith chart, circles of constant noise figure are formed. The noise factor, F, with an arbitrary source impedance Γ_s can be expressed as:

$$F = F_{min} + 4R_n \frac{|\Gamma_s - \Gamma_{opt}|^2}{(1 - |\Gamma_s|^2)|1 + \Gamma_{opt}|^2} \tag{5.31}$$

where R_n is known as the equivalent noise resistance of the device. It can be seen that R_n determines how rapidly the noise figure degrades as Γ_s moves away from Γ_{opt}: A small R_n means that the noise figure is less sensitive to the source impedance, making the circuit easier to design.

Fig. 5.45 shows the noise circles at 1.6 GHz for a 300 μm gate-width MESFET with 1 μm gate length. Note that Γ_{opt} is considerably different from $S_{11}*$ (and hence from the simultaneous match input reflection coefficient). This implies that an amplifier designed for minimum noise figure must be mismatched to some extent from the maximum power transfer condition, unless the device has $S_{11}*$ and Γ_{opt} very close or they are brought close together by using feedback. This poor input match problem has traditionally been overcome by using a balanced amplifier topology, and this has been described earlier.

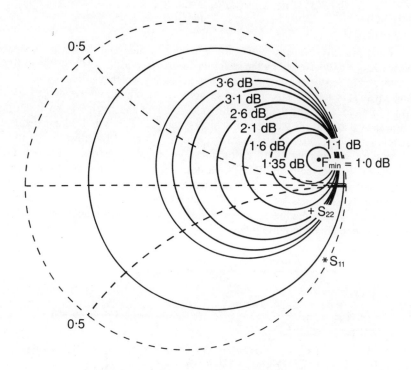

Fig. 5.45 Noise figure circles for a 300μm MESFET at 1.6 GHz

5.11.2 Simultaneous match LNA

For an MMIC low noise amplifier, the balanced amplifier solution is rather unattractive because it doubles the chip area at least. A technique was proposed many years ago for using common lead inductance to improve the signal-to-noise ratio of triode circuits [45]. Later, Engberg [46] showed that a combination of series and shunt feedback, with suitable output loading, could be used to make the power matching and noise matching reflection coefficients converge. He reported that the noise measure was unaffected by these lossless feedback elements. In 1985, Lehmann and Heston [47, 48] first demonstrated that this series feedback technique could be applied to monolithic low noise amplifier design.

When applied to an FET, the principle of the technique is that a carefully controlled value of source inductance can be used to move $S_{11}*$ and Γ_{opt} closer together, resulting in an LNA with simultaneous noise and power match. With reference to Fig. 5.46, it can be shown that:

$$Z_{in} = (R_{ch} + R_{ds} + \frac{g_m L_s}{C_{gs}}) + j\omega L_s + \frac{g_m R_s}{j\omega C_{gs}} + \frac{1}{j\omega C_{gs}} \qquad (5.32)$$

It is found that S_{11}^* varies with both the output loading and the series feedback, whereas Z_{opt} is only affected by the series feedback. As a result, it is possible to find a combination of feedback inductance and output loading which result in S_{11}^* and Γ_{opt} being equal. It is actually possible that the noise figure might improve at the expense of the gain. In a practical design, the gate-width of the device is an important parameter which controls the proximity of S_{11}^* and Γ_{opt}. For each different frequency of application, a different gate-width is optimum. In any LNA design, the use of resistors for biasing or stabilisation must be avoided because of their noise contribution. Furthermore, in MMIC LNA design it is often found that the resistance of spiral inductors can introduce a significant noise penalty. Unfortunately, at low frequencies the simultaneous match LNA may require a spiral inductor of one or two turns for the series feedback, and the resistance of this inductor can cause a significant degradation in noise figure.

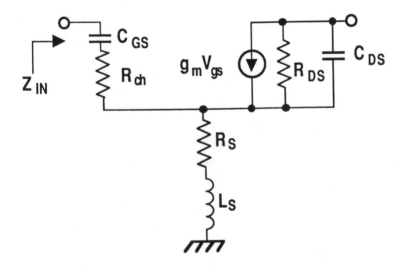

Fig. 5.46 Equivalent circuit of the series feedback FET LNA

Amplifier type	Advantages	Disadvantages	Typical bandwidth
Reactively matched	Best gain per stage, noise figure and output power	Difficult to control the input and output VSWR: Poor gain flatness when cascaded. Hard to achieve unconditional stability. Very sensitive to process variations	Typically 20 % bandwidths at any centre frequency from 1 GHz to over 100 GHz (device permitting). Octave bandwidths just feasible
Lossy matched	Good VSWRs and gain flatness: Broadband design easier, and readily cascadable	Reduced gain per stage compared with the reactive match. Some degradation in noise figure and output power	Octave bandwidths readily achievable. Frequency range again determined by device performance
Feedback	Excellent gain flatness over a wide frequency range, good stability, and less sensitive to transistor spreads	Noise figure is mediocre and gain is reduced significantly. Tends to need large gate-widths, leading to high DC power	0.1 to 6 GHz (MESFET) 6 to 18 GHz
Distributed	Ultra-wide bandwidths achievable (more than one decade). Good VSWRs, easy to cascade	Low gain, mediocre noise figure and limited output power. High DC power. Large total gate-width required	1 to 20 GHz (MESFET) 2 to >50 GHz (pHEMT)
Darlington pair	Very simple topology with small chip size. Broadband, with quite high gain per stage and good VSWRs. Can be DC coupled	Mediocre noise figure and fairly high DC power	DC to 5 GHz (Si bipolar) DC to 20 GHz (HBT)
Actively matched	A simple topology for broadband amplifiers at lower frequencies. Small chip size	Poor noise figure and high DC power requirement	1 to 6 GHz (MESFET)
DC coupled (OP-AMP style)	The highest packing density: No passive elements required. Ideally suited to BiCMOS technology. Highly competitive for mobile radio applications	Limited noise figure and power handling performance compared with traditional techniques	DC to 5 GHz This technique is generally best suited to bipolar transistors

Table 5.1 Comparison of the key MMIC amplifier topologies

5.12 Summary

The first part of this chapter reviewed the theory and analysis of microwave amplifiers in terms of the device's two-port S-parameters. This provides the foundations for the understanding of the problems of impedance matching and stability, and demonstrates that amplifier design is far more involved than simply matching to S_{11} and S_{22}. The practical implications of the theoretical analysis have been identified, and a wide range of matching techniques and DC biasing circuits have been described. The key MMIC amplifier topologies have been studied, and the design of high-power and low-noise amplifiers has been discussed. The remaining references [49-107] provide specific details of a number of MMIC amplifier designs, as well as some further general information on amplifiers. To conclude the chapter, Table 5.1 presents a comparison of the features of each of the major MMIC amplifier topologies: the general advantages and disadvantages of each amplifier type are listed, along with an indication of the typical frequency range which they can cover. This summary table should assist in the first and most important step in MMIC amplifier design, which is choosing the most suitable topology.

5.13 References

1. VENDELIN, G. D., PAVIO, A. M., and ROHDE, U. L.: *Microwave circuit design using linear and nonlinear techniques*, John Wiley and Sons, 1990
2. PENGELLY, R. S.: *Microwave field effect transistors–Theory, design and applications*, Research Studies Press, 1986
3. LIAO, S.: *Microwave circuit analysis and amplifier design*, Prentice Hall, 1987
4. SOARES, R. (*ed.*):*GaAs MESFET Circuit Design*, Artech House, 1988
5. YIP, P. C. L: *High-frequency circuit design and measurements*, Chapman and Hall, 1990
6. POZAR, D. M: *Microwave Engineering*, Addison Wesley, 1990
7. YOUNG, G. P., and O' SCANLAN, S.: 'Matching network design studies for microwave transistor amplifiers', *IEEE Trans. Microwave Theory Tech.*, 1981, pp. 1027-1034
8. VILLAR, J. C., and PEREZ, F.: 'Graphic design of matching and interstage lossy networks for microwave transistor amplifier', *IEEE Trans. Microwave Theory Tech.*, 1985, pp. 210-215
9. KHATIBZADEH, A, *et al.*: 'High-efficiency X-band HBT power amplifier', IEEE Microwave and Millimeter-wave Monolithic Circuits Symposium, 1994, pp. 117-120
10. SCHUMACHER, H., ERBEN, U., and GRUHLE, A.: 'Low-noise performance of SiGe heterojunction bipolar transistors', IEEE Microwave and Millimeter-wave Monolithic Circuits Symposium, 1994, pp. 213-216
11. ROLLETT, J.: 'Stability and power gain invariants of linear two-ports, *IRE Trans. on Circuit Theory*, CT-9, March 1962, pp. 29-32

12. ROBERTSON, I. D., and AGHVAMI, A. H.: 'Ultrawideband biasing of MMIC distributed amplifiers using an improved active load', *Electron. Lett.*, vol. 27 no. 21, 1991, pp. 1907-1908
13. NICLAS, K. B.: 'On the design and performance of lossy match GaAs MESFET amplifiers', *IEEE Trans. Microwave Theory Tech.*, 1982, pp. 1900-1906
14. ULRICH, E.: 'Use negative feedback to slash wideband VSWR', Microwave Journal, October 1978, pp. 66-70
15. NICLAS, K. B., WILSER, W. T., GOLD, R. B., and HITCHENS, W. R.: 'The matched feedback amplifier: ultra-wideband microwave amplification with GaAs MESFETs', *IEEE Trans. Microwave Theory Tech.*, 1980, pp. 285-294
16. RIGBY, P. N., SUFFOLK, J. R., and PENGELLY, R. S.: 'Broadband monolithic low-noise feedback amplifiers', IEEE Microwave and Millimeter-wave Monolithic Circuits Symposium, 1983, pp. 71-75
17. BEALL, J. M., NELSON, S. R., and WILLIAM, R. E.: 'Design and process sensitivity of a two-stage 6-18 GHz monolithic feedback amplifier', *IEEE Trans. Microwave Theory Tech.*, MTT-33, 1985, pp. 1567-1571
18. PERCIVAL, W.S.: 'Thermionic valve circuits', British Patent 460562, July 1936
19. STRID, E. W., and GLEESON, K. R.: 'A DC-12 GHz Monolithic GaAs FET Distributed Amplifier', *IEEE Trans. Microwave Theory Tech.*, MTT-30, no. 7, July 1982, pp. 969-975
20. AYASLI, Y., MOZZI, R. L., VORHAUS, J. L., REYNOLDS, L. D., and PUCEL, R. A.: 'A monolithic GaAs 1-13 GHz traveling-wave amplifier," *IEEE Trans. Microwave Theory Tech.*, MTT-30, 1982, pp. 976-980
21. BEYER, J. B., PRASAD, S. N., BECKER, R. C., NORDMAN, J. E., and HOHENWARTER, G. K.: 'MESFET distributed amplifier design guidelines', *IEEE Trans. Microwave Theory Tech.*, MTT-32, 1984, pp. 268-275
22. CHASE, E. M., and KENNAN, W.: 'A power distributed amplifier using constant-R networks', IEEE MTT-S Int. Microwave Symp. Dig., 1986, pp. 811-815
23. KENNAN, W., ANDRADE, T., and HUANG, C. C.: 'A 2-18 GHz monolithic distributed amplifier using dual-gate GaAs FETs', *IEEE Trans. Microwave Theory Tech.*, 1984, pp. 1693-1697
24. MINNIS, B. J.: 'The traveling wave matching technique for cascadable MMIC amplifiers', *IEEE Trans. Microwave Theory Tech.*, MTT-42, 1994, pp. 690-692
25. NICLAS, K. B., and PEREIRA, R. R.: 'The matrix amplifier: A high-gain module for multioctave frequency bands', *IEEE Trans. Microwave Theory Tech.*, MTT-35, 1987, pp. 296-306
26. NICLAS, K. B., PEREIRA, R. R., and CHANG, A. P.: 'A 2-18 GHz low-noise/high-gain amplifier module', *IEEE Trans. Microwave Theory Tech.*, MTT-37, 1989, pp. 198-207
27. CHANG, A. P., NICLAS, K. B., CANTOS, B. D., and STRIFLER, W. A.: 'Design and performance of a 2-18 GHz monolithic matrix amplifier', IEEE MTT-S Int. Microwave Symp. Dig., 1989, pp. 287-289

28. CHANG, A. P., NICLAS, K. B., and CANTOS, B. D.: 'Monolithic 2-18 GHz matrix amplifier', *IEEE Trans. Microwave Theory Tech.*, MTT-37, 1989, pp. 2159-2162
29. CHU, S. L., TAJIMA, Y., COLE, J. B., PLATZKER, A., and SCHINDLER, M. J.: 'A novel 4-18 GHz monolithic matrix distributed amplifier', IEEE MTT-S Int. Microwave Symp. Dig., 1989, pp. 291-295
30. SIMON, K. M., CHU, S. L. G. and WOLHERT, R.: 'A millimeter-wave monolithic matrix distributed amplifier', IEEE Microwave and Millimeter-wave Monolithic Circuits Symposium, 1992, pp. 73-75
31. KOBAYASHI, K. W., *et al.*: 'GaAs HBT MMIC broadband amplifiers from DC to 20 GHz', IEEE Microwave and Millimeter-wave Monolithic Circuits Symposium, 1990, pp. 19-22
32. NAGANO, N., *et al.*: 'Monolithic ultra-broadband transimpedance amplifiers using AlGaAs/GaAs heterojunction bipolar transistors', *IEEE Trans. Microwave Theory Tech.*, MTT-42, 1994, pp. 2-9
33. HORNBUCKLE, D. P.: 'GaAs IC direct-coupled amplifiers', IEEE Int Microwave Symposium, 1980, pp. 387-389
34. HORNBUCKLE, D. P., and VAN TUYL, R. L.: 'Monolithic GaAs direct-coupled amplifiers', *IEEE Trans. Electron Devices*, 1981, pp. 175-182
35. ESTREICH, D. B.: 'A monolithic wide-band GaAs IC amplifier', *IEEE Journal of Solid-State Circuits*, SC-17, no. 6, Dec. 1982, pp. 1166-1173
36. WALKER, J. L. B.: *High-power GaAs FET amplifiers*, Artech House, 1993
37. SCHELLENBERG, J. M., and YAMASAKI, H.: 'A new approach to FET power amplifiers', Microwave Journal, March 1982, pp. 51-66
38. DEGENFORD, J. E., FREITAG, R. G., BOIRE, D. C., and COHN, M.: 'Broadband monolithic MIC power amplifier development', Microwave Journal, March 1982, pp. 89-97
39. PAVLIDIS, D., *et al.*: 'A new specifically monolithic approach to microwave power amplifiers', IEEE Microwave and Millimeter-wave Monolithic Circuits Symposium, 1983, pp. 54-58
40. LANE, J. R., *et al.*: 'High efficiency 1-, 2-, and 4-W class-B FET power amplifiers', *IEEE Trans. Microwave Theory Tech.*, MTT-34, no. 12, Dec. 1986, pp. 1318-1326
41. HENRY, H. G., *et al.*: 'A compact 3W X-band GaAs MMIC amplifier based on a novel multi-push-pull circuit concept', IEEE GaAs IC Symp. Dig., 1991, pp. 327-330
42. AYASLI, Y., MILLER, S. W., MOZZI, R. L., and HANES, L. K.: 'Capacitively coupled traveling-wave power amplifier', *IEEE Trans. Microwave Theory Tech.*, 1984, pp. 1704-1709
43. AYASLI, Y., REYNOLDS, L. D., MOZZI, R. L., and HANES, L. K.: '2-20 GHz GaAs traveling-wave power amplifier', *IEEE Trans. Microwave Theory Tech.*, MTT-32, no. 3, March 1984, pp. 290-294
44. JONES, K. E., BARTA, G. S., and HERRICK, G. C.: 'A 1 to 10 GHz tapered distributed amplifier in a hermetic surface mount package', IEEE Int. Microwave Symp. Dig., 1985, pp. 137-140

45. STRUTT, M. J. O., and VAN DER ZIEL, A.: 'Suppression of spontaneous fluctuations in amplifiers and receivers for electrical communication and for measuring devices', *Physica*, vol. IX, no. 6, June 1942, pp. 513-538
46. ENGBERG, J.: 'Simultaneous input power match and noise optimization using feedback', European Microwave Conference, 1974, pp. 385-389
47. LEHMAN, R. E., and HESTON, D. D.: 'X-band monolithic series feedback LNA', IEEE MTT-S Int. Microwave Symp. Dig., 1985, pp. 51-54
48. LEHMAN, R. E., and HESTON, D. D.: 'X-band monolithic series feedback LNA', *IEEE Trans. Microwave Theory Tech.*, MTT-33, 1985, pp. 1560-1566
49. NICLAS, K. B.: 'Reflective match, lossy match, feedback and distributed amplifiers: A comparison of multi-octave performance characteristics', IEEE MTT-S Int. Microwave Symp. Dig., 1984, pp. 215-217
50. NICLAS, K. B.: 'Multi-octave performance of single-ended microwave solid-state amplifiers', *IEEE Trans. Microwave Theory Tech.*, MTT-32, 1984, pp. 896-908
51. NICLAS, K. B.: 'Noise in broad-band GaAs MESFET amplifiers with parallel feedback', *IEEE Trans. Microwave Theory Tech.*, 1982, pp. 63- 70
52. NICLAS, K. B.: "The exact noise figure of amplifiers with parallel feedback and lossy matching circuits', *IEEE Trans. Microwave Theory Tech.*, MTT-30, 1982, pp. 832-834
53. NICLAS, K. B., WINSER, W. T., KRITZER, T. R., and PEREIRA, R. R.: 'On theory and performance of solid-state microwave distributed amplifiers', *IEEE Trans. Microwave Theory Tech.*, MTT-31, 1983, pp. 447-456
54. KURDOGHLIAN, A., *et al.*: 'High-efficiency InP-based HEMT MMIC power amplifier', IEEE GaAs IC Symp. Dig., 1993, pp. 375-377
55. FUNABASHI, M., and HOSOYA, K.: 'High gain V-band heterojunction FET MMIC power amplifiers', IEEE GaAs IC Symp. Dig., 1993, pp. 379-382
56. ZDEBEL, P. J.: 'Current status of high performance silicon bipolar technology', IEEE GaAs IC Symp. Dig., 1992, pp. 15-18
57. TAN, K. L., *et al.*: 'A manufacturable high performance 0.1-µm pseudomorphic AlGaAs/InGaAs HEMT process for W-band MMICs', IEEE GaAs IC Symp. Dig., 1992, pp. 251-253
58. WU, C. S., *et al.*: 'High efficiency X-band power HBTs', IEEE GaAs IC Symp. Dig., 1992, pp. 259-262
59. HO, W. J., WANG, N. L., and CHANG, M. F.: 'Producibility and performance of the microwave power HBT', IEEE GaAs IC Symp. Dig., 1992, pp. 263-266
60. LE, H. M., *et al.*: 'A Ku-band high efficiency ion-implanted amplifier', IEEE GaAs IC Symp. Dig., 1991, pp. 335-338
61. MOGHE, S. B., *et al.*: 'A monolithic direct-coupled GaAs IC amplifier with 12 GHz bandwidth', *IEEE Trans. Microwave Theory Tech.*, 1984, pp. 1698-1703
62. PLATZKER, A.: 'Monolithic broadband power amplifier at X-band', IEEE Microwave and Millimeter-wave Monolithic Circuits Symposium, 1983, pp. 59-61

63. NICLAS, K. B., and TUCKER, B. A.: 'On noise in distributed amplifiers at microwave frequencies', *IEEE Trans. Microwave Theory Tech.*, 1983, pp. 661-668
64. LIU, L. C., MAKI, D. W., and FENG, M.: 'Single and dual stage monolithic low noise amplifier', IEEE GaAs IC Symp. Dig., 1982, pp. 94-97
65. HO, W. J., and WANG, N. L.: 'Producibility and performance of the microwave power HBT', *IEEE Trans. Microwave Theory Tech.*, MTT-30, 1982, pp. 263-266
66. TSERNG, H. Q., MACKSEY, H. M., and NELSON, S. R.: 'Design, fabrication, and characterization of monolithic microwave GaAs power FET amplifiers', *IEEE Trans. Electron Devices*, 1981, pp. 183-190
67. SECHI, F. N.: 'Design procedure for high-efficiency linear microwave power amplifiers', *IEEE Trans. Microwave Theory Tech.*, 1980, pp. 1157-1163
68. MARTINES, G., and SANNINO, M.: 'The determination of the noise, gain and scattering parameters of microwave transistors (HEMT's) using only an automatic noise figure test-set', *IEEE Trans. Microwave Theory Tech.*, MTT-42, 1994, pp. 1105-1113
69. CROUCH, M. A., and HILTON, K. P.: 'An introduction to GaAs/GaAlAs HBTs for power applications', Microelectronics Journal, vol. 24, 1993, pp. 779-794
70. WINSLOW, T. A., and TREW, R. J.: 'Principles of large-signal MESFET operation', *IEEE Trans. Microwave Theory Tech.*, MTT-42, 1994, pp. 935-942
71. ALI, F.: 'Design considerations for high efficiency GaAs HBT MMIC power amplifiers', European Microwave Conference, 1994, pp. 156-176
72. KASODY, R. E., and DOW, G. S.: 'A high efficiency V-band monolithic HEMT power amplifier, IEEE Microwave and Guided Wave Letters, vol. 4, 1994, pp. 303-304
73. LIU, W., *et al.*: 'First demonstration of high-power GaInP/GaAs HBT MMIC power amplifier with 9.9 W output power at X-band', IEEE Microwave and Guided Wave Letters, vol. 4, 1994, pp. 293-295
74. ASBECK, P. M., *et al.*: 'GaAs-based heterojunction bipolar transistors for very high performance electronic circuits', *Proc. IEEE*, vol. 81, 1993, pp. 1709-1725
75. BAYRAKTAROGLU, B.: 'GaAs HBT's for microwave integrated circuits', *Proc. IEEE*, vol. 81, 1993, pp. 1762-1785
76. PLATZKER, A., HETZLER, K., and COLE, J. B.: 'High density dual-channel C-X-Ku and 6-18 GHz MMIC power amplifiers', IEEE GaAs IC Symp. Dig., 1991, pp. 339-342
77. YUEN, C., *et al.*: '5-60 GHz high-gain distributed amplifier utilizing InP cascode HEMTs', IEEE GaAs IC Symp. Dig., 1991, pp. 319-322
78. HIGGINS, J. A.: 'GaAs heterojunction bipolar transistors: A second generation microwave power amplifier transistor', Microwave Journal, May 1991, pp. 176-194
79. HUANG, J. C., *et al.*: 'An AlGaAs/InGaAs pseudomorphic high electron mobility transistor (PHEMT) for X-and Ku-band power applications', IEEE MTT-S Int. Microwave Symp. Dig., 1991, pp. 713-716

80. MATLOUBIAN, M., *et al.*: 'High power and high efficiency AlInAs/GaInAs on InP HEMTs', IEEE MTT-S Int. Microwave Symp. Dig., 1991, pp. 721-724
81. GAT, M., *et al.*: 'A 3.0 Watt high efficiency C-band power MMIC', IEEE GaAs IC Symp. Dig., 1991, pp. 331-334
82. OHMURO, K., FUJISHIRO, H. I., and ITOH, M.: 'Enhanced-mode pseudomorphic inverted HEMT for low noise amplifier', IEEE MTT-S Int. Microwave Symp. Dig., 1991, pp. 709-712
83. D'AGOSTINO, S., *et al.*: 'A 0.5-12 GHz hybrid matrix distributed amplifier using commercially available FETs', IEEE MTT-S Int. Microwave Symp. Dig., 1991, pp. 289-292
84. YUEN, C., *et al.*: 'Monolithic InP cascode HEMT distributed amplifier from 5 to 40 GHz', *Electron. Lett.*, vol. 26, 1990, pp. 1411-1412
85. MAJIDI-AHY, R., *et al.*: '94 GHz InP MMIC five-section distributed amplifier', *Electron. Lett.*, vol. 26, 1990, pp. 91-92
86. HO, W. J., *et al.*: 'A multifunctional HBT technology', IEEE GaAs IC Symp. Dig., 1990, pp. 67-70
87. RIAZIAT, M., *et al.*: 'HEMT millimetre wave monolithic amplifier on InP', *Electron. Lett.*, vol. 25, 1989, pp. 1328-1329
88. PERDOMO, *et al.*: 'A monolithic 0.5 to 50 GHz MODFET distributed amplifier with 6dB gain', IEEE GaAs IC Symp. Dig., 1989, pp. 91-94
89. BAYRAKTAROGLU, B., KHATIBZADEH, M. A., and HUDGENS, R. D.: 'Monolithic X-band heterojunction bipolar transistor power amplifiers', IEEE GaAs IC Symp. Dig., 1989, pp. 271-274
90. AUST, M., *et al.*: 'A family of InGaAs/AlGaAs V-band monolithic HEMT LNA's', IEEE GaAs IC Symp. Dig., 1989, pp. 95-98
91. AVASARALA, *et al.*: 'A 1.6-Watt high efficiency X-band power MMIC', IEEE GaAs IC Symp. Dig., 1989, pp. 263-266
92. AYAKI, N., *et al.*: 'A 12 GHz-band monolithic HEMT low-noise amplifier', IEEE GaAs IC Symp. Dig., 1988, pp. 101-104
93. YUEN, C., *et al.*: 'A monolithic 3 to 40 GHz HEMT distributed amplifier', IEEE GaAs IC Symp. Dig., 1988, pp. 105-108
94. GAMAND, P., *et al.*: '2 to 42 GHz flat gain monolithic HEMT distributed amplifiers', IEEE GaAs IC Symp. Dig., 1988, pp. 109-111
95. KOMIAK, J. J.: 'S-band eight Watt power amplifier MMICs', IEEE GaAs IC Symp. Dig., 1988, pp. 45-48
96. LIU, C. S., WANG, K. G., and CHANG, C. D.: 'A 6-18 GHz monolithic low noise amplifier', IEEE GaAs IC Symp. Dig., 1987, pp. 211-214
97. ERON, M., TAYLOR, G., and MENNA, R.: 'X-band MMIC amplifier on GaAs/Si', IEEE GaAs IC Symp. Dig., 1987, pp. 171-173
98. WANG, S. K., *et al.*: 'Producibility of GaAs monolithic microwave integrated circuits', Microwave Journal, June 1986, pp. 121-133
99. BINGHAM, S. D., MCCARTER, S. D., and PAVIO, A. M.: 'A 6.5-16 GHz monolithic power amplifier module', *IEEE Trans. Microwave Theory Tech.*, MTT-33, 1985, pp. 1555-1558

100. SHIBATA, K., ABE, B., and KAWASAKI, H.: 'Broadband HEMT and GaAs FET amplifiers for 18-26.5 GHz', IEEE MTT-S Int. Microwave Symp. Dig., 1985, pp. 547-550
101. WATKINS, E. T., SCHELLENBERG, J. M., and YAMASAKI, H.: 'A 30 GHz low noise FET amplifier', IEEE MTT-S Int. Microwave Symp. Dig., 1985, pp. 321-323
102. AITCHISON, C. S.: 'The intrinsic noise figure of the MESFET distributed amplifier', *IEEE Trans. Microwave Theory Tech.*, MTT-33, 1985, pp. 460-466
103. BERENZ, J. J., NAKANO, K., and WELLER, P.: 'Low noise high electron mobility transistors', IEEE Int. Microwave Symp. Dig., 1984, pp. 98-101
104. YODER, M. N.: 'Distributed, traveling wave monolithic amplifiers', IEEE GaAs IC Symp. Dig., 1984, pp. 73-76
105. SCHELLENBERG, J. M., YAMASAKI, H., and ASHER, P. G.: '2 to 30 GHz monolithic distributed amplifier', IEEE GaAs IC Symp. Dig., 1984, pp. 77-79
106. CALANDRA, E., MARTINES, G., and SANNINO, M.: 'Characterization of GaAs FET's in terms of noise, gain, and scattering parameters through a noise parameter test set', *IEEE Trans. Microwave Theory Tech.*, MTT-32, 1984, pp. 231-242
107. TAJIMA, Y., and MILLER, P. D.: 'Design of broad-band power GaAs FET amplifiers', *IEEE Trans. Microwave Theory Tech.*, MTT-32, 1984, pp. 261-267

Chapter 6

Mixers

S. J. Nightingale

6.1 Introduction

A mixer, or frequency converter, has the prime function of converting a signal from one frequency to another with minimum loss of the signal and minimum noise performance degradation. However, as will be discussed later, the requirements for low loss and noise performance together with low spurious signal levels (mixed and intermodulation products) are often conflicting and an appropriate compromise must be made. The mixer consists of a 'pump' or 'local oscillator' and a network containing a non-linear device with one or more non-linear elements and/or non-linear current or voltage generators. The network also contains a circuit to couple the pump and RF signals to the non-linear device(s) together with a means to extract the required output signal. The output signal is formed by the fundamental or a harmonic component of the pump mixing with the RF signal. It is usually at a lower frequency than the pump or RF and is often referred to as the intermediate frequency (IF). The impedances presented across the non-linear device(s) by the surrounding network, at the discrete frequencies occurring within the mixer, are referred to as embedding impedances. In general, the pump voltage or current has a much larger amplitude than the input signal voltage or current and its sole function is to drive the non-linear device(s) to provide time-varying circuit parameters in the mixer. Under this condition, it can be shown that the frequency conversion process can be analysed by considering the mixer as a linear circuit with a time-varying impedance(s) or current or voltage generator(s). This type of operation in which mixed products of the form $nf_p \pm f_s$ are produced (n is an integer and f_p and f_s are the pump and signal RF frequencies respectively) is commonly referred to as linear mixing.

The non-linear device can be a non-linear passive element or a non-linear current or voltage generator. An example of the former is a non-linear resistor, capacitor or inductor or a combination of these such as found in a PIN or Schottky diode. An example of the latter is the transconductance, g_m, of a FET which varies as a function of gate bias. A network using a device which is predominantly resistive is called a mixer, whereas one using a device which is predominantly reactive is called a parametric converter. However, strictly speaking, they are both mixers.

6.1.1 The three basic types of mixer

6.1.1.1 Single-ended mixers (SEMs)
These mixers have no inherent isolation between pump, signal or IF due to the circuit geometry. This isolation is only achieved by using filters.

6.1.1.2 Single-balanced mixers (SBMs)
In these mixers, two signals are mutually isolated due to the circuit geometry rather than the use of filters. Usually, the pump and RF signal are mutually isolated and separated from the IF port by means of a filter. There is potential for pump AM noise suppression.

6.1.1.3 Double-balanced mixers (DBMs)
In these mixers all three signals are mutually isolated due to the balanced nature of the circuit geometry rather than the use of filters. All these mixers have pump AM noise suppression.

The spectrum generated in a linear mixer is shown in Fig. 6.1. When the signal is large compared with the pump, then non-linear mixing occurs and intermodulation products of the general form $nf_p \pm mf_s$ (m and n both integers) are generated. In order to analyse this general case, a full non-linear analysis must be performed where both signals are present.

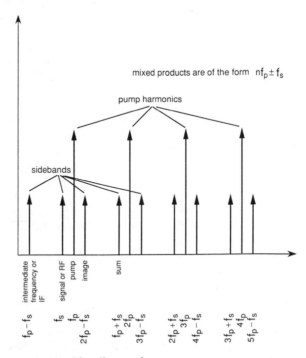

Fig. 6.1 Spectrum generated in a linear mixer

6.2 Mixer terminology

The following are some commonly used definitions:

6.2.1 Mixer bandwidth

The bandwidth of a mixer is defined by the range of frequencies which can be fed into the pump and RF ports and taken from the output port (IF), while maintaining acceptable performance. It is often specified for pump, RF and IF ports separately.

6.2.2 Narrowband mixer

A mixer of narrow operating bandwidth for pump and RF signal, typically 10-15%.

6.2.3 Wideband mixer

A mixer of wide operating bandwidth for pump and RF signal, typically 100% or more, e.g. 6-18 GHz.

6.2.4 Broadband mixer

A mixer is commonly referred to as being broadband or having a broadband input when the signal and image frequencies 'see' the same termination. A broadband mixer can be narrow- or wideband.

6.2.5 Harmonic mixer

Most mixers use the fundamental of the pump to produce an IF. Harmonic mixers deliberately use a harmonic ($n>1$) of the pump and are used in synthesisers and sampling systems or at millimetric frequencies where the fundamental may be difficult to generate.

6.2.6 Image recovery mixer (IRM)

An IRM is a circuit in which the internally generated image signal is fed back into the mixer for remixing to enhance the overall conversion loss/gain. This mixer also rejects the externally generated image frequency.

6.2.7 Image rejection mixer

This mixer is also referred to as a single sideband mixer. In this mixer, the internally or externally generated image frequency is terminated in a resistive load. This does not have as good a conversion loss/gain as the image recovery mixer, but usually has greater bandwidth.

N.B. Image recovery mixers exhibit image rejection, but the reverse is not true.

6.3 Mixer analysis

6.3.1 General analysis

In this analysis no assumptions are made about the relative magnitudes of pump or input signal(s). Theoretical implementation is difficult and such analyses are computationally slow compared with the restricted case. This is due to having to develop a robust convergence algorithm using time domain or harmonic balance methods which makes use of a special type of Fourier transform for multiple non-harmonically related signals. Such a general analysis can be simulated with a range of different CAD software packages as given in Table 6.1. Mixer products are produced of the form $nf_p \pm mf_s$. (f_p = pump frequency, f_s = signal frequency (RF) and n and m are integers).

6.3.2 Restricted analysis

In this analysis the assumption is made that the pump is much larger than the signal and is solely responsible for the time-varying impedance(s) and/or generators. The analysis is made in two parts: firstly a large signal analysis is performed using only the pump, and secondly a linear analysis is performed using the time-varying impedances and/or generators derived from the pump analysis. The former is performed using time domain or harmonic balance techniques. The latter is usually performed in the frequency domain using the Fourier components of the time-varying impedance(s) and/or generators to set up a conversion matrix. Mixed products are produced of the form $nf_p \pm f_s$ (f_p = pump frequency, f_s = signal frequency and n is an integer). When mixed products of the above form are produced this is referred to as linear mixing.

6.4 Background reading

There have been numerous publications over the past 40 years relating to the theoretical and practical aspects of mixer design. One of the most useful books, in the author's opinion, is the one by Tucker [1]. This book treats mixers, modulators and parametric converters and amplifiers purely in circuit terms and lays down the basic principles in a clear and concise way. Although the examples given are for two terminal devices (diodes), the analytical techniques can also be applied to three (single-gate FET and HEMT) or four (dual-gate FET and HEMT) terminal devices.

In 1971, Saleh [2] wrote a research monograph on mixers using non-linear resistors (ideal diodes). This work was a milestone in classifying circuits into different types such as the Z, Y, G, and H mixers and increased the general understanding of mixer operation. His classification was based on the parameters and matrix representation required to perform the conversion loss analysis. The book by Maas [3], published in 1986, presents an up-to-date treatment of mixers covering both diode and FET based circuits. Another book [4] by the same author, published in 1988, covers the main techniques in use today for solving for the large signal performance of mixers or power amplifiers. The book covers solid-state device modelling, harmonic balance and large-signal-small-signal analysis, Volterra-series and power-series analysis together with applications to diode and FET circuits such as mixers, multipliers and power amplifiers.

Table 6.1 A selection of available large signal non-linear analysis packages suitable for mixer analysis

Name	Supplier	Comments
MICROWAVE HARMONICA	Compact Software	Harmonic balance program, uses SUPERCOMPACT models for linear circuit analysis, has circuit optimisation
LIBRA	HP-EEsof	Harmonic balance program, uses TOUCHSTONE models for linear circuit analysis, has circuit optimisation
MICROWAVE SPICE	HP-EEsof	Time-domain analysis similar to SPICE with microwave models, no circuit optimisation
ANAMIC	Sobhy and Jastrzebski Microwave Consultants	Time-domain, uses state space techniques, includes optimisation
CIRCEC	Thomson	Time domain, limited microwave models
SPECTRE (HARMONICA)	K. Kundert, U.C. Berkeley	Harmonic balance similar to MICROWAVE HARMONICA
ASTEC 3	SIA	Time domain, no microwave models
GISSMIX	P. Siegel, NASA	Restricted to single loop diode circuits
DIODEMX	S. Maas	Uses large signal pump analysis followed by small signal conversion loss and noise analysis. No optimisation

6.5 Analysis of mixer circuits

In order to give some understanding of mixer operation, a simple single loop circuit will now be analysed.

6.5.1 Analysis of a simple single loop circuit

Consider the simple single loop circuit shown in Fig. 6.2. This represents a single-ended mixer design where all mixed products and intermodulation products are terminated in the source and load resistances, i.e. a wideband mixer design. This circuit will now be examined using both a general non-linear analysis and the method of linear circuits with time-varying parameters. The latter method is also referred to as the conversion loss matrix method.

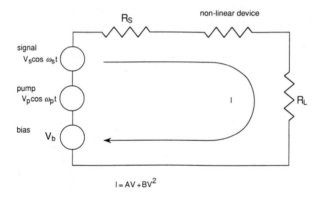

Fig. 6.2 A simple single-ended mixer

6.5.2 General non-linear analysis

We begin by combining the source and load resistances with the non-linear device to form a new non-linear equation:

$$I = AV + BV^2 \qquad (6.1)$$

After substituting for V and some trivial manipulation we have:

$$I = AV + BV^2$$

$$= AV_b + B(V_b^2 + \frac{1}{2}(V_p^2 + V_s^2)) \qquad \text{- DC}$$

$$+ (A + 2BV_b) V_s\cos\omega_s t \qquad \text{-Signal}$$

$+ \frac{1}{2}BV_s^2 \cos 2\omega_s t$ -Signal second harmonic

$+ (A + 2BV_b)V_p \cos \omega_p t$ -Pump

$+ \frac{1}{2}BV_p^2 \cos 2\omega_p t$ -Pump second harmonic

$+ BV_p V_s \cos(\omega_p + \omega_s)t$ -Sum frequency

$+ BV_p V_s \cos(\omega_p - \omega_s)t$ -Difference or intermediate frequency   (6.2)

6.5.2.1 Conversion loss
The power available from the signal, P_{av}, is given by:

$$P_{av} = \frac{V^2}{8R_s} \qquad (V = V_p \text{ or } V_s) \tag{6.3}$$

Power into the load at any frequency is given by:

$$P_{out} = \frac{I^2 R_L}{2} \tag{6.4}$$

Conversion loss (CL) is therefore given by:

$$CL = \frac{V^2}{4I^2 R_s R_L} \qquad (V = V_p \text{ or } V_s) \tag{6.5}$$

f_s to $f_p - f_s$ and $f_p + f_s$ $$CL = \frac{1}{4R_s R_L (BV_p)^2} \tag{6.6}$$

f_s to $2f_s$ and f_p to $2f_p$ $$CL = \frac{1}{2 R_s R_L B^2} \tag{6.7}$$

6.5.2.2 Input and output impedance
The input or output impedances of the circuit at a particular frequency are obtained by exciting the circuit with a voltage generator at this frequency and finding the current which flows at this frequency. In any practical measurement of this parameter it must be possible to separate the required current from others which are flowing in the circuit.

$$Z_{in} = \frac{V \text{ at excitation frequency}}{I \text{ at excitation frequency}} - R_S \qquad (6.8)$$

$$Z_{out} = \frac{V \text{ at excitation frequency}}{I \text{ at excitation frequency}} - R_L \qquad (6.9)$$

f_s and f_p $\qquad Z_{in} = \dfrac{1}{(A + 2BV_b)} - R_S \qquad (6.10)$

$f_p\text{-}f_s$ and $f_p\text{+}f_s$ $\qquad Z_{out} = \dfrac{1}{(A + 2BV_b)} - R_L \qquad (6.11)$

6.5.2.3 Isolation

The isolation (I) between ports is given by the ratio of the power available from the source to the power dissipated in the load at the same frequency:

f_s and f_p $\qquad I = \dfrac{1}{4R_S R_L (A + 2BV_b)^2} \qquad (6.12)$

6.5.3 Conversion loss matrix method

In this approach the local oscillator or pump is assumed to be solely responsible for the mixing. Under these conditions, the mixer can be analysed in two stages:

(1) A non-linear analysis to determine the steady-state time-varying conductance waveform.

(2) A linear analysis to determine the small signal performance.

The current waveform is given by:

$$I = AV + BV^2 \qquad (6.13)$$

Therefore the conductance waveform is given by:

$$G = \frac{dI}{dV} = A + 2BV \qquad (6.14)$$

therefore $\qquad G = A + 2B(V_p \cos \omega_p t + V_b) \qquad (6.15)$

The currents flowing in this time-varying conductance waveform are now given by:

$$I(t) = V(t)G(t) \qquad (6.16)$$

$$I = V_s \cos\omega_s t \left(A + 2B(V_p \cos\omega_p t + V_b)\right) \qquad (6.17)$$

$$= (A + 2BV_b)V_s \cos\omega_s t \qquad \text{- signal}$$

$$+ BV_p V_s \cos(\omega_p + \omega_s)t \qquad \text{- sum}$$

$$+ BV_p V_s \cos(\omega_p - \omega_s)t \qquad \text{- difference frequency (IF)}$$

6.5.3.1 Conversion loss
The conversion loss is now given by:

$$\begin{array}{c} f_s \text{ to } f_p + f_s \\ \text{and } f_s \text{ to } f_p - f_s \end{array} \qquad CL = \frac{1}{4R_S R_L (BV_p)^2} \qquad (6.18)$$

which is identical to the result given in eqn. 6.6. The results for the input and output impedance and signal isolation will also be found to be identical.

The above is a very simple example to illustrate how the mixer performance can be determined when the device non-linearity is expressed as the first two terms of a power series. Such a device does not exist in practice because, although some two-terminal devices have a square law characteristic in the forward direction, they have very little conduction in reverse. The characteristic can then be represented by either a much higher order power series [5] or by splitting the forward and reverse parts of the characteristic and determining the Fourier series for the conductance waveform separately [6]. When the diode is being driven hard it can be considered as a sampler switching between two states. This bilinear or switch model has been used extensively to analyse the performance of diode mixers and will be referred to in the forthcoming sections.

The references are divided into six areas: those giving background information on mixer operation and performance [1-15], diode MMIC mixers [16-57], subharmonic MMIC mixers [58-59], FET and HEMT MMIC mixers [60-86], balun design [87-98] and miscellaneous references [99-105]. References from number 16 onwards are given in chronological order for ease of identification.

Further details on the operation of diode and FET or HEMT mixers will now be discussed separately.

6.6 Diode mixers

6.6.1 Introduction

In order to achieve minimum conversion loss in a mixer, there are two fundamental conditions which must be satisfied:

(1) Minimum power from the input signal and mixed product frequencies must be dissipated in the non-linear element or elements

(2) All out of band mixed product frequencies must be terminated reactively

However, it should be noted this is in general contrary to the requirements for low out of band mixed and intermodulation product levels where it is 'good mixer practice' to terminate out of band products resistively [7]. An appropriate compromise must be made which depends on the system application. In a wideband ESM receiver, for example, low spurious signal levels are more important than low loss and noise figure whereas these are not so critical in a narrowband Satcom receiver, where they can be removed by filtering.

Condition (1) is obvious and, ideally, would be satisfied by a device which at no time during the pump cycle assumed an impedance with a finite resistive component. Although this condition is satisfied by a non-linear capacitor, the Manley-Rowe relationships [8, 9] show that a parametric down converter has a power loss which is equal to the ratio between the input and output frequencies in the unconditionally stable state. Although it is theoretically possible to make a down converter with gain, it is not desirable since it would be potentially unstable. These points suggest that the optimum device would be one switching between two resistance values which, for a zero loss mixer, tend to 0 and ∞, i.e. a short and open circuit. Saleh [2] presented a rigorous mathematical proof for this and showed that the optimum time-varying conductance or resistance waveform was a rectangular pulse or group of pulses.

Although it is possible to simulate this ideal situation at low frequencies using, for example, a relay or reed switch, at RF and above it is necessary to utilise the non-linear properties of a semiconductor or thermionic device. A device which is used almost exclusively nowadays in low loss mixers at microwave frequencies is the Schottky barrier diode or a related structure. This device has largely superseded its forerunner – the point contact diode – particularly at lower microwave frequencies (i.e. <40 GHz). An equivalent circuit for a typical microwave diode is given in Fig. 6.3 together with the key equations defining the non-linearities R_j and C_j. The elements L_s, r_s and C_m are the device parasitics. In order to obtain low loss, it is necessary to pump the diode strongly so that the junction closely approximates a fully on or off state and, therefore, dissipates minimum power. Under high drive conditions, a typical diode impedance is limited by the series resistance in the forward direction, and the junction capacitance in parallel with a high resistance, representing the leakage, in reverse. The time-varying impedance or admittance then tends to a rectangular pulse waveform and a particular value of pump voltage or current and DC bias will determine how long the device conducts for each pump cycle. This two-state or bilinear model has been used extensively in the past to determine the optimum conditions for minimum loss in a specified embedding structure as well as to investigate other mixer phenomena. Although such a model does provide considerable insight into optimum operating conditions, it has some limitations at low pump power levels. The main advantage of this model is the ease with which a solution for the conversion loss performance may be obtained for a specified embedding structure. Once the device parameters and embedding structure have been defined, the only variables are the source and load impedances and the conduction time per pump cycle or pulse duty ratio (PDR).

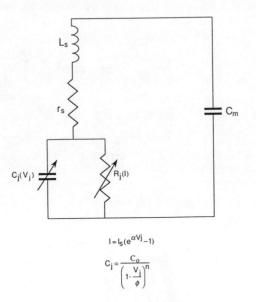

Fig. 6.3 A Schottky barrier diode equivalent circuit

Condition (2) must be satisfied in order to avoid power loss at out of band frequencies, although there are cases where the difference between total and selective reactive termination is small. Since a reactive termination can vary from an open to a short circuit, it is informative to note the general effects of these. Fig. 6.4, taken from a study by the author [10], shows how the conversion loss of a mixer can be improved by progressively terminating more higher-order mixed products in a short circuit. The main conclusions of this work on single loop circuits were as follows:-

(a) As more currents are terminated in a short circuit (current idler) the minimum conversion loss attainable decreases, the optimum pulse duty ratio decreases and the optimum source and load impedances increase. In the limit, where all higher order mixed products are terminated in a short circuit, we have a Y mixer. In this circuit, with a high Q device, the optimum operating conditions are very high source and load impedances and a very small pulse duty ratio, tending to 0; i.e. impulse conditions.

(b) The reverse of (a) is also true. In the limit, where all higher order mixed products are terminated in an open circuit (voltage idler), we have a Z mixer. In this circuit, with a high Q device, the optimum operating conditions are very low source and load impedances and a very large pulse duty ratio tending to 1; i.e. inverse impulse conditions.

212 *Mixers*

$R_{on} = 2\ \Omega$
$R_{off} = 10^6\ \Omega$

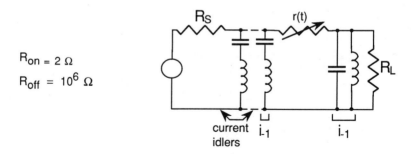

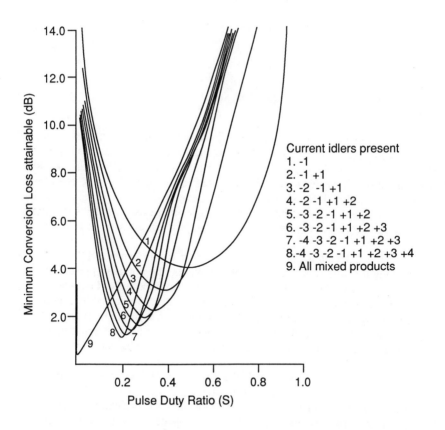

Current idlers present
1. -1
2. -1 +1
3. -2 -1 +1
4. -2 -1 +1 +2
5. -3 -2 -1 +1 +2
6. -3 -2 -1 +1 +2 +3
7. -4 -3 -2 -1 +1 +2 +3
8. -4 -3 -2 -1 +1 +2 +3 +4
9. All mixed products

Fig. 6.4 Conversion loss against pulse duty ratio S for different current idlers present in a mixer

(c) Following from (a) and (b) it is possible to select a pulse duty ratio and terminate the mixed products reactively but in different ways (i.e. s/c, o/c or in between) so that the minimum conversion loss is realised. Specific cases of this are the G(H) mixers where odd order products are terminated in an open (short) circuit and even order products are terminated in a short (open) circuit. For these two mixers the optimum pulse duty ratio is 0·5.

(d) In general, once reactive terminations have been obtained by correctly terminating the sum and image frequencies, there is little to be gained by attempting to recover energy from other mixed products unless a high Q medium for the embedding structure and a high Q device are available. Consideration of several microstrip structures shows that the extra potential improvement in conversion loss achieved by terminating higher order mixed products reactively is usually more than cancelled out by the additional loss of the embedding circuit (filter).

(e) In most practical cases, the non-linear device will have parasitic capacitance, inductance and resistance. The main parasitics in a Schottky barrier diode are the 'on' resistance, which is primarily due to the device series and junction resistance, and the 'off' capacitance which is due to the junction. The latter tends to progressively 'short circuit' higher order mixed products across the diode junction. Therefore the Y mixer is the single most commonly used circuit at microwave frequencies. Consideration of the noise performance shows that it is preferable to use short circuit terminations to avoid parametric amplification of noise from mixed products to the IF.

(f) In mixers where high Q embedding circuits cannot be realised (e.g. microstrip), the device zero bias junction capacitance should have a value such that the reactance at the signal frequency is typically twice the signal source impedance. Once this has been achieved, the series resistance should be made as small as possible.

6.6.2 Equivalences

There are several important equivalences which enable the lattice or ring mixer and shunt and series mounted diode structures to be analysed from a common circuit. These are summarised in Fig. 6.5. It should be noted that in this case odd and even order mixed products are defined by the integer n in the frequency $nf_p \pm f_s$. This definition is different from that used by Saleh [2].

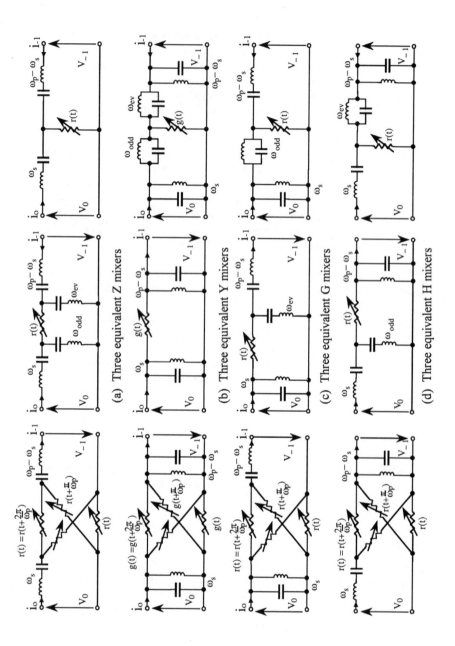

Fig. 6.5 Equivalences between different mixer types
(where $V_o = V_s \cos \omega_s t$ and $V_{-1} = V_{IF} \cos(\omega_p - \omega_s)t$)

6.6.3 Noise figure

The noise figure of a mixer is quite difficult to calculate for the general case. However, if it can be assumed that the dominant noise mechanisms are shot and thermal noise, i.e. pump noise and other device noise such as $1/f$ noise are assumed to be zero, the situation is simplified considerably.

It was shown by the author [10, 11], developing the work of Nemlikher and Strukov [12, 13], that the noise figure, F_m, of a mixer using Schottky barrier diodes could be represented by the following equations:

$$t_{eq} = \frac{F_m - 1}{L_i - 1} \qquad (6.19)$$

$$= \frac{p_r + \frac{n}{2} p_j}{p_r + p_j} \qquad (6.20)$$

$$L_m = L_i L_{mm} \qquad (6.21)$$

where: t_{eq} = the homogeneous equivalent temperature ratio of the mixer
F_m = the mixer noise figure
L_i = the mixer intrinsic conversion loss
p_r, p_j = the normalised power coefficients which account for the proportions of power dissipated in the circuit resistances and device junction(s), respectively.
n = the ideality factor of the diode (typically 1·1)
L_m = the mixer conversion loss
L_{mm} = the mixer mismatch loss at the input

The above equation assumes the mixer is at 290 K.

It should be noted that an alternative definition for the noise temperature ratio, t_m, of a mixer with total loss L_m is given by:

$$t_m = \frac{F_m}{L_m} \qquad (6.22)$$

This definition is commonly used because it is compatible with Friis' formula for cascaded networks. Although it is one way to represent the noise performance of a mixer or, generally, any lossy network, unfortunately it does not help the analysis or understanding of the problem because both types of loss are lumped together.

When the diode is weakly driven, i.e. the pump drive power is low, and the signal and IF are perfectly matched and all out of band mixed products are terminated reactively:

$$p_j \gg p_n$$
$$t_{eq} \to n/2$$

$$\therefore \quad F_m \to nL_i/2 + (2-n)/2 \quad (6.23)$$
$$\approx nL_i/2 \text{ when } L_i \text{ is large}$$

This situation is seldom of practical interest unless the resistive losses are very low, because the conversion loss will be very high.

In most mixers operating under practical pumping conditions p_r is much greater than p_j [10]. Indeed, under strong pumping conditions where the junction is made to switch hard between an 'on' and an 'off' state, close to zero power is dissipated in the junction, hence p_j tends to zero and t_{eq} tends to 1.

$$p_j \to 0$$
$$t_{eq} \to 1$$

$$\therefore \quad F_m = L_i \quad (6.23)$$

The overall noise figure, F_t, of a mixer in cascade with an IF amplifier of noise figure F_{if}, can be found using the Friis formula for cascaded networks, *viz*:

$$F_t = F_m + (F_{if} - 1) L_m$$

$$= F_{if} L_m \quad (6.24)$$

6.6.4 A simple single-ended mixer

One of the most basic forms of mixer circuit is the single-ended mixer and an example of this is shown in Fig. 6.6. The single-ended mixer described here is often used to form the basis of a single-balanced mixer. The lumped-element circuit shows the diode represented by a time-varying resistance ($r(t)$), with the pump and RF signals being fed from a common input. A parallel circuit tuned to the IF at the output provides a short circuit to all pump harmonics and mixed product frequencies other than the IF. A current idler is provided at the input which provides a short circuit to the IF at the signal port, thus preventing IF power being dissipated in the signal source resistance. On the right hand side of the lumped-element circuit, the approximate distributed equivalent is given in two-wire transmission line form. In this circuit, the assumption is made that the IF is low and therefore, assuming linear mixing, the frequency spectrum is as shown in Fig. 6.1 where, apart from the IF, all other sideband pairs are closely spaced either side of the pump fundamental and its

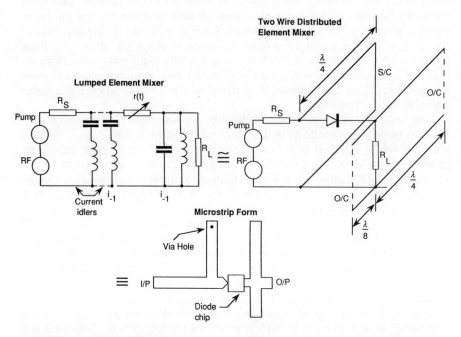

Fig. 6.6 A single-ended mixer

harmonics. The assumption is made that the characteristic impedances of the stubs are sufficiently low that each resonant element presents approximately the same impedance to a given pump harmonic and its associated sidebands. The IF output has two open circuit stubs which are $\lambda/4$ and $\lambda/8$ at the pump fundamental. The $\lambda/4$ stub provides a short circuit at the plane of the diode for the odd order multiples of the pump frequency and the associated sidebands, whereas the $\lambda/8$ stub provides a short circuit for the pump second harmonic and odd order multiples of it and associated sidebands. Thus, in combination they approximate the parallel tuned circuit of the lumped-element form. The short circuit stub on the pump and RF side of the diode is $\lambda/4$ at the pump fundamental and its associated sidebands (the RF and the image frequency). Therefore, the RF, image and pump frequencies reach the diode, but the second harmonic of the pump and the sum frequency see a short circuit at the diode plane. In addition, the short circuit stub provides a DC return for the rectified current generated by the diode. A DC short circuit is also required across the IF output, but this is not shown and could in practice be provided by a shunt inductor. Any DC bias supplied to the diode to define its operating point is also best applied via the IF port. Below the two idealised forms of the lumped element and two-wire distributed circuits is a typical microstrip representation which is self explanatory. In practice it may be necessary to add additional filter elements to the IF port to provide adequate rejection of pump harmonics and higher order mixed product frequencies. Generally, the pump fundamental and harmonics cause the main problem because the pump power is much larger than typical RF levels. A simple matching section may also be included on the pump and RF side of the diode to match the RF to the diode impedance. It is not possible to match the pump and RF simultaneously but, since the object of frequency conversion is usually to minimise the conversion loss and noise figure, the matching circuit should be designed to match the RF to the diode under typical pumping conditions.

The single-ended mixer is not normally used in practice because it has no mutual isolation between the pump, RF and IF. However, in some designs, where the IF is high, this can be achieved by the use of filters. A second reason is that there is no pump AM noise suppression. Usually, two single-ended mixers are connected to the outputs of a 3 dB coupler, or equivalent form of four-port circuit, where the pump and RF input signals are mutually isolated due to the properties of the coupler. It should be noted that good isolation between pump and RF is only achieved when the coupler's 3 dB outputs are well matched in amplitude and the coupler provides a high directivity when operated under matched conditions. Pump AM noise suppression is achieved with a mixer of this type and it is commonly referred to as a single-balanced mixer because of the inherent balance or mutual isolation between the pump and signal. There is no balance between the pump and IF or the RF and IF, so the required isolation is only achieved by filtering.

A considerable number of publications exist on the design of diode mixers and a selection of those realised as MMICs is given in References 16-59. Tables 6.2 and 6.3 provide a summary of these. A diode mixer here has been considered as one which uses diodes or a FET or HEMT configured as a diode to perform the switched sampling or commutation frequency conversion process. Therefore, a mixer which uses FETs to form active baluns has been included. The references in the tables cover only mixers where performance was given in terms of conversion loss or noise figure for the mixer alone. Therefore, references have been excluded which do not allow the performance of the mixer alone to be calculated or only give a loss/gain figure for a mixer and amplifier combination.

6.6.5 Design of a 94 GHz mixer

6.6.5.1 Introduction

The following paragraphs describe some of the steps involved in designing a 94 GHz mixer. Although the work described below makes rather ideal assumptions for some of the local oscillator and mixed product terminations, the overall procedure and method of design can be followed for other designs at different frequencies. The analyses presented here were made with the program DIODEMX which is a public domain program written by Stephen Maas and is listed at the back of the first edition of Reference 3.

Mixers 219

Ref.	Company	RF (GHz)	IF (GHz)	CL (dB)	DSB NF (dB)	SSB NF (dB)	Diode f_t (GHz)	Comments
16	Siemens	15	70 (MHz)	8.5	[5.5]	10 inc. 1.5 dB IF [8.5]	130	Microstrip SBM with branchline coupler.
17	Honeywell	30-40	5-500 (MHz)	[7.5]	6.0 (31-39GHz) inc. 1.5dB IF [4.5]	[7.5]	660	Suspended stripline SBM in WG, pump power 6mW, no dc bias.
18	GEC	8-12	1-500 (MHz)	4.95	[1.95]	[4.95]	-	Theoretical figure. Microstrip SBM with Lange couplers.
19	Honeywell	93-97	5-500 (MHz)	[7.8]	6.3 inc. 1.5dB IF [4.8]	[7.8]	600	Cross bar SBM in WG.
20	Hughes	75-110	1-9	9.0 av. 7.5 @76GHz	[6.0] [4.5]	[9.0] [7.5]	3000	Cross bar SBM in WG.
21	Honeywell	80-105	-	[7.1]	5.6 @ 94.5GHz inc. 1.5dB IF [4.1]	[7.1]	700	Microstrip SBM with rat race coupler.
23	Hughes	73.6-83.6	8-18 (GHz)	8.0, 4.6 @ 91.1GHz	[5.0] [1.6]	[8.0] [4.6]	-	CPW & slotline SBM.
24	GE	26-40	1	6.0	[3.0]	[6.0]	640	CPW & CPS SBM with bow tie transition.
25	MIT Lincoln Labs.	34-36	1-2	6.5 +/- 0.5 (34-36GHz)	[3.5 +/- 0.5]	[6.5 +/- 0.5]	-	Microstrip SBM with branchline coupler.
27	GE	30	1	6.0	[3.0]	[6.0]	637	CPW and CPS SBM using Mott diodes.
28	Hughes	27-30	3.5-6.5	10.5	[7.5]	[10.5]	550	Microstrip SBM with Lange coupler. Part of complete RX.
29	Honeywell	27-30	5.5-8.0	8.0 @ 30GHz	[5.0]	[8.0]	>1000	Microstrip SBM with modified rat race coupler.
30	Pacific Monolithics	3.7-4.2	0.1-1.8	-	-	-	-	Complete RX with RF & IF LNAs. DBM. NF=4.0dB.
33	Hughes	75-88	4-17	6.8-10.0	[3.8-7.0]	[6.8-10.0]	450-560	Microstrip SBM with Lange coupler.
34	TI	2-18	500 (MHz)	12.0 @ 10GHz	[9.0]	12.0-15.0	-	DBM with FET baluns.
35	AEG	50-70	-	6 @ 60GHz	3.3 @ 60GHz	[6.3]	2300	Microstrip SBM with modified branceline coupler.
36	Thomson	7.47	30 (MHz)	6.7	[3.7]	[6.7]	-	Lumped element rat race coupler.
37	Telefunken Sys'technik	94	1	8 @ 94GHz	6 @ 94GHz	[9.0]	2000	Microstrip SBM with branchline coupler.
38	TRW	7-10	5	10.0	[7.0]	[10.0]	-	MESFET technology DBM using lumped element high & low pass networks for balun. High IP3.
39	TRW & ESG	30-40	20-100 (MHz)	5.0 @ 35GHz	[2.0]	[5.0]	-	Complete downconverter in HEMT technology. Microstrip SBM with Lange coupler.

Table 6.2 A comparison of different monolithic mixers using diodes

220 Mixers

Table 6.2 continued

#	Company	Col1	Col2	Col3	Col4	Col5	Col6	Comments
41	Martin Marietta & GAMMA	85-100	-	-	-	-	-	HEMT technology. Microstrip SBM with branchline coupler.
42	TRW & ESG	92-96	1-8	<9.0 (90-98GHz) 7.5 @ 94GHz	<6.0 [4.5]	[<9.0] [7.5]	370	PHEMT technology downconverter with 2 stage RF LNA. Microstrip SBM with rat race coupler. 5.3dB gain & 6.8dB NF @ 94GHz.
43	Varian	6-20	2-7	6.2-9.8	[3.2-6.8]	[6.2-9.8]	-	CPW, CPS & slotline SBM.
45	Telefunken Sys'technik	94	0.1-1.8	6.0 @ 94GHz	4.0	[7.0]	-	MESFET technology downconverter. Microstrip SBM with Lange coupler.
46	Raytheon et al.	35	-	19 @ 35GHz	8.7 @ 35GHz	[11.7]	-	Quasi-optical millimeter-wave monopulse RX with slot ring SBMs.
47	THORN EMI & Alpha	35	100 (MHz)	4.6	[1.6]	[4.6]	1058	Microstrip SBM with sum enhancement. Rat race mixer with vertical diode.
47	THORN EMI & Alpha	35	100 (MHz)	5.7	[2.7]	[5.7]	1058	Microstrip SBM with Lange coupler with vertical diode.
48	Nonlinear Consulting & TRW	26-40	DC-10	5.0-10.0	[2.0-7.0]	[5.0-10.0]	235	Planar Marchand baluns. DBM with diodes in star configuration.
49	Northrop	16-21	200 (MHz)	8.0-11.0	-	-	-	PHEMT technology. Microstrip DBM.
50	TRW	10-18	-	-	-	-	-	HEMT technology image rejection RX. Microstrip SBM with Lange couplers. 7-10dB gain.
51	COMSAT	91-96	50 (MHz)	8.5	-	-	-	LNA SBM combination. CG=7dB & NF=7.3dB.
52	TRW, Hughes & Hexwave	20.2-21.2	1.55	-	-	6.0	-	LNA SBM combination. CG=15dB & NF=6.0dB.
53	TI	6-18	500 (MHz)	<8.0	[<5.0]	[<5.0]	-	DBM with novel form of Marchnd balun.
54	TRW	18-22	6-10	5.0-6.0	[2.0-3.0]	[5.0-6.0]	-	HBT diode DBM.
55	TRW	43-46	16-21	<6.0	[<3.0]	[<6.0]	-	HEMT diode DBM with Lange couplers.
56	TRW	60-63	8.1	-	-	-	1.02THz	HEMT diode SBMs using modified rat race couplers. RF & IF LNAs. CG >10dB.
57	Hittite	5-20 6-11 4-8 1.8-5 1-2	dc-3.0 dc-2.0 dc-2.0 dc-2.0 dc-1.0	10.0 10.0 10.0 11.0 11.0	- 7.0 7.0 7.0 8.0 8.0	[10.0] [10.0] [10.0] [11.0] [11.0]	-	DBMs with folded passive baluns or planar transformers. 25-35dB pump-RF isolation. P_p=13-20dbm. P_{1dB}=10dbm. IP3=20dbm.

Mixers 221

Ref.	Company	RF (GHz)	IF (GHz)	CG (dB)	DSB NF (dB)	SSB NF (dB)	Gate Length (µm)	Comments
60	CISE & Alcatel	11.7-12.4	-	-7.5	-	-	-	Dual gate FET SEM.
61	Matsushita	100-800 (MHz)	100MHz	6.0-8.0	-	-	1.0	DBM using dual gate FETs. P_o=10dBm.
62	GE	6-18	0.1-2.0	-8.0	-	-	0.5	DBM using single gate FETs. Theoretical figure based on switching model.
64	TI	2-18	0.2-1.4	0 - -6.5 (0.1-1.4GHz IF)	15.0-18.0	-	0.5	4 stage distributed SEM realised with dual gate FETs.
65	TRW	6-10	3	7+/-1.5 (6-10GHz)	13.0	-	0.5	Dual gate SBM. Complete RX with RF & IF LNAs. P_o=10dBm.
66	Plessey	6-4	2-7	5.0	13.0	-	1.0	Dual gate common source SBM. Complete downconverter with RF & IF LNAs.
67	Avantek	0.1-6	70MHz	8-15 (RF=2GHz, IF=250MHz)	-	15.5-16 (RF=2GHz & IF=250MHz)	-	Silicon bipolar active mixer using a Gilbert cell.
73	Sharp	11.7-12.2	1.0-1.5	>0 (11.7-12.2GHz)	<6.5 (11.7-12.2GHz)	-	-	Drain pump injection SBM. Downconverter for broadcast satellite applications. P_o=-9dBm.
75	TRW & UCLA	DC-5	8-10	-4.0 @ 2.5GHz	-	-	0.5	DBM using dual gate FETs. Upconverter chip with RF & IF LNAs.
76	TRW	3-18	-	-7.9-9.0 (4-14GHz) -7.5--11(3-18GHz)	-	-	0.2-0.25	HEMT DBM. Active pump balun, passive RF & IF baluns. HEMTS used as switches in mixer.
79	Pacific Monolithics	34.1-36	0.1-2	-4.0 @3GHz	-	-	-	PHEMT mixer fed via drain. Complete RX chip with RF & IF LNA. P_o=-7dBm.
80	Alcatel	12.7-15.5	50 (MHz)	-6.2-6.9	-	-	0.25	HEMT process - cold shunt HEMT mixer cell SBM.
81	Comms. Res. Centre.	19.5-22	4	<-3.0	-	-	0.25	PHEMT process. SBM. Pump and RF to gate via external diplexer.
82	Comms. Res. Centre.	28.5-31	4	1-3.7	-	-	0.25	PHEMT process. Same as above.
83	Fujitsu	58-62	140 (MHz)	-7.5 @ 59.86GHz	-	-	0.15	SBM active drain mixer with Lange couplers in an image rejection configuration.
84	M/A COM & Anadigics	2.5	100 (MHz)	0.9	-	15.0	-	Silicon MMIC Gilbert cell mixer with 20dB improvement in dynamic range.
86	Raytheon	4-18	1-2	<12.0	<12.0	-	-	Switched PHEMT SBM with active pump balun & passive RF balun.

Table 6.3 A comparison of different monolithic mixers using FETs, HEMTs and pHEMTs

6.6.5.2 General background to a 94 GHz mixer design

The design of a mixer with low conversion loss and noise figure involves consideration of both the small and large signal circuits. We begin by considering the small signal circuit and associated conditions.

The two basic conditions which must be fulfilled in order to minimise the conversion loss were described earlier in this section. Firstly, the device must ideally be driven by the pump such that at no time during the pump period does it assume an impedance with a finite real value and secondly, all out of band mixed products must be terminated reactively. In the case of a Schottky barrier diode, the first condition requires that the diode is driven with sufficient pump power such that the current waveform and hence the conductance waveform meets this requirement. The second condition to minimise the conversion loss is that the input and output signals must be matched and out of band mixed products must be terminated reactively across the diode. This is obvious from conservation of energy. Clearly, as stated previously, such a condition can be met by terminating out of band mixed products in open or short circuits or anything in between. However, in any practical mixer, such as one in microstrip, it is unrealistic to attempt reactive termination for all the out of band mixed products since the loss associated with the complex filters or embedding circuits required to do this would more than offset any potential conversion loss improvement. Therefore, in the practical case, one normally attempts to terminate reactively (i.e. recover energy from) those mixed products which have significant power transferred to them during the mixing process. In general, one can say that the amount of power in the mixed products falls off with increasing product order, although the fall off is by no means proportional to the product order (the order of a mixed product is given by the associated pump harmonic which produces it during the basic mixing process, e.g. $3f_p + f_s$ is a third order product). Analysis of a large number of circuits has shown that the greatest improvement to the conversion loss is obtained by correctly terminating the sum ($f_p + f_s$) and image ($2f_p - f_s$) frequencies. Indeed, in many practical cases, the difference in conversion loss between partial reactive termination where only energy is recovered from the sum and image compared with total reactive termination is very small. This is assuming that, in both cases, the source and load, i.e. RF and IF signals, are conjugately matched. We now consider how to provide these reactive terminations.

Undoubtedly the most difficult frequency to deal with is the image frequency, because of its close proximity to the signal. In this 94 GHz mixer, the signal (RF) is at 94 GHz and the pump is at 93 GHz to give a 1 GHz IF. Therefore, the image is at 92 GHz and it is not possible, bearing in mind the finite Qs of microstrip filters, to pass the signal and capture or block the image frequency. The only practical alternative is to provide an effective reactive termination for the image frequency by using two mixers which are pumped in phase quadrature. This technique is well known and will not be discussed further here except to say that the overall structure is twice as large and doubles the pump power requirement. It is worth noting at this point that the maximum potential improvement to the conversion loss using image recovery is 3 dB. However, this oft quoted figure is for the case of a mixer using a pure non-linear resistance and the figure is less when linear or non-linear capacitance is present. In fact, the potential performance improvement is sometimes as little as

1dB which is not worth pursuing unless it is also desired to use the mixer for single channel or single sideband reception, i.e. the incoming signal is converted to the IF while the conversion loss for the image channel is so high it is effectively rejected. Bearing in mind the above points, the mixer was designed to make use only of sum recovery.

It was mentioned earlier that, in principle, any form of reactive termination could be used ranging from an open circuit to a short circuit. However, in practice, there are two effects that arise from the junction capacitance which restrict our choice: firstly, the very presence of junction capacitance tends to progressively short circuit higher order mixed products directly across the junction, therefore making it impossible to realise an open circuit condition, and secondly, the junction capacitance is non-linear and increasing with forward bias voltage. This latter effect gives increased conversion loss when the diode is operated with high forward bias due to signal shunting effects and also gives rise to parametric effects which increase the overall noise figure when open circuit terminations are used. This effect has been reported by a number of authors including Torrey and Whitmer, Pound and more recently Liechti [15]. The effect occurs because the non-linear junction capacitance makes the mixer behave like a negative resistance parametric amplifier producing negative impedances across the junction at some or all of the mixed product frequencies. This amplifies the noise voltages at those frequencies generated both from the diode junction (shot) and the diode series resistance and the real part of the embedding impedance (thermal or Johnson) which in turn is mixed down to the IF to emerge from the IF port and degrade the overall noise figure. Clearly, the approach to adopt to prevent this effect is to design an embedding structure which provides a short circuit to the noise voltages at the major mixed product frequencies. It should be noted that this effect exists with standard Schottky barrier diodes, but can be drastically reduced or eliminated by using a Mott diode where the capacitance-voltage characteristic is virtually flat.

It should now be clear from the above discussion that, for the small signal circuit, a short circuit termination or a close approximation to this should be provided for the sum frequency. This is most easily achieved, in the case of a microstrip mixer, by using an open circuit shunt stub which is a quarter wavelength long at the sum frequency. In view of the close proximity of the pump second harmonic and the sideband $3f_p-f_s$, these will also be terminated in a short circuit.

We will now consider what happens to the large signal waveform. This is, in general, quite complex, but several effects can be considered qualitatively: firstly, for a given pump power, it is desirable to minimise the conversion loss by turning the diode on and off as rapidly as possible. To a first approximation, turn-on-time or rise time is related to the frequency response by the simple approximation Bandwidth × Risetime ≈ 0·35. Therefore, to decrease the rise time we must increase the bandwidth. In the case of the mixer diode, this means ensuring that reasonably large currents flow at the pump harmonics so as to give a short rise and fall time to the current waveform. A few analyses with DIODEMX readily show this effect. It is very important to terminate the second harmonic correctly. Higher harmonics will be short circuited across the junction directly by the junction capacitance. A short circuit

224 Mixers

second harmonic will ensure a high current and thus help to decrease the rise and fall time. It is interesting to note that an open circuit condition at the second harmonic across the junction can give rise to a double pulse condition for the current and make it impossible to realise narrow pulses with sharp risetimes.

6.6.5.3 Details of a 94 GHz single-balanced mixer design

A single diode mixer was designed first using the program DIODEMX. The general aim was to design a single-balanced mixer at 94 GHz operating between source and load impedances of 50 Ω. Therefore, a single-ended mixer was designed operating from a source impedance of 50 Ω driving an IF load impedance of 100 Ω (these two outputs are combined at the IF to give 50 Ω).

One of the fundamental problems in mixer design is that there are so many variables that it is difficult to see how to optimise the circuit. In this design approach, the concept of equivalent pulse duty ratio for the pump waveform is used. This is described in Reference 10 and the key equations are given in Fig. 6.7.

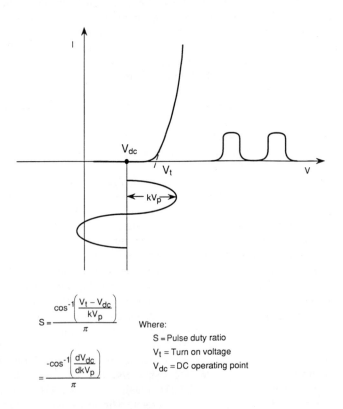

$$S = \frac{\cos^{-1}\left(\frac{V_t - V_{dc}}{kV_p}\right)}{\pi}$$

$$= \frac{-\cos^{-1}\left(\frac{dV_{dc}}{dkV_p}\right)}{\pi}$$

Where:
S = Pulse duty ratio
V_t = Turn on voltage
V_{dc} = DC operating point

Fig. 6.7 Approximation of pulse duty ratio

The basic assumptions made here are that the diode is voltage driven by a pump waveform of amplitude kV_p and it operates at a point V_{dc} on the DC characteristic. This operating point must be derived from the values used in DIODEMX by subtracting the voltage drop across the series resistance; i.e. V_{dc} as defined here is V_{dc} (given by DIODEMX) $- I_{dc} \times$ diode series resistance. The parameter k is a voltage tapping factor and accounts for the fact that only a portion of the pump voltage appears across the diode junction. When the voltage kV_p plus the diode bias voltage V_{dc} exceeds the turn on voltage, the diode turns on and when this combination is less than V_t, the diode turns off. The on time per pump cycle is the pulse duty ratio denoted by S. We now have to determine the turn-on voltage for a device with a VI characteristic which is a continuous function. This can be determined by a series of calculations on the selected mixer circuit.

Examination of the equation for pulse duty ratio S given in Fig. 6.8 shows that, given that V_t is fixed for a given device, there are a range of values for kV_p and V_{dc} which give the same value of S. Further investigation shows that, if we plot kV_p against V_{dc} for a constant pulse duty ratio, we have a straight line with a negative slope. The slope of this line gives the pulse duty ratio and the point at which it intersects the DC bias voltage axis gives V_t. In general k will not be known, therefore, we plot V_p against V_{dc}. Since a plot of kV_p against V_{dc} has the same slope and intersects the DC bias voltage axis at V_t, k can be found if the value of V_t is known. This is given approximately by the junction voltage which causes a current to flow in the junction such that the small signal resistance is equal to the real part of the RF source impedance. However, since the pulse duty ratio is given by the slope of the plot, it is not necessary to determine the value of k, a plot of V_p against V_{dc} is sufficient. We know that a given embedding structure requires a particular value of S to achieve minimum conversion loss. We can now compute a series of conversion loss curves for different DC operating points (n.b. this is the operating point on the VI characteristic and not the external bias voltage). We know that the minimum value of conversion loss on each curve corresponds to the optimum pulse duty ratio S_{opt}. Therefore, if we calculate the generator voltage magnitude from the pump power level at this minimum we can plot this value of V_p together with the operating point voltage V_{dc}. Plotting a series of these results gives a straight line as mentioned above.

In the first stages of this work, an attempt was made to find these minima in conversion loss by plotting conversion loss against pump power for fixed external bias voltages. A typical result for the mixer under consideration is given in Fig. 6.8. One problem in using this method is that the minimum is very poorly defined. The reason for this is that, although the external bias voltage is fixed, the operating voltage on the IV characteristic decreases as pump power is increased due to the opposing voltage developed across the diode series resistance caused by the increasing rectified current. What is happening in this case is that the operating point voltage is maintained at approximately the optimum value for a wide range of pump powers (this phenomenon can be deliberately exploited in some mixer designs where it is required to maintain a low conversion loss over a wide range of pump powers. In this case the total series resistance in the DC bias circuit would be carefully selected to keep the operating point constant over the range of pump power levels required).

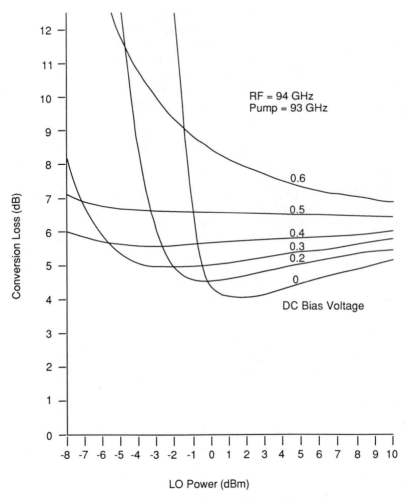

Fig. 6.8 Conversion loss against pump power for different bias voltages

In view of this problem, conversion loss was plotted against the operating point voltage. Conversion loss curves were plotted for pump power levels corresponding to pump generator voltages of 0.4, 0.6, 0.8, 1.0 and 1.2 volts which correspond to power levels of -3.98, -0.46, 2.04, 3.98 and 5.56 dBm, respectively, as shown in Fig. 6.9. The minima of these curves enabled five conversion loss minima to be obtained. These minima were obtained for the optimum value of pulse duty ratio. Therefore, five values of V_p against V_{dc} were plotted in Fig. 6.10. These points showed a very good fit to a straight line and enabled the value for V_t/k and the optimum pulse duty ratio S_{opt} to be determined as previously described. In this case they were found to be 0.65 and 0.1609, respectively.

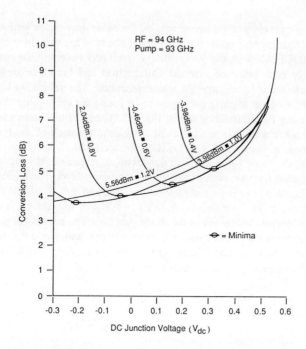

Fig. 6.9 Conversion loss against DC junction voltage for different power levels

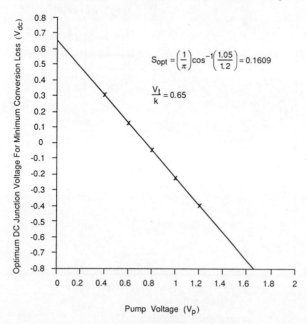

Fig. 6.10 Optimum DC junction voltage against pump voltage

Following computation of the above results, the pulse duty ratios were calculated for each of the four highest power settings given above. This was done by running the program DIODEMX with the pump voltage fixed and sweeping the external DC voltage bias. For each value of external voltage bias and the calculated rectified current, the operating point voltage V_{dc} was determined. The values for V_{dc} and the fixed V_p enabled a value of pulse duty ratio to be found for each value of operating point voltage using the equation given in Fig. 6.7. These values were used to plot four curves of conversion loss against pulse duty ratio for each of the four largest pump power levels as shown in Fig. 6.11. It can be seen that all four conversion loss minima occur at the same value of pulse duty ratio, as expected, but that the actual values of the minima increase with decreasing pump power level. The reasons for this are that as the pump power is decreased, the time taken to turn the diode on and off increases and that, in the on-state, the device has a higher impedance thus causing increased signal power dissipation in the diode junction. The reason for the latter is that in the on-state, the diode is passing a lower current, and, since the differential resistance is inversely proportional to the current, the 'on' junction resistance is higher and hence dissipates more power and raises the conversion loss.

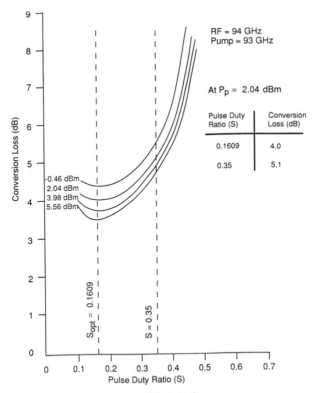

Fig. 6.11 Conversion loss against pulse duty ratio

Mixers 229

Having determined the equivalent pulse duty ratios to plot the conversion loss curves, the input and output impedance were plotted against pulse duty ratio as shown in Figs. 6.12 and 6.13. The impedance plots were drawn this way rather than on a Smith chart as it is easier to interpret and interpolate values. The curves given in Figs. 6.11, 6.12 and 6.13 completely characterise the mixer circuit within the pump power range -0.46 to 5.56 dBm. The use of these figures to determine an optimum design will now be discussed.

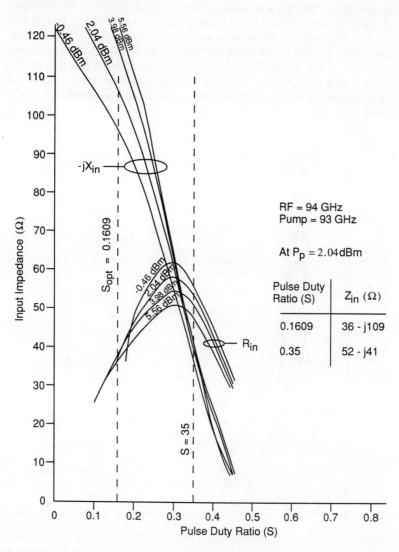

Fig. 6.12 Input impedance against pulse duty ratio for different pump power levels

230 Mixers

The first point to be resolved in the mixer design is how much pump power will be available. Clearly, the higher the pump power level, the lower the conversion loss and noise figure, but beyond a certain point the improvement in performance becomes negligible. Examination of Fig. 6.11 suggests that a pump power level of 2 dBm would give a theoretical conversion loss of around 4 dB. However, from Figs. 6.12 and 6.13, we see that for a pulse duty ratio corresponding to $S_{opt} = 0.1609$, the input impedance is $36-j109.5$ Ω and the output impedance is $482+j105$ Ω. The output impedance is clearly too high and, although this value will be halved when two single-ended mixers are combined in the balanced design, i.e. $241-j52.5$ Ω, it is still about twice the value required. However, these three figures enable a trade-off to be readily established. We first set the constraint that the output impedance must have a real part which is about 100 Ω. Examination of Fig. 6.13 shows that this can be achieved with a pulse duty ratio of 0.35 which gives an input impedance of $52-j41$ Ω and an output impedance of $117+j12.5$ and that, at this point, the conversion loss is only degraded to 5.1 dB.

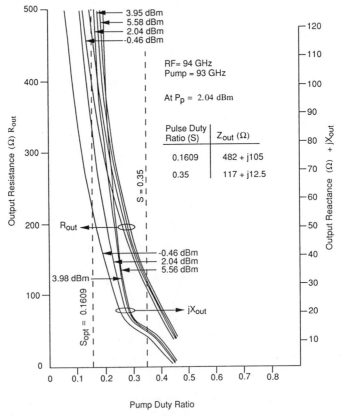

Fig. 6.13 Output impedance against pulse duty ratio for different pump powers

Finally, we now wish to do a check on the mixer performance. The first step in this process is to determine the optimum value of external voltage bias. This is readily determined using the equation in Fig. 6.7 rewritten to calculate V_{dc} from which we find that for $V_p = 0.8$, $V_j/k = 0.65$, $S = 0.35$, so $V_{dc} = 0.2868$. We can now sweep through this voltage using DIODEMX to produce a set of results either side of the optimum bias voltage. The analysis shows input and output impedances of 51.8-j41.2 Ω and 117+j12.5 Ω thus confirming the values selected from Figs. 6.12 and 6.13. It should be remembered that these are the input and output impedances which require a conjugate match to achieve the conversion loss figure of 5.22 dB, therefore the embedding impedance round the diode should be 51.8+j41.2 Ω and 117-j12.5 Ω at the signal and IF, respectively. Having found the optimum operating conditions for this application, it is now necessary to design the matching circuit for the RF input. This can be realised by an additional open circuit stub. However, after insertion of this stub, it is necessary to iterate the design process to ensure that the operating conditions are still optimum. The reason for this is that when the matching stub is placed in the circuit, the terminations at other frequencies are also affected thus making the arrival at an optimum condition an iterative process. The IF is matched by a lumped element at the IF output.

In this example, we are assuming a pump power of 2 dBm will be available. It should be noted that when the two SEMs are combined with a 3 dB coupler to form an SBM, the power required will be 5 dBm plus an extra amount to compensate for the loss in the coupler.

6.7 FET mixers

6.7.1 Design considerations for FET mixers

In general, there are three different ways a FET can be used to perform the frequency conversion process. These are shown in Fig. 6.14 and are: (a) in an active mode, (b) as a source-drain zero current switch and (c) as a diode switch. In the active mode, the device is used in a range of circuit configurations resembling those for small signal amplifier design. When operating as a zero current switch, the source-drain voltage is zero and the gate is used to switch the source-drain small signal conductance between two values; usually between a gate-source voltage equal to zero and pinch-off (a negative value for a depletion mode FET). Ideally, no current from the switching signal, applied to the gate, flows in the drain. In the diode switch, the FET has the source and drain connected together to form the cathode of a diode and the gate forms the anode. The small signal conductance is proportional to the current flowing through it. Generally, diodes made in this way have rather low cut-off frequencies of around 200-300 GHz compared with those realised with a planar diode process using a vertical structure which can achieve cut-off frequencies of 1000-1500 GHz without recourse to sub-micron geometries.

A FET mixer operating in the active mode can have a number of advantages over the diode mixer depending on its configuration. These include: (1) the possibility of conversion gain, (2) lower pump power, (3) the potential for pump signal

- AS AN AMPLIFIER (VARIABLE gm)

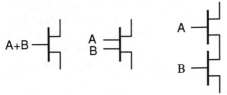

- AS A SWITCH (SWITCHABLE SMALL SIGNAL CONDUCTANCE)

- AS A DIODE

Fig. 6.14 The three ways a FET or HEMT can be used for frequency conversion

isolation in a dual-gate FET by applying pump and signal to different gates, and (4) high reverse isolation, i.e. IF-signal and IF-pump, due to the unilateral properties of the FET.

A study by Fairburn *et al.* [14] identified four sets of mixing actions using the active mode in a single-gate FET.

6.7.2 Pump-drain, RF-gate

The non-linearity at the knee of the I/V characteristic is used for this mixing action. Therefore, the optimum drain-source bias is when $V_{ds}=V_{knee}$. The dominant contribution to frequency conversion is the variation of the drain-source resistance.

6.7.3 Pump-gate, RF-drain

To utilise this part of the V/I characteristics the drain-source voltage is zero. The large pump voltage applied to the gate varies the channel resistance. The properties of this type of mixing are: (1) low $1/f$ noise, and (2) low unwanted intermodulation products for high RF drive levels. When a FET is used as a source-drain zero current switch or as a diode, the embedding circuits used to achieve a given mixer performance will be identical to those described for diodes in the previous sections. However, when the FET is used as a zero current switch there is one major advantage that is often not fully understood or exploited. This is the fact that switching is achieved without having to pass current as in a diode. When driving a diode, it is necessary to drive the device with a sufficiently large signal that it switches between high and low impedance states. Since these states are defined by the current which flows through the junction and the small signal conductance is proportional to the current, currents of several mA have to pass to minimise the RF energy lost in the junction and hence minimise the conversion loss. This in turn leads to high pump fundamental and harmonic currents flowing through the diode which can cause a filtering problem on the IF port, excessive pump power leaking to the RF port etc. These problems do not exist with the FET switch. Furthermore, difficulties with device stability are avoided because the FET has no source-drain voltage to make it capable of giving any gain.

6.7.4 Pump-gate, RF-drain

The pump voltage is applied to the gate and the time-varying transconductance is responsible for frequency conversion. The gate is biased near pinch-off and the drain biased at V_{dd} to provide the best conversion response. Under these conditions the instantaneous values of the other non-linear elements in the model are approximately constant.

6.7.5 Pump-source, RF-gate

The FET is now being used under common-drain configuration and the pump voltage modulates both V_{gs} and V_{ds}. This is the most complicated mode of mixing from the point of view of analysis since all the non-linear elements within the FET model are time-varying.

6.7.6 General comments on FET and HEMT mixer design

The pump and RF can be applied to the gate of a single-gate FET or to separate gates in a dual-gate FET to provide a simple means of pump to RF isolation. Some examples of single-ended and balanced FET mixers are given in Fig. 6.15 and these are described in more detail in Reference 3.

Many of the techniques used in amplifier designs can be applied to mixer design. The aim of FET mixer design is to achieve optimum RF performance. In most receiver applications this is achieved when minimum noise figure is obtained with adequate conversion gain. This is different from the diode mixer where maximum conversion efficiency is usually sought. It is also important that the mixer is a 'well-behaved' component. 'Well-behaved' is a concept not as precisely defined as one might like. It means that the mixer does not have any unusual properties that compromise its practical use in systems. Specifically, it must be stable under small-signal and large-signal conditions, its performance parameters must vary gracefully with pump level and environmental temperature, and it should be approximately unilateral, i.e. have high enough reverse loss that its performance is not unduly sensitive to source or load impedance.

These properties are best realised by a transconductance mixer; one in which the time-varying transconductance is the dominant contributor to frequency conversion, and the variation of other parameters is minimal. The effect of harmonically varying gate-drain and gate-source capacitance must be minimised, as must that of R_{ds}. Additionally, the transconductance swing must be maximised for a given device, in order to minimise noise figure, to optimise gain for a given set of source-load impedances, and to minimise gain variation with pump level. The most important requirement in achieving these conditions is that the FET is kept in its current saturation region throughout the pump cycle. Full saturation can be achieved by ensuring that the variation of $V_{ds}(t)$ under pumping is minimal; it is best if it remains constant at its DC value. This condition is achieved by short circuiting the drain at all pump harmonics; then the drain current, which may have a fairly high peak value, cannot cause any drain-source voltage variation. The RF voltage across the gate-drain capacitance is then minimal, so feedback in minimal and large-signal stability is good.

A typical circuit configuration for a microstrip GaAs FET mixer [102] is shown in Fig. 6.16. This circuit is a single-ended circuit and therefore has no pump AM noise suppression as with diode SEMs. However there is inherent isolation from IF to RF and IF to pump due to the unilateral properties of the FET. This is further enhanced by the use of filters. In general, this type of mixer should be designed to have an RF and pump filter on the drain which essentially provides a short circuit for these frequencies. Also, a short circuit for the IF needs to be provided at the input to assist with stability. In general, no attempt is made to enhance the performance by feeding back higher order mixed products as with the diode mixer as this can lead to instabilities and oscillation unless it is very carefully controlled. The circuit shown in Fig. 6.16 can form the basis of an SEM if two identical circuits are connected to the outputs of a 3dB coupler or equivalent structure. Possible coupling structures are described in Section 6.8. A novel form of FET mixer has been investigated by Robertson and Aghvami [72] which makes use of a Wilkinson power divider. Although this circuit can be made compact, particularly by realising it in lumped element form, is has an inherent loss due to the insertion loss of the Wilkinson splitter. The RF loss can be minimised by using an unequal split such that the RF loss is lower, but this is only achieved at the expense of higher pump losses and consequently the requirement for a higher pump power level at the input.

The dual-gate FET has been investigated extensively by a number of authors and appears to provide an answer to the problem of pump-RF isolation due to the inherent isolation between the two gates. Unfortunately this is often not large enough for many applications and, although the gate stripes can be placed further apart to minimise the mutual capacitance, this compromises the performance of the device as a mixer. The dual-gate FET mixer is quite difficult to analyse as a mixer because of the changing operating conditions. The dual-gate FET can be considered as being equivalent to two single-gate FETs connected in series. This configuration is very

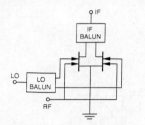
(a) Dual gate single balanced mixer

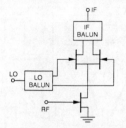
(b) Single gate single balanced mixer

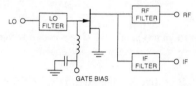
(c) Single ended resistive switch mixer

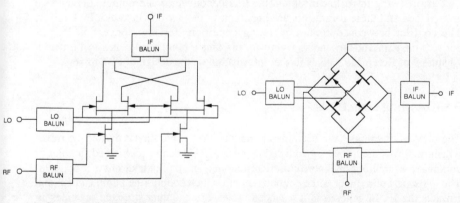
(d) Single gate double balanced mixer

(e) Double balanced resistive switch mixer

Fig. 6.15 Examples of single and double-balanced FET and HEMT mixers

236 Mixers

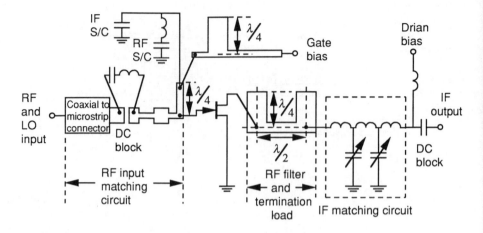

Fig. 6.16 Schematic diagram of a typical FET single-ended mixer

similar to a cascode amplifier, but the operation is very different. In this circuit, the pumped FET controls the bias conditions of the other FET. This is discussed in detail in Reference 3. The FET performing the mixing is in saturation over a large part of the pump cycle and this is the primary reason why dual-gate FETs tend to exhibit lower conversion loss and higher noise figure than single-gate devices. When the FET is operating in its linear region, the transconductance and output conductance are reduced and these parameters are important to the performance of a FET mixer. The conflict between optimum operating conditions for low conversion loss or conversion gain and low noise figure means that a compromise has to be made. Chapter 9 of Reference 3 gives further information on the design of FET mixers.

6.8 Coupling structures

The previous sections have been mainly devoted to the non-linear part of the mixer which performs the frequency conversion process. This can be a switched sampler or commutator circuit or one more closely resembling an amplitude or phase modulator. The other part of a mixer is the coupling circuit which couples the pump and RF and extracts the IF. In general, in a balanced mixer, the coupling structure provides an equal amplitude in-phase split for one input signal and an equal amplitude anti-phase split for the other input signal. This enables mutual isolation between the two input signals to be achieved together with pump AM noise suppression. The structure itself may combine both these features, such as in the rat race or ring hybrid 180° coupler, or there may be separate circuits for the pump and RF, such as a Wilkinson in-phase splitter and a second Wilkinson splitter with the in-phase outputs put through separate circuits which have an insertion phase difference of 180°. In single-balanced mixers, the balance is usually between the pump and the signal and the IF is extracted via a low pass filter. This filter prevents the pump and RF reaching the IF amplifier connected to the IF output. The coupling structures described in this section are equally applicable to both diode and FET or HEMT mixers.

In MMICs it is desirable to make the overall mixer circuit as small as possible. Generally, the coupling structures are the circuits which require the most area. This is particularly true for distributed circuits at low frequencies. Therefore, it is common practice to use either a folded version of the distributed circuit or a lumped approximation in order to reduce the overall size. At millimetre wavelengths, distributed circuits are already quite small and are commonly used. An alternative to the passive distributed and lumped element coupling circuits is an active circuit. Active coupling circuits have the potential to be much smaller than the distributed or lumped circuits, but, although they can have gain, the noise performance of most circuits investigated to date is poorer. Often the asymmetry of an active balun or splitter means that the noise figure to each output is not the same even though the gains may be equally balanced. This can be quite acceptable in some receiver architectures where the mixer is preceded by a low noise amplifier. However, although insertion of a low noise amplifier in front of a mixer can decrease the overall noise figure of the combination, it also decreases the dynamic range.

6.8.1 Passive coupling structures

One very important feature to understand in passive coupling structures is the significance of in-phase, quadrature and anti-phase splitting. This is best illustrated by use of the diagram in Fig. 6.17.

6.8.2 180° 3 dB coupler

In this coupler shown in Fig. 6.17(a), a signal applied to port 1 gives equal amplitude anti-phase signals at ports 3 and 4; one applied to port 2 gives equal amplitude in-phase signals to ports 3 and 4. When this coupler is matched at all ports, there is mutual isolation between ports 1 and 2. If this coupler is used in a mixer, the pump would be connected to port 1 and the RF to port 2 or vice versa. The two output ports would be connected to the diodes with lines of equal length. When the diodes present perfect loads, there is mutual isolation between ports 1 and 2. When the diodes are not matched, but have equal reflection coefficients, a signal fed into port 1(2) splits between ports 3 and 4 and the two reflected parts return to the coupler to add constructively at port 1(2) and cancel each other at port 2(1). This means that when this circuit is used for a single-balanced mixer, any diode mismatch results in pump power being returned to the pump input port and RF power being returned to the RF input port. An example of a 180° coupler is the rat race or ring hybrid coupler shown in Fig. 6.18(a). Since the IF outputs from the diodes are combined, it is necessary to mount the diodes in opposite directions on the two outputs of the coupler to ensure the IFs produced from the diodes are in phase.

238 Mixers

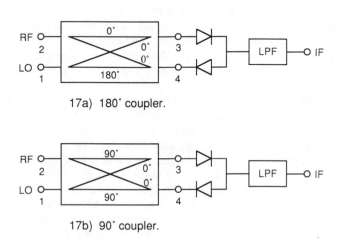

Fig. 6.17 Single-balanced mixers using 180° and 90° 3dB couplers

The bandwidth of the rat race is 10-15%, but when the effects of junction discontinuities and transmission line dispersion are considered, it may be even lower. The coupler's bandwidth is limited mainly by the fact that the line between ports 1 and 4 must be $3\lambda/4$ in length in order to achieve the 180° difference between ports 4 and 3. Developments of the basic rat race coupler are given in Figs. 6.18(b) and 6.18(c). The bandwidth of the basic coupler can be improved significantly to about an octave by using a frequency independent phase inverter between ports 1 and 4 as proposed by March [87] and shown in Fig. 6.18(b). This does, however, have the disadvantage of the need for ground connections and that the required odd- and even-mode impedances are such that an interdigital coupled line section will be required. A further improvement to this design has been proposed by Walker [98] who developed a compensated coupler using two $\lambda/4$ short circuited stubs, with a line impedance equal to the even-mode impedance of the coupled line section, connected in shunt with ports 1 and 2 as shown in Fig. 6.18(c). This design has perfect anti-phase outputs over the operating bandwidth of the coupler.

Examples of lumped element equivalent circuits of the basic rat race coupler are given in Fig. 6.19. These make use of a 'pi' or 'tee' representation of a $\lambda/4$ length of line. It should be noted that, by replacing the low pass 'pi' or 'tee' section by its high pass equivalent, an additional 180° of phase shift is achieved, hence the $3\lambda/4$ section can be simplified as shown in Fig. 6.19(b). The design equation is:

$$\omega L = \frac{1}{\omega C}\sqrt{2}\,R \tag{6.25}$$

Where R is the impedance of the port terminations.

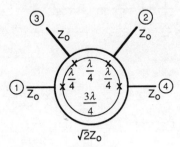

(a) Basic rat-race coupler

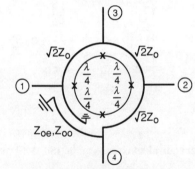

(b) Modified rat-race coupler

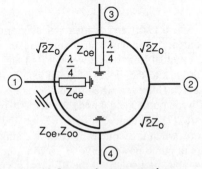

(c) Improved rat-race coupler

Fig. 6.18 The distributed 180° 3 dB rat race coupler

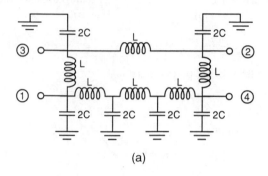

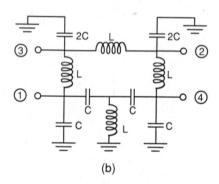

Fig. 6.19 Lumped element equivalent circuits of the 180° 3 dB rat race coupler

6.8.3 90° 3dB coupler

This coupler is shown in Fig. 6.17(b) and is identical diagramatically to the one above except that the two outputs are in phase quadrature. Again, when the coupler is matched at all ports, there is mutual isolation between ports 1 and 2. If this coupler is used in a mixer, again with lines to the diodes of equal length, there is also mutual isolation between ports 1 and 2. However, when the diodes are not matched, a signal fed into port 1(2) splits between ports 3 and 4 and the two parts return to the coupler to add constructively at port 2(1) and cancel at port 1(2), i.e. the opposite of the rat race 180° coupler. Therefore, when this coupler is used in a single-balanced mixer, any diode mismatch results in pump power being returned to the RF input and vice versa. This is often not desirable, particularly if the RF input is connected directly to an antenna, as a significant amount of pump power could be radiated. Fortunately, this situation can be easily corrected, by inserting an extra $\lambda/4$ section of line between the coupler output and one of the diodes to turn it into a 180° coupler.

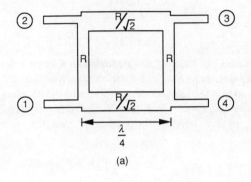

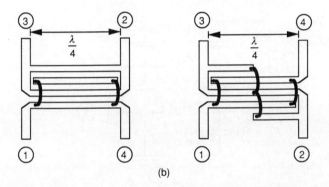

Fig. 6.20 Distributed 90° 3 dB couplers: (a) Branch-line or quadrature hybrid coupler; (b) Lange couplers

Two examples of 90° or quadrature couplers are the branch-line or quadrature hybrid and the interdigital or Lange coupler. These are shown in Fig. 6.20. The latter is basically a standard λ/4 proximity coupler with interdigital line sections to increase the mutual capacitance in order to realise the necessary odd- and even-mode impedances for tight coupling. An example of the lumped element form of the branch-line and interdigital or Lange coupler is given in Fig. 6.21. The design equations for the lumped element coupler are:

$$C_1 = \frac{1}{\omega R \sqrt{K}} \qquad (6.26)$$

242 Mixers

$$L = \frac{1}{\omega\sqrt{1+\omega C_1 R}} \quad (6.27)$$

$$C_o = \frac{1}{\omega^2 L} - C_1 \quad (6.28)$$

where R is the impedance of the port terminations and K is the ratio of powers at the output and coupled ports, therefore, for a 3 dB coupler, $K=1$.

It should be noted that the lumped element forms of the couplers described above have a much narrower bandwidth than the distributed forms. Staudinger and Seely [92] proposed an octave bandwidth quadrature coupler suitable for MMIC implementation.

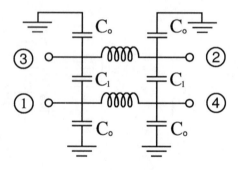

Fig. 6.21 Lumped element equivalent circuit of a 90° 3 dB coupler

6.8.4 Wilkinson with high-pass and low-pass sections

An alternative way to construct an in-phase or anti-phase equal amplitude split is to use the Wilkinson power splitter. The in-phase splitter is a standard Wilkinson splitter which can be realised in distributed or lumped form as shown in Fig. 6.22. A lumped element anti-phase splitter can be realised by following one output with a high-pass filter structure and the other by a low-pass filter structure as shown in Fig. 6.23. The distributed form (see Section 7.3.5.1) can, for homogeneous dielectrics, have exact anti-phase tracking over infinite bandwidth and the inhomogeneous microstrip form can have similar performance. This circuit can be in folded form to reduce the size. The lumped element splitter would be realised using spiral inductors and MIM capacitors, but would tend to have narrower bandwidth than the distributed form. These structures have been investigated by a number of authors [62, 89, 91].

Mixers 243

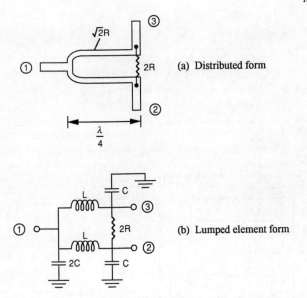

Fig. 6.22 Distributed and lumped element form of the Wilkinson power splitter

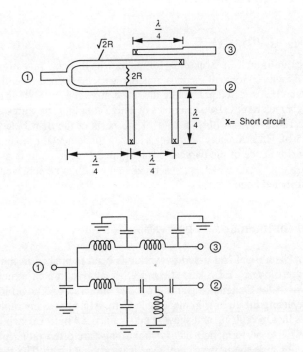

Fig. 6.23 Distributed and lumped element forms of an anti-phase power splitter

6.8.5 Planar transformers

Another type of coupler is the centre-tapped transformer, shown in Fig. 6.24, which can also be realised as a planar spiral. This structure, however, tends to be only suitable for low frequency operation because of the high parasitic inter-turn capacitance.

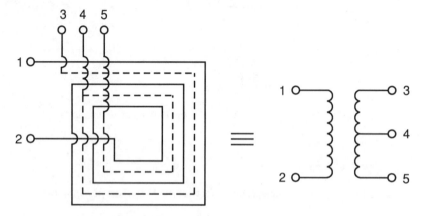

Fig. 6.24 A lumped element centre-tapped transformer

6.8.6 Active coupling structures

A number of active coupling structures for mixers have been investigated and published in the literature. Some of the most notable work performed recently was by Robertson [96] who investigated practically the noise properties as well as the small and large signal characteristics. Two examples of active coupling structures are given in Fig. 6.25. The form shown in Fig. 6.25(a) is the basis of the distributed version given in Fig. 6.25(b). Unfortunately, this circuit has the fundamental problem that the gain and noise performance of the two halves are quite different and it is difficult to design a compensation circuit which enables broad bandwidth to be achieved without resorting to a distributed structure.

6.9 Some final comments and observations

A vast amount of theoretical and practical work has been published in the field of mixer design using diodes, FETs, HEMTs and other more complex structures. In recent years, much of the experience gained, together with new and novel ideas, has been applied to realising the circuits in monolithic form. Although many of the active forms using FETs look promising and show good small signal performance it should be noted that the noise performance and dynamic range are often rather poor. If a publication does not give the noise figure, then it is probably bad! The best mixer conversion performances and noise figures appear to be obtained when switching

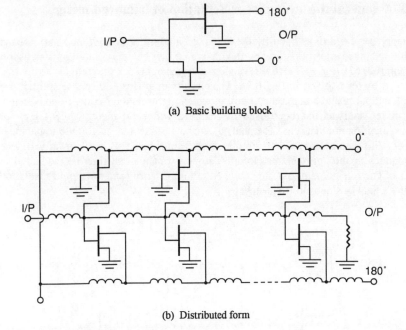

(a) Basic building block

(b) Distributed form

Fig. 6.25 Active coupling structures

devices are used together with distributed (sometimes folded to reduce the size) or lumped element coupling structures. In this type of circuit one must then consider carefully whether it would be more cost effective to put the coupling structure on a separate MIC substrate rather than use costly GaAs real estate for a purely passive circuit.

The key parameters of the most notable recently reported diode mixers are summarised in Table 6.2. Note that the figures in square brackets are not given in the references and are based on the assumption that the single sideband noise figure is equal to the conversion loss and that the single sideband noise figure is 3 dB higher than the double sideband noise figure. They are given for comparison purposes only. This assumption can only be made when the baluns or coupling circuits are passive. The mixers in this table are all diode mixers. A mixer is considered to be a diode mixer when the part which is switching and performs the frequency conversion process is a diode.

Recently, the Hittite Microwave Corporation [57] has developed some very small double-balanced MMIC mixers. A 5.0–20.0 GHz DBM with an IF of DC–3.0 GHz occupies a die size of 1.58 × 1.53 mm and shows a conversion loss of 10 dB over that frequency range with pump-RF and pump-IF isolation of 25 dB. Table 6.3 gives a summary of some of the more notable mixer designs that have been reported using FETs and HEMTs.

6.10 A general theory and representation of balanced mixers

A theory has been developed by the author to analyse both wide and narrowband double-balanced mixers using any device which can be represented as a two-port element [62, 63]; e.g. FET, HEMT, diode etc. A general block diagram of the double-balanced mixer is given in Fig. 6.26. The circuit comprises 180° phase splitters for signal and pump and a commutator circuit which performs the frequency conversion. The mixer analysis has been used to design a wideband mixer (6-18 GHz) for communication applications. The analysis enables phase and magnitude imbalances in both the pump and signal circuits to be considered as well as any device imbalances in the commutator circuit. The performance can then be evaluated in terms of conversion gain/loss, input and output impedances and port to port isolation for any signal or sideband frequency.

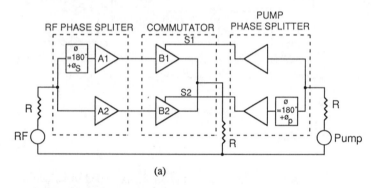

(a)

NOMINAL PARAMETER VALUES
RF phase splitter gains A1 = A2 = 3 dB
RF phase splitter phase error $\phi_s = 0°$
Pump pulse duty ratios S1 = S2 = 0.5
Pump phase splitter phase error $\phi_p = 0°$
Commutator gains B1 = B2 = 0 dB (on) = -∞ dB (off)

Under the above conditions the above circuit is equivalent to the lattice mixer shown below.

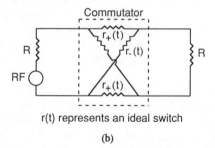

r(t) represents an ideal switch

(b)

Fig. 6.26 Block diagram of a wideband balanced mixer

The mixer analysis has been developed using the theory of linear circuits with time-varying parameters. This analysis technique relies on the assumption that the pump is much larger than the input signal and is, therefore, solely responsible for changing the impedance of a non-linear device in a time-varying periodic manner. Accepting this assumption, it is then possible to analyse the circuit in two stages. Firstly, a large signal non-linear analysis is performed to determine the harmonic voltages and currents present in the circuit and hence the time-varying impedances. Secondly, using these time-varying impedances, a small signal analysis is performed to calculate the conversion gain, input and output impedances and port to port isolation.

The analysis developed here used a bilinear model for the commutator circuit. This model assumes that the device switches between two levels at the local oscillator frequency. It has already been shown from work with diodes that the bilinear or pulse approximation is valid under medium to strong pumping conditions. However, the analysis could readily be modified to incorporate the time-varying impedance derived from a more general treatment if required. The pumped commutator is defined by 5 real and 16 complex parameters for each signal or mixed product frequency under investigation.

Each half can be specified by the S-parameters in the 'on' and 'off' states (8 complex), a pulse duty ratio (1 real) which defines the time spent in the on state and a rise time (1 real) which defines the transition between the on and off states. A phase error term (1 real) defines the departure from anti-phase pumping of the two halves. The signal and pump splitters are completely defined by three-port S-parameters (12 complex each).

A technique was developed to analyse the mixer using S, $ABCD$ and Y parameters to determine four overall admittance functions. These functions enabled the currents to be found in the source and load impedances when the mixer was driven from the signal and IF port. Deriving each of these admittance functions for the four commutator states enabled four new variables to be found which were used to define the Fourier series admittance waveforms. Four sets of Fourier series were found corresponding to each of the conditions mentioned above for each frequency in the commutator. Appropriate cross-multiples of the terms of the Fourier series enabled the currents at the signal and different mixed product frequencies to be calculated to determine conversion gain/loss, input and output impedances and port isolation for any signal or sideband frequency. The overall procedure is quite complex and a computer program was written which incorporates all of the analysis theory.

This program was used to investigate the sensitivity of the performance of a double-balanced mixer to variations in different circuit parameters and is described in the next section.

6.10.1 Diode lattice mixer

The ring or lattice mixer circuit given in Fig. 6.26(b) was analysed using the above program. The circuit considered has infinite bandwidth and the phase splitter has 3 dB gain. A perfect switch (represented in the program by a very high or a very low impedance value) provides the time-varying admittance which performs the low frequency conversion process. Since any part of the circuit has constant gain with frequency, the analysis for this simple circuit predicts a somewhat higher value for the higher order mixed products than would be the case in any practical circuit. However, the circuit given provides a valuable insight into what happens when the different circuit parameters are varied. Figs. 6.27 to 6.30 show the variation in conversion gain and RF/IF mutual isolation as a function of pump phase splitter error and pulse duty ratio imbalance, RF phase splitter error, pump phase splitter error and RF phase splitter gain imbalance. These results show that the conversion loss is relatively insensitive to amplitude and phase errors in the pump and RF splitters, and only degrades by 6 dB if one half of the commutator circuit fails in an open or short circuit. The parameter which is most sensitive is the RF-IF mutual isolation. This requires amplitude and phase balance between channels of better than 0.5 dB and 10° in both the pump and RF splitter.

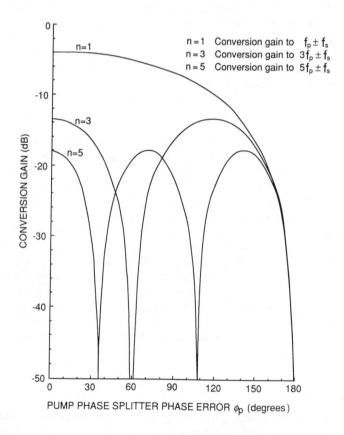

Fig. 6.27 Conversion gain against pump phase splitter error

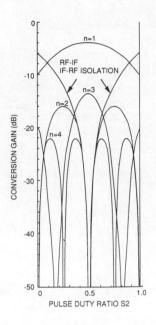

Fig. 6.28 Conversion gain against pump pulse duty ratio

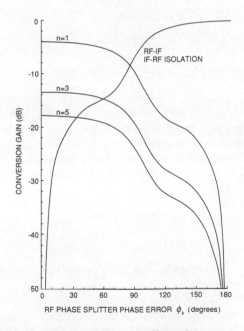

Fig. 6.29 Conversion gain against RF phase splitter phase error

250 Mixers

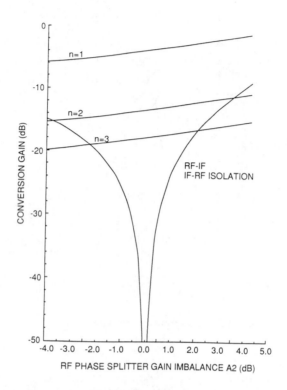

Fig. 6.30 Conversion gain against RF phase splitter gain imbalance

Although the results are given for the simple case of a switched mixer using diodes or a zero current FET or HEMT switch, they give some insight into the performance of a mixer using devices operating in an active mode. In this case, the IF-RF isolation would be enhanced due to the unilateral properties of the devices, but the RF-IF figure would be similar after allowing for the gain in the devices.

6.11 References

6.11.1 General

1. TUCKER, D. G.: *Circuits with periodically-varying parameters*, Macdonald and Co. Ltd., London, 1964
2. SALEH, A. A. M.: *Theory of resistive mixers*, Research Monograph No. 64 MIT Press, Cambridge Massachusetts and London, 1971
3. MAAS, S. A.: *Microwave mixers*, Artech House, Dedham, Massachusetts, 1986

4. MAAS, S. A.: *Non-linear microwave circuits*, Artech House, Norwood, Massachusetts, 1988.
5. FAYOS, J. R., and NIGHTINGALE, S. J.: 'Determination of the transfer function of a non-linear circuit for prediction of intermodulation characteristics', IEEE MTT-S International Microwave Symposium, Baltimore, June 1986, pp. 499-502
6. HOWSON, D. P., and GARDINER, J. G.: 'High-frequency mixers using square-law diodes', *The Radio and Electronic Engineer*, Nov. 1968, pp. 311-316
7. WILL, P.: 'Reactive loads -the big mixer menace', *Microwaves*, April 1971, pp. 38-42
8. MANLEY, J. M., and ROWE, H. E.: 'Some general properties of non-linear elements: Part (1) General energy relations', *Proc. IRE*, vol. 44, no. 7, July 1956, pp. 904-913
9. ROWE, H. E.: 'Some general properties of non-linear elements: Part (2) Small signal theory', *Proc. IRE*, vol. 46, No. 5, May 1958, pp. 850-860
10. NIGHTINGALE, S. J.: 'Loss and noise characteristics of microwave mixers', PhD thesis, University of Canterbury, Kent, 1980. Copy available from the Electronics Laboratory Library
11. NIGHTINGALE, S. J., and SCHIEK, B.: 'Direct and accurate measurements of mixer equivalent noise temperatures', Proc. 8th EuMC, Paris, Sept. 1978, pp. 712-716
12. NEMLIKHER, Y. A., and STRUKOV, I. A.: 'The impedance and transfer characteristics of resistive frequency converters using Schottky barrier diodes', *Radio Engineering and Electronic Physics*, vol. 19, no. 1, 1974, pp. 111-118
13. NEMLIKHER, Y. A., and STRUCKOV, I. A.: 'Noise characteristics of frequency converters using Schottky barrier diodes', *Radio Engineering and Electronic Physics*, vol. 19, no. 1, 1974, pp. 119-126
14. FAIRBURN, M., MINNIS, B. J., and NEALE, J.: 'A novel monolithic distributed mixer design', IEE Colloquium on Microwave and Millimetre Wave Monolithic Integrated Circuits, Digest no. 1988/117, Nov. 11th, 1988
15. LIECHTI, C. A.: 'Down converters using Schottky barrier diodes', *IEEE Trans. on Electron Devices*, ED-17, no. 11, Nov. 1970, pp. 975-983

6.11.2 Diode mixers

16. RISTOW, D., ENDERS, N., and KNIEPKAMP.: 'A monolithic GaAs Schottky barrier diode mixer for 15 GHz', Proc. EuMC, Paris, Sept. 1978, pp. 707-711
17. CONTOLATIS, A., CHAO, C., JAMISON, S., and BUTTER, C.: 'Ka-band monolithic GaAs balanced mixers', IEEE Microwave and Millimeter Wave Monolithic Circuits Symposium, Dallas, June 1982, pp. 28-30
18. JACOMB-HOOD, W. W., SMITH, J. R., and HIPWOOD, L. G.: 'High quality X-band monolithic diode mixer', IEEE Microwave and Millimeter Wave Monolithic Circuits Symposium, Boston, May 1983, pp. 1-4

19. CHAO, C., CONTOLATIS, A., JAMISON, S., and BUTTER, C.: '94 GHz monolithic GaAs balanced mixers', *ibid.*, pp. 50-53
20. YUAN, L. T.: 'A W-band monolithic GaAs crossbar mixer', IEEE Microwave and Millimeter Wave Monolithic Circuits Symposium, San Francisco, May 1984, pp. 67-69
21. BAUHAN, P., CONTOLATIS, T., ABROKWAK, J., and CHAO, C.: '94 GHz planar GaAs monolithic balanced mixer', *ibid.*, pp. 70-73
22. KERMARREC, C., FAGUET, J., VANION, B., MAZOUSSE, C., COLLET, A., KAIKATI, P., and BEAUFORT, D.: 'The first GaAs fully integrated microwave receiver for DBS applications at 12 GHz', Proc. EuMC, Liège, Sept. 1984, pp. 749-754
23. YUAN, L. T., and ASHER, P. G.: 'A W-band monolithic balanced mixer', IEEE Microwave and Millimeter Wave Monolithic Circuits Symposium, St Louis, June 1985, pp. 71-73
24. NIGHTINGALE, S. J., UPTON, M. A. G., MISHRA, U. K., PALMATEER, S. C., and SMITH, P. M.: 'A 30 GHz monolithic single balanced mixer with integrated dipole receiving element', *ibid.*, pp. 74-77
25. CHU, A., COURTNEY, W. E., MAHONEY, L. J., MANFRA, M. J., and CALAWA, A. R.: 'Dual function mixer circuit for millimeter wave transceiver applications', *ibid.*, pp. 78-81
26. KERMARREC, C.: 'GaAs MMICs for mass production 12 GHz down converters', Proc. EuMC, Paris, Sept. 1985, pp. 87-95
27. NIGHTINGALE, S. J., UPTON, M. A. G., MISHRA, U. K., PALMATEER, S. C., and SMITH, P. M.: 'An EHF coplanar monolithic single balanced mixer using Mott diodes', *ibid.*, pp. 635-640
28. LIU, L. C. T, LIU, C. S., KESSLET, J. R., and WANG, S. K.: 'A 30 GHz monolithic receiver', IEEE Microwave and Millimeter Wave Monolithic Circuits Symposium, Baltimore, June 1986, pp. 41-44
29. BAUHAN, P., CONTOLATIS, A., SOKOLOV, V., and CHAO, C.: '30 GHz balanced mixers using an ion-implanted FET-compatible 3-inch GaAs wafer process technology', *ibid.*, pp. 45-49
30. PODELL, A. F., and NELSON, W. W.: 'High volume, low cost, MMIC receiver front end', *ibid.*, 1986, pp. 57-59
31. GELLER, B. D., and ABITA, J. L.: 'An integrated DBS receiver front-end using a monolithic frequency converter', Proc. EuMC, Dublin, Sept. 1986, pp. 613-618
32. RAMACHANDRAN, R., MOGHE, S., HO, P., and PODELL, A.: 'An 8-15 GHz GaAs monolithic frequency converter', IEEE Microwave and Millimeter Wave Monolithic Circuits Symposium, Las Vegas, June 1987, pp. 31-34
33. TRINH, T. N., WONG, W. S., LI, D., and KESSLER, J. R.: 'Ion implanted W-band monolithic balanced mixers for broadband applications', *ibid.*, pp. 89-92
34. PAVIO, A. M., HALLADAY, R. H., BINGHAM, S. D., and SAPASHE, C. A.: 'Broadband monolithic single and double ring active/passive mixers', IEEE Microwave and Millimeter Wave Monolithic Circuits Symposium, New York, May 1988, pp. 71-74

35. ADELSEK, B., COLQUHOUN, A., DIEUDONNÉ, J.-M., EBERT, G., SELDERS, J., SCHMEGNER, K. E., and SCHWAB, W.: 'A monolithic 60 GHz diode mixer in FET compatible technology', IEEE Microwave and Millimeter Wave Monolithic Circuits Symposium, Long Beach, June 1989, pp. 91-94
36. TELLIEZ, I., CHAUMAS, P., RUMELHARD, C., and PATAUT, G.: 'GaAs monolithic balanced mixer for C-band direct demodulation receiver', Proc. EuMC, Wembley, Sept. 1989, pp. 731-736
37. ADELSECK, B., DIEUDONNE, J.-M., SCHMEGNER, K. E., COLQUHOUN, A., EBERT, G., and SELDERS, J.: 'A monolithic 94 GHz balanced mixer', IEEE Microwave and Millimeter Wave Monolithic Circuits Symposium, Dallas, May 1990, pp. 111-114
38. TON, T. N., DOW, G. S., CHEN, T. H., LACON, M., LIN, T. S., BUI, S., and YANG, D.: 'An X-band monolithic double double-balanced mixer for high dynamic range receiver application', *ibid.*, pp. 115-118
39. MAOZ, B., REYNOLDS, L. R., and OKI, A.: 'FM-CW radar on a single GaAs/AlGaAs HBT MMIC chip', IEEE Microwave and Millimeter Wave Monolithic Circuits Symposium, Boston, June 1991, pp. 3-6
40. YONAKI, J., CARANDANG, R., ALLEN, B. A., HOPPE, M., JONES, W. L., YANG, D. C., and BRUNNENMEYER, C. L.: '35 GHz InGaAs HEMT MMIC downconverter', *ibid.*, pp. 47-50
41. TRIPPE, M. W., WEINREB, S., DUNCAN, S. W., ESKANDARIAN, A., GOLJA, B. A., MANTEL, D. C., MENDENILLA, G., POWER, B., SEQUEIRA, H. B., SOUTHWICK, S. B., SVENSSON, S. P., TU, D. W., and BYER, N. E.: 'mm-wave MIMIC receiver components', *ibid.*, pp. 51-54
42. CHANG, K. W., WANG, H., CHEN, T. H., JAN, K., BERENZ, J., DOW, G. S., HAN, A. C., GARSKE, D., and LIU, L. C. T.: 'A W-band monolithic pseudomorphic InGaAs HEMT downconverter', *ibid.*, pp. 55-58
43. EISENBERG, J., PANELLI, J., and OU, W.: 'A new planar double-double balanced MMIC mixer structure', *ibid.*, pp. 69-72
44. DÄMBKES, H., and SCHMIDT, L. P.: 'MMIC technology in Europe for millimetre-wave applications', Proc. of GaAs '92 Symposium, Noordwijk, session 1B, paper 1, April 1992
45. DIEUDONNE, J.-M., RITTMEYER, R., ADELSEK, B., and COLQUHOUN, A.: 'A 94 GHz monolithic downconverter in a MESFET technology', IEEE Microwave and Millimeter Wave Monolithic Circuits Symposium, Alburquerque, June 1992, pp. 69-72
46. GINGRAS, R. L., DRUBIN, C., COLE, B., STACEY, W., PAVIO, R., WOLVERTON, J., YNGVESSON, K., and CARDIASMENOS, A.: 'Millimeter-wave slot ring mixer array', *ibid.*, pp. 105-107
47. LANG, R. J., BHUMBRA, S. S., NIGHTINGALE, S. J., and McDERMOTT, M. G.: 'Low conversion loss Ka-band GaAs monolithic diode mixers', Proc. EuMC, Helsinki, Aug. 1992, pp. 735-740

48. MAAS, S. A., and CHANG, K. W.: 'A broadband planar doubly balanced monolithic Ka-band mixer', IEEE Microwave and Millimeter Wave Monolithic Circuits Symposium, Atlanta, June 1993, pp. 53-55
49. FUDEM, H., MOGHE, S., DIETZ, G., and CONSOLAZIO, S.: 'A highly integrated MMIC K-band transmit/receive chip', *ibid.*, pp. 119-122
50. KATZ, R., AUST, M. V., KASODY, R., WANG, H., ALLEN, B., DOW, G. S. TANK, K., LIN, S., and MEYERS, R.: 'A hightly compact wideband GaAs HEMT X-Ku band image-reject receiver MMIC', *ibid.*, pp. 131-134
51. HUANG, H. C., LAUX, P., BASS, J. F., CHEN, S. W., LEE, T. T., TADAYON, S., SINGER, J. L., KEARNEY, J., and AINA, O. A.: 'A W-band multifunction MMIC', IEEE Microwave and Millimeter Wave Monolithic Circuits Symposium, San Diego, May 1994, pp. 37-40
52. LESTER, J. A., HUANG, P., AHMADI, M., DUFAULT, M., INGRAM, D., CHEN, T. H., GARSHE, D., and CHOW, P. D.: 'HEMT MMIC chip set for low cost miniaturised EHF satcom transceiver', *ibid.*, pp. 85-88
53. BRINLEE, W. R., PAVIO, A. M., and VARIAN, K. R.: 'Novel planar double-balanced 6-18 GHz MMIC mixer', *ibid.*, pp. 139-142
54. KABAYASHI, K. W., KASODY, R., OKI, A. K., DOW, S., ALLEN, B., and STREIT, D. C.: 'K-band double balanced mixer using GaAs HBT THz Schottky diodes', *ibid.*, pp. 209-212
55. AUST, M. V., ALLEN, B., DOW, G. S., KASODY, R., BIEDENBENDER, M., and WANG, N.: 'A low noise high gain Q-band monolithic HEMT receiver', *ibid.*, pp. 217-220
56. MINOT, K., AUST, M., KASODY, R., KATZ, R., WANG, H., ROGERS, P., SMITH, D., SHAW, L., TAN, K., WANG, N., DOW, S., and ALLEN, B.: 'A MMIC chip set for V-band crosslink communication systems', *ibid.*, pp. 221-224
57. Hittite Microwave Corporation, 21 Cabot Road, Woburn, MA 01801, USA

6.11.3 Subharmonic mixers

58. BEMKOPF, P., and TAJIMA, Y.: 'A monolithic Ka-band sub-harmonically pumped frequency converter', IEEE Microwave and Millimeter Wave Monolithic Circuits Symposium, Boston, May 1991, pp. 43-46
59. PAYNE, D., HICKS, R., LAMPORT, R., STEWART, E., VAN LEEUWEN, B., and RAFFAELLI, L.: 'A GaAs monolithic V-band receiver for space communications', European GaAs Applications Symposium, session 1B, paper 2, April 1992

6.11.4 FET mixers

60. BASTIDA, E. M., CAVALIERI D'ORO, E., DONZELLI, G. P., FANELLI, N., FAZZINI, G., and SIMONETTI, G.: 'A monolithic 800 MHz bandwidth DBS front end receiver for mass production', Proc. EuMC. Liège, Sept. 1984, pp. 755-760

61. KANAZAWA, K., KAZUMURA, M., NAMBU, S., KANO, G., and TERAMOTO, I.: 'A GaAs double-balanced dual-gate FET mixer IC for UHF receiver front-end applications', IEEE Microwave and Millimeter Wave Monolithic Circuits Symposium, St. Louis, June 1985, pp. 63-65
62. NIGHTINGALE, S. J., UPTON, M. A. G., and DANDEKAR, N. V.: 'The design and performance of a double balanced mixer using FETs', IEEE MTT-S Symposium, St. Louis, June 1985, pp. 717-720
63. NIGHTINGALE, S. J., UPTON, M. A. G., DANDEKAR, N. V., and KONG, W. M.: 'The design and performance of a wideband monolithic double balanced mixer using FETs', Proc. EuMC, Paris, Sept. 1985, pp. 940-945
64. HOWARD, T. S., and PAVIO, A. M.: 'A distributed monolithic 2-18 GHz dual-gate FET mixer', IEEE Microwave and Millimeter Wave Monolithic Circuits Symposium, Las Vegas, June 1987, pp. 27-30
65. YANG, D. C., ESFANDIARI, R., LIN, T. S., and O'NEILL, T.: 'Wideband GaAs MMIC receiver', *ibid.*, pp. 101-103
66. HARVEY, A. R., COTTON, F. J., PENNINGTON, D. C., EATON, R. M., and COOPER, P. D.: 'A low noise GaAs MMIC satellite downconverter for the 6-4 GHz band', *ibid.*, pp.139-142
67. WHOLEY, J., KIPNIS, I., and SNAPP, C.: 'Silicon bipolar double balanced active mixer MMICs for RF and microwave applications up to 6 GHz', IEEE Microwave and Millimeter Wave Monolithic Circuits Symposium, Long Beach, June 1989, pp. 133-137
68. ROBERTSON, I. D., and AGHVAMI, A. H.: 'A practical distributed FET mixer for MMIC applications', IEEE MTT-S Symposium, Long Beach, June 1989, pp. 1031-1032
69. ROBERTSON, I. D.: 'Mixer designs for GaAs technology', Chapter 9 of *GaAs technology and its impact on circuits and systems*, eds. Haigh, D., and Everard, J., Peter Peregrinus Ltd., on behalf of the Institution of Electrical Engineers, September 1989
70. PHILIPPE, P., and PERTUS, M.: 'A 2 GHz enhancement mode GaAs down converter IC', IEEE Microwave and Millimeter Wave Monolithic Circuits Symposium, Boston, June 1991, pp. 61-64
71. ROBERTSON, I. D., and AGHVAMI, A. H.: 'A novel 1-15 GHz matrix distributed mixer', Proc. 21st European Microwave Conference, Stuttgart, Sept. 1991, pp. 489-494
72. ROBERTSON, I. D., and AGHVAMI, A. H.: 'A compact X-band GaAs monolithic balanced FET mixer', European GaAs Applications Symposium, session 3A, paper 2, April 1992
73. SAKUNO, K., YOSHIMASU, T., MATSUMOTO, N., TSUKAO, T., NAKAGAWA, Y., SUEMATSU, E., and TOMITA, T.: 'A miniature low current GaAs MMIC downconverter for Ku-band broadcast satellite applications' IEEE Microwave and Millimeter Wave Monolithic Circuits Symposium, Albuquerque, June 1992, pp. 101-104
74. BONATO, G. L., and BÓVEDA, A.: 'GaAs monolithic image rejection down-converter for point-to-multipoint communication systems', *ibid.*, pp. 131-134

75. NEILSON, D., ALLEN, B., KINTIS, M., HOPPE, M., and MAAS, S.: 'A broadband upconverter IC', *ibid.*, pp. 163-165
76. CHEN, T. H., CHANG, K. W., BUI, S. B. T., LIU, L. C. T., and PAK, S.: 'A double balanced 3-18 GHz resistive HEMT monolithic mixer', *ibid.*, pp. 167-170
77. DEVLIN, L. M., BUCK, B. J., CLIFTON, J. C., DEARN, A. W., and LONG, A. P.: 'A 2.4 GHz single chip transceiver', IEEE Microwave and Millimeter Wave Monolithic Circuits Symposium, Atlanta, June 1993, pp. 23-26
78. KAMITSUNA, H., and OGAWA, H.: 'Monolithic image rejection optoelectronic up-converters that employ the MMIC process', *ibid.*, pp. 75-78
79. RAMACHANDRAN R., WOO C., NIJJAR, M., and FISHER D.: 'A fully integrated 35 GHz MMIC receiver with on-chip LO', *ibid.*, pp. 93-95
80. HÜBNER, M., and RATTAY.: 'K- and Ku-band MMICs for radio link communications', Proc. of GaAs '94, Torino, April 1994, pp. 59-62
81. FEUDALE, M. A., SURIANI, A. M., McPARTLIN, M. J., and OLSEN, E. A.: 'MMICs for satellite Ku band TLC repeaters', *ibid.*, pp. 229-232
82. VERVER, C. J., and STUBBS, M. G.: 'Development of 30 and 20 GHz mixers for miniaturised personal communications systems', *ibid.*, pp. 269-272
83. SAITO, T., HIDAKA, N., OHASHI, Y., SHIMURA, T., and AOKI, Y.: '60 GHz MMIC downconverter using an image-rejection active HEMT mixer', IEEE Microwave and Millimeter Wave Monolithic Circuits Symposium, San Diego, May 1994, pp. 77-80
84. MONIZ, J. M., and MAOZ, B.: 'Improving the dynamic range of Si MMIC Gilbert cell mixers for homodyne receivers', *ibid.*, pp. 103-106
85. WOO, C., RAMACHANDRAN, R., BURNS, L., MARASCO, K., and FISHER, D.: 'A frequency agile X-band homodyne GaAs MMIC transceiver with a synthesized phase locked source for automotive collision avoidance radar', *ibid.*, pp. 129-132
86. TSAI, M. C., SCHINDLER, M. J., STRUBLE, W., VENTRESIA, M., BINDER, R., WATERMAN R., and DANZILIO D.: 'A compact wideband balanced mixer', *ibid.*, pp.135-138

6.11.5 Baluns

87. MARCH, S.: 'A wideband stripline ring hybrid', *IEEE Transactions Microwave Theory Tech.*, MTT-16, 1968, p. 361
88. CLOETE, J. H.: 'Exact design of the Marchand balun', *Microwave Journal*, May 1980, pp. 99-102
89. BHARJ, S. S., TAN, S. P., and THOMPSON, B.: 'A 2-18 GHz 180 degree phase splitter network', IEEE MTT-S Symposium Digest, Long Beach, June 1989, pp. 959-962
90. TOKUMITSU, T., HARA, S., and AIKAWA, M.: 'Very small ultra-wideband MMIC magic-T and application to combiners and dividers', *ibid.*, June 1989, pp. 963-966

91. BHARJ, S. S., THOMPSON, B., and TAN, S. P.: 'Integrated circuit 2-18 GHz balun', *Applied Microwave Magazine*, Nov./Dec. 1989, pp. 110-119
92. STAUDINGER, J., and SEELY, W.: 'An octave bandwidth 90-degree coupler topology suitable for MMICs', *Microwave Journal*, vol. 33, no. 9, 1990, p. 117
93. MINNIS, B. J., and HEALY, M.: 'New broadband balun structure for monolithic microwave integrated circuits', IEEE MTT-S Symposium Digest, Boston, June 1991, pp. 425-428
94. CHEN, T. H., CHANG, K. W., WONG, H., DOW, G. S., LIU, L. C. T., BUI, S. B. T., and LIN, T.S.: 'Broadband monolithic passive baluns and monolithic double-balanced mixer', *ibid.*, pp. 861-864
95. ROGERS, J., and BHARTIA, R.: 'A 6-20 GHz planar balun using a Wilkinson divider and Lange couplers', *ibid.*, pp. 865-868
96. ROBERTSON, I. D.: 'Novel techniques for GaAs MMIC receivers', PhD Thesis, Kings College, London, June 1991
97. TRIFUNOVI , V., and JOKANOVI , B.: 'Four decade bandwidth uniplanar balun', *Electron. Lett.*, vol. 28, no. 6, 12th March 1992, pp. 521-604
98. WALKER, J. L. B.: 'Improvements to the design of the 0-180 rat race coupler and its application to the design of balanced mixers with high LO to RF isolation', private communication, 1994

6.11.6 Miscellaneous MMIC and MIC references

99. PUCEL, R. A.: *Monolithic microwave integrated circuits*, IEEE Press (selection of reprints) 1985
100. TANG, O. S. A., and AITCHISON, C. S.: 'A practical microwave travelling-wave MESFET gate mixer' , IEEE MTT-S Symposium, St. Louis, June 1985, pp. 605-608
101. TANG, O. S. A., and AITCHISON, C. S.: 'A very wide-band microwave MESFET mixer using the distributed mixing principle', *IEEE Trans. Microwave Theory and Tech.*, MTT-33, no. 12, Dec. 1985, pp. 1470-1478
102. PENGELLY, R.S.: *Microwave field-effect transistors -theory, design and applications*, Research Studios Press, John Wiley and Sons Inc., 1986.
103. CAHANA, D.: 'A new coplanar waveguide/slotline double-balanced mixer', IEEE MTT-S, Long Beach, June 1989, pp. 967-968
104. MAAS, S.: 'The star mixer', *Microwave Journal*, July 1993, pp. 36-46
105. BASU, S., and MAAS, S. A.: 'Design and performance of a planar star mixer', *IEEE Trans. on Microwave Theory and Tech.*, MTT-41, no. 11, Nov. 1993, pp. 2028-2030

Chapter 7

Phase shifters

S. Lucyszyn and J. S. Joshi

7.1 Introduction

A phase shifter is a control device found in many microwave communication, radar and measurement systems. This chapter describes many of the design techniques for MMIC phase shifters, implemented with either analogue or digital control. In order to understand the subtle differences between the two main generic types of phase shifters, both the true phase shifter and the true delay line must first be defined.

A true phase shifter can be defined as a control device that has a flat group delay frequency response within its defined bandwidth of operation, the level of which does not change as the insertion phase is varied. Its two characteristics features are:

(1) a flat relative phase shift frequency response, at all levels of relative phase shift

(2) constant group delay, resulting in no change in the timing of an input RF pulse envelope.

The frequency characteristics of a true phase shifter are illustrated in Fig. 7.1. True phase shifters can be employed in multiple space diversity receiver-combiners for aligning RF signals within a pulse envelope without changing the timing of the pulse edges. However, they should not be employed in wideband beam-forming networks for large aperture phased-array antennas, in order to avoid the effects of 'phase squinting' and 'pulse stretching'.

A time shifter can be defined as a control device that has a flat group delay frequency response with a level that changes as its insertion phase varies, within a defined bandwidth of operation. Its two characteristic features are:

(1) a linear relative phase shift frequency response, with a gradient that changes as the value of relative phase shift varies

(2) a flat but varying group delay frequency response, resulting in a change in the timing of an input RF pulse envelope.

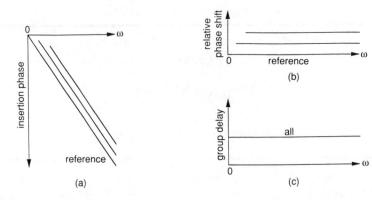

Fig. 7.1 Frequency characteristics of a true phase shifter: (a) Insertion phase; (b) Relative phase shift; (c) Group delay

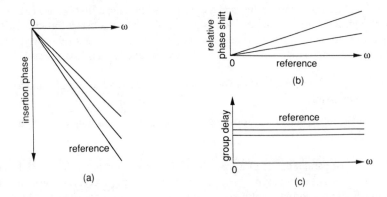

Fig. 7.2 Frequency characteristics of a true delay line: (a) Insertion phase; (b) Relative phase shift; (c) Group delay

A special case of a time shifter is a true delay line, which is defined by the following expression:

$$\Delta\tau = \frac{\Delta\angle S_{21}(\omega)}{\omega} \qquad (7.1)$$

where $\Delta\tau$ = relative shift in group delay, $\Delta\angle S_{21}(\omega)$ = relative phase shift at ω, and ω = angular frequency.

The frequency characteristics of the true delay line are illustrated in Fig. 7.2. True delay lines find many applications in general wideband microwave signal processing applications, including beam-forming networks for large aperture phased-array antennas. It must be noted that when the non-idealities of the circuit elements are included in a design, an ideal true phase shifter circuit may have the frequency characteristics of a true delay line, and vice versa.

7.2 Analogue implementations

When compared with a purely digital implementation, analogue control devices have the following advantages:

(1) they require only one control wire per device
(2) no special foundry processing is required to realise high quality switches
(3) they require almost no control power with a passive reflection topology incorporating varactor diodes and cold-FETs
(4) if using a reflection topology with active circulators, they can make much more efficient use of chip space
(5) they do not suffer from quantisation errors
(6) since the phase variation is continuous, any degradation in performance attributed to fabrication process variations can easily be corrected
(7) calibration corrections can be performed after integration into a subsystem
(8) any degradation in performance attributed to adverse operating conditions can be corrected.

As a result of these significant advantages, analogue control devices appear ideal for large adaptive phased arrays and other high performance microwave signal processing applications. However, it should also be pointed out that a purely analogue implementation is not inherently compatible with digital system architectures and may not provide the best overall performance across wide bandwidths

7.2.1 Introduction to analogue phase shifters

Today's spread spectrum communication, radar, EW, remote sensing, high data rate digital transmission and state-of-the-art microwave measurement systems all require very large RF bandwidths. While the scientific community has addressed the problem of bandwidth for most microwave devices, high performance analogue phase shifters have only recently received such attention. The principle reason for this is the multi-dimensional nature of the problem. It is a relatively straightforward task to design a microwave phase shifter to meet a specified level of phase shift at a specific frequency. However, this task becomes more difficult if the appropriate level of phase

262 Phase shifters

shift has to be maintained across a very large frequency range. This difficulty is further compounded when this wideband relative phase shift performance is required to be continuously tuned over a large relative phase shift range. This three-dimensional scenario has assumed that all the other device specifications, such as insertion loss and return loss, are kept within acceptable limits over the wide bandwidth and large tuning range. It will be shown that effective solutions to the three-dimensional problem can be implemented in monolithic technology as well as in conventional hybrid technology.

7.2.2 Single-stage reflection-type phase shifters

The basic topology of a single-stage reflection-type phase shifter (RTPS) is shown in Fig. 7.3. For simplicity, the directional coupler is assumed to be lossless, perfectly matched and having infinite isolation. With reference to Fig. 7.3, a proportion of the input power emerges from the direct port, where the resulting incident voltage wave is reflected from its terminating impedance, Z_T. A proportion of the reflected signal is transmitted back to the input port and a proportion is coupled into the isolated port. Similarly, the remainder of the input power emerges from the coupled port where it is reflected and sent to both the input and isolated ports.

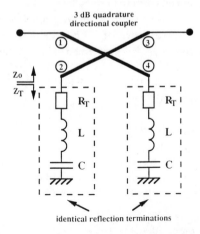

Fig. 7.3 Basic topology of a single-stage reflection-type phase shifter

It can be deduced that the voltage wave emerging from the input port will have zero amplitude when equal power split and phase quadrature, between the coupled and direct ports, are maintained–resulting in a perfect input and output impedance match. Also, with a 3 dB directional coupler and purely reactive reflection terminations, the voltage wave emerging from the isolation port will have an amplitude equal to that entering the input port–resulting in zero insertion loss.

7.2.2.1 General design approaches

The RTPS was first introduced more than three decades ago by Hardin et al. [1]. Hardin employed a directional coupler and simple series-tuned circuit reflection terminations, incorporating a single varactor diode in each. The resulting frequency characteristic of the relative phase shift resembles a single hump. This hump provides a maximum bandwidth of approximately 5-10%, for a maximum phase error of ±5°, within a maximum relative phase shift of 180°. This approach was further developed by Searing [2] and Garver [3] without much improvement in the bandwidth. Attempts to widen the bandwidth were proposed by Henoch and Tamm [4] and Ulriksson [5], by synthesising a parallel resonance mode in the reflection terminations. Impedance transformers were used to effectively produce a double humped frequency response. This technique increased the bandwidth by a factor of two. Here, the component values of the reflection terminations could only be determined by intensive optimisation and human judgement. In recent years, a number of variations to the basic analogue RTPS have been reported in monolithic form without significant advances in their overall performance [6-11].

7.2.2.2 Optimum design approaches

For a reflection-type phase shifter having single reactive element reflection terminations, two optimum design approaches have been recently identified to provide the maximum bandwidth for a defined level of maximum phase error. The first design approach gives the *minimum group delay variation* at centre frequency. The second design approach gives the *maximum relative phase shift* at centre frequency. With both approaches, bandwidth is directly proportional to the level of phase error. When compared with the first design approach, the second approach provides an increase of 25% in the level of maximum relative phase shift, however, this is at the expense of a doubling of the relative shift in group delay.

A rigorous theoretical investigation, complete design examples and the first proof-of-concept measured performance for an X-band MMIC implementation are given in Reference 12. The GaAs MMIC was fabricated at the GEC-Marconi Material Technology Ltd (GMMT) foundry, using their standard F20 foundry process. A microphotograph of the phase shifter is shown in Fig. 7.4. Standard 0.5 µm MESFETs are used to realise interdigitated planar Schottky varactor diodes (IPSVDs) [13-15]. A back-to-back arrangement was adopted for the varactor diodes. This increases the maximum RF power level of the devices and reduces the errors in the resulting relative phase shift levels when operating at higher RF power levels [6]. As RF voltage increases, the capacitance of one of the capacitors increases and that of the other capacitor decreases. Therefore, the net change in capacitance, due to excess RF power, should ideally be zero. With the use of a folded Lange coupler [16], the X-band MMIC phase shifter has an active area of only 2.4×1.0 mm^2.

With the phase shifter operating in the *maximum relative phase shift* condition, the measured relative phase shift performance, with its associated tuning characteristic, is shown in Fig. 7.5(a). Here, the *rms* phase error increases almost linearly from zero at centre frequency, f_o, to ±2.5° at the band edges. The corresponding group delay performance, with its associated variation in group delay,

264 *Phase shifters*

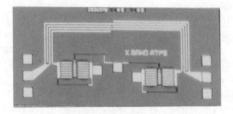

Fig. 7.4 Microphotograph of the optimum design MMIC analogue phase shifter

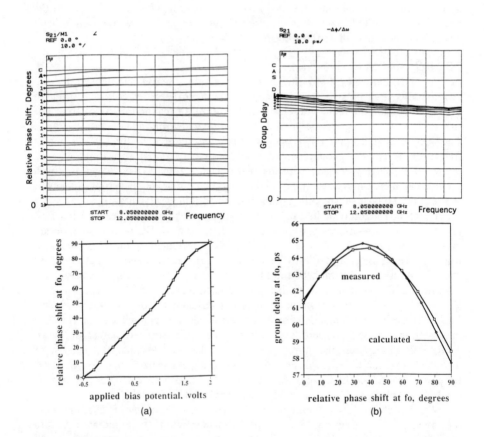

Fig. 7.5 Frequency response of the optimum design MMIC phase shifter: (a) Relative phase shift; (b) Group delay

is shown in Fig. 7.5(b). The calculated variation in group delay has also been superimposed onto the measured results–demonstrating an almost perfect agreement. The mean level of insertion loss is 1.5 dB, with an *rms* amplitude error of ±0.7 dB maintained at all levels of relative phase shift and at all frequencies. The worst-case return loss decreases almost linearly from 33 dB at centre frequency to 18 dB at the band edges.

7.2.3 Single-stage reflection-type delay lines

Variable delay lines are required in adaptive beam-forming networks of wideband phased-array antennas. For large aperture arrays, containing hundreds or thousands of elements, the delay lines should ideally have the following characteristics:

(1) at the lowest frequency of operation, ω_1, $\Delta\angle S_{21}(\omega_1)|_{max}=360°$
(2) a large number of phase states, with high accuracy
(3) the group delay, $\Delta\tau,=(\Delta\angle S_{21})/\omega$, at all operational frequencies
(4) negligible PM/AM conversion
(5) very low insertion loss
(6) good return loss
(7) low control power.

Common classes of general time shifters that are compatible with both hybrid and monolithic microwave integrated circuit technologies are switched-line delay lines [17] and digital reflection-type delay lines [18].

It has been found that the single-stage reflection-type phase shifter, which has a single reactive element reflection termination, can exhibit the time shifter frequency characteristics [19]. However, a true delay line characteristic can only be exhibited over a very narrow frequency range. In order to achieve a wideband performance, the reflection terminations must consist of a variable capacitor in series with a fixed inductor [20]. The non-idealities of the Lange coupler compensate for the non-linear behaviour of the group delay frequency response [12]. Quick empirical design equations for an octave bandwidth variable delay line, employing a traditional four-finger Lange coupler, are as follows:

$$L = \frac{4.0}{f_o} \text{ nH}, \quad C_{max} = \frac{3.0}{f_o} \text{ pF}, \text{ and } R_T \leq \frac{Zo}{10} \text{ }\Omega$$

where f_o = centre frequency, having units of GHz

With an ideal 3 dB quadrature coupler, the relative phase shift is equal to the amount of change in the phase angle of voltage reflection coefficient of the reflection termination. Therefore, to a good approximation, the tuning ratio, *m*, required by the variable capacitor can be determined from the following, for $\omega_1 \leq \omega \leq \omega_2$:

$$m = \frac{1}{1 - \frac{\Delta \angle S_{21}(\omega)|_{max}}{180°} \cdot \left(\frac{\omega_{os}|_{min}}{\omega}\right)} \tag{7.2}$$

where $\omega_{os}|_{min} = \frac{1}{\sqrt{LC_{max}}}$.

The first proof-of-concept analogue reflection-type delay line, employing a Lange coupler, was demonstrated using hybrid technology [21]. Here, the centre frequency was 750 MHz and a high performance was measured over a 40% bandwidth. An MMIC analogue delay line has recently been demonstrated at X-band [12]. The GaAs MMIC was fabricated at GMMT, using their F20 process. A microphotograph of the analogue delay line is shown in Fig. 7.6. As with the optimally designed MMIC phase shifter, standard 0.5 μm MESFETs were used to realise the IPSVDs, which were then arranged back-to-back. With a folded Lange coupler, the X-band MMIC delay line has an active area of only 2.4 × 1.0 mm².

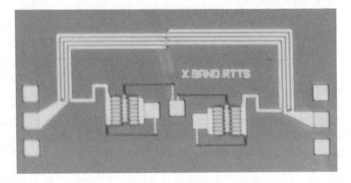

Fig. 7.6 Microphotograph of the MMIC analogue delay line

The measured relative phase shift performance for the delay line, and its associated tuning characteristic, are shown in Fig. 7.7(a). The corresponding group delay performance, and its associated variation in group delay, are shown in Fig. 7.7(b). The calculated variation in group delay for an ideal variable delay line has been superimposed onto the measured results–again, an almost perfect agreement is demonstrated. The variation in group delay for the phase shifter operating in the *minimum group delay variation* condition has also been included for comparison. The mean level of insertion loss is 1.6 dB, with an *rms* amplitude error of ±0.6 dB maintained at all levels of relative phase shift and at all frequencies. The worst-case return loss decreases almost linearly from 28 dB at centre frequency to 18 dB at the band edges.

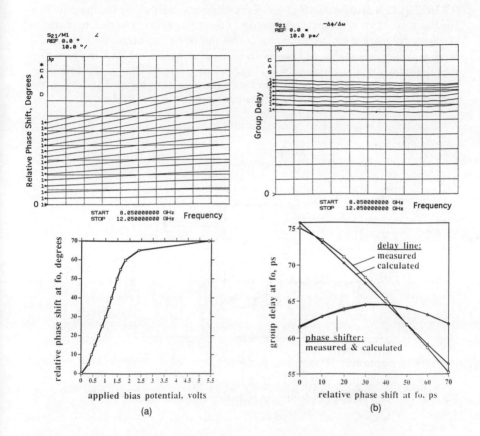

Fig. 7.7 Frequency responses of the MMIC delay line: (a) Relative phase shift; (b) Group delay

7.2.4 Cascaded-match reflection-type phase shifters

Single-stage RTPSs have either a narrow bandwidth, with a good phase error performance at all phase states, or a wide bandwidth, with a poor phase error performance at most phase states. Also, because single-stage RTPSs have an inherent variation in the their group delay response, they technically only approximate true phase shifters.

A novel method for implementing a true wideband analogue phase shifter, with a low phase error performance and constant group delay, is to combine the simplicity of Hardin's approach with the type of rippled frequency response produced by Henoch and Ulriksson. Here, the phase shifter is split into two matched cascaded stages. The topology of an ideal cascaded-match reflection-type phase shifter (CMRTPS) [22] is shown in Fig. 7.8. Each stage is similar to the one proposed by Hardin, except that the resonant frequency of the series-tuned reflection termination of the second stage is always much greater than that of the first stage. With ideal 3 dB quadrature coupling maintained across the desired bandwidth, the relative phase shift would simply be the superposition of the relative phase shifts in the reflection coefficients produced by the reflection terminations of both stages. In practice, the non-ideal effects of the couplers have to be corrected for.

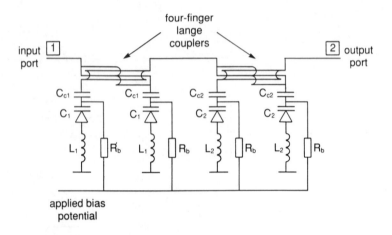

Fig. 7.8 Topology of the cascaded-match reflection-type phase shifter

With the CMRTPS, the non-linear relative phase shift behaviour of one stage complements that of the other to produce a high performance true wideband phase shifter. In order to maintain a low phase error over more than an octave bandwidth, at all levels of relative phase shift up to 180°, it has been found that the following conditions must be satisfied by the reflection terminations [22]:

(1) The resonant frequency of the second stage reflection termination must correctly scale that of the first stage, at the reference bias potential (usually zero volts).

(2) The change in the resonant frequency of the second stage reflection termination must correctly track the change in that of the first stage.

(3) The reflection coefficient of the second stage reflection termination must have a phase angle whose gradient, with respect to frequency, correctly scales that of the first stage, at their respective frequency of resonance.

The result is a phase shifter that has the same bandwidth performance as the directional coupler or circulator employed. Quick empirical design equations for an octave bandwidth CMRTPS, to be realised with an alumina substrate and employing traditional four-finger Lange couplers, are given below:

$$L_1 = \frac{3.9}{f_o} \text{ nH}; \quad C_1(0) = \frac{48}{f_o} \text{ pF}; \quad C_{C1} > \frac{960}{f_o} \text{ pF};$$

$$L_2 = \frac{2.55}{f_o} \text{ nH}; \quad C_2(0) = \frac{8.55}{f_o} \text{ pF}; \quad C_{C2} = \frac{13.5}{f_o} \text{ pF};$$

and $f_C = 1.087 \cdot f_o$

where f_C = directional couplers' frequency of maximum coupling

A rigorous theoretical investigation and a simulated design example for a Ku-band hybrid MIC phase shifter is given in Reference 22. The simulated results for this CMRTPS design demonstrate that a very low maximum *rms* phase error and maximum *rms* amplitude error of ±1.7° and ±0.15 dB, respectively, can be achieved over a full octave bandwidth and at all levels of relative phase shift up to 180°. For an MMIC implementation, component values determined from these design equations can be used as a starting point in the modelling process.

7.2.4.1 One octave bandwidth (6.5-13 GHz) design example

Since the traditional four-finger Lange coupler is common-place in MMICs, it was chosen for the first proof-of-concept monolithic CMRTPS design [23]. With GMMT's F20 processing specifications, Jansen Microwave's LINMIC+ CAD software was used to design the coupler (courtesy of Ron Arnold at GMMT). It was found that for equal power coupling, the finger width and separation distance should be 19 µm and 11 µm, respectively. For a frequency of maximum coupling of 10.9 GHz, the total length of the coupler is 2.7 mm. In practice, the two couplers had to be folded, with a single bend in each, in order to reduce the aspect ratio so that the design would fit into its allotted space in the wafer array. Also, the length of the couplers had to be shortened by 9.3%, again because of the limited available chip space. In the first stage reflection termination, an effective zero bias junction capacitance of 4.62 pF was implemented with a 12 × 300 µm IPSVD, resulting in a total gate width of 3.6 mm. The first stage coupling capacitor, C_{C1}, was omitted from the design. In the second stage reflection termination, an effective zero bias junction

270 Phase shifters

capacitance of 0.77 pF was implemented using a 4×150 μm IPSVD, resulting in a total gate width of 0.6 mm. A second stage coupling capacitance of 1.5 pF was realised using 55×55 μm^2 silicon nitride MIM capacitors. All inductors were realised using 20 μm wide microstrip lines, having a characteristic impedance of 100 Ω at 10 GHz. High value mesa etched resistors (~6 KΩ) were used to prevent RF leakage and as forward bias current limiters. A microphotograph of the monolithic CMRTPS, with its maximum dimensions of 3.0×2.0 mm^2 is shown in Fig. 7.9.

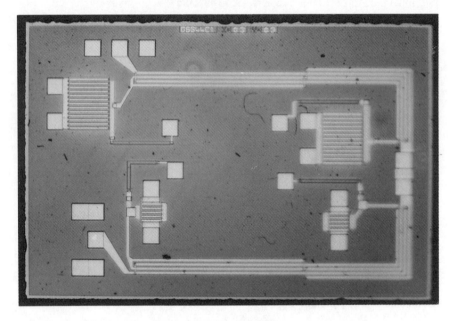

Fig. 7.9 Microphotograph of the one octave bandwidth monolithic CMRTPS

Fig. 7.10(a) shows the relative phase shift frequency responses. The tuning curve (also known as the phase modulation (PM) characteristic) for this MMIC, at centre frequency, is given in Fig. 7.10(b). A maximum relative phase shift of 90° is achieved with a total capacitance ratio of C(0)/C(-7.5 V)=4.4, using the F20 process. If the bias potential is decreased to -0.7 V or increased from 7.5 V to 17 V, there is only a 3.5° increase in the level of maximum relative phase shift. Therefore, the absolute maximum relative phase shift that can be safely introduced with this first phase shifter design is ~95°. Fig. 7.10(c) shows the corresponding phase error performance, with the reference taken at centre frequency. A maximum *rms* phase error of ±2.6° is maintained across the octave bandwidth and for all levels of relative phase shift. It can be seen that if the required bandwidth and/or the maximum relative phase shift level is reduced, this phase error will be significantly reduced.

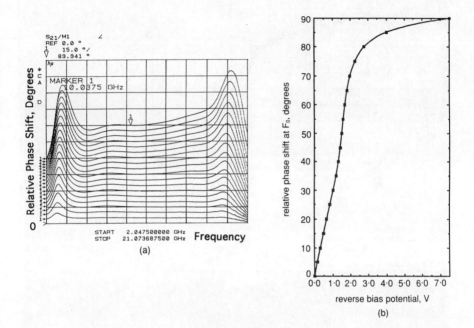

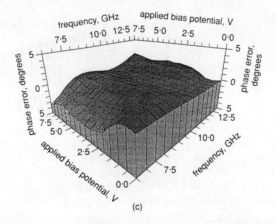

Fig. 7.10 Measured relative phase shift responses: (a) Frequency responses; (b) Tuning curve; (c) Phase error performance

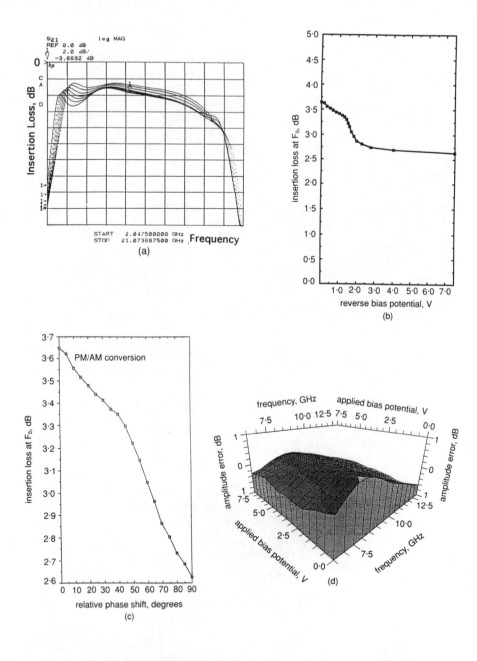

Fig. 7.11 Measured insertion loss response: (a) Frequency responses; (b) Bias dependency; (c) PM/AM conversion; (d) Amplitude error performance

Fig. 7.11(a) shows the insertion loss frequency responses. The 3.14 dB level of mean insertion loss is attributed to both the couplers and IPSVDs. Within the couplers, the fingers lie on a lossy polyimide interlayer dielectric. Therefore, a significant reduction in the insertion loss would be achieved if this lossy dielectric layer was removed. As for the IPSVDs, their effective Q-factors can be increased by changing their topographies [15]. The bias dependency of insertion loss (also known as the amplitude modulation (AM) characteristic) for this MMIC, at centre frequency, is shown in Fig. 7.11(b). It can be seen that insertion loss decreases with an increase in the reverse bias potential. This is contrary to what would normally be expected, since the series resistance of GMMT's IPSVDs increases with an increase in the reverse bias potential [13-14]. However, at most frequencies, the overall loss of the reflection termination decreases with an increase in the reverse bias potential, due to a greater increase in the corresponding reactance of the junction capacitance; i.e. the Q-factor of the IPSVD increases with an increase in the reverse bias potential. Fig. 7.11(c) shows the corresponding PM/AM conversion. The amplitude error performance is shown in Fig. 7.11(d) (with the reference taken at centre frequency). An *rms* amplitude error of ±0.45 dB is maintained across the octave bandwidth and for all levels of relative phase shift.

The input return loss performance is shown in Fig. 7.12. A good input match is achieved, with a corresponding return loss of approximately 20 dB across most of the octave bandwidth. This level of input match is very difficult to achieve with digital switched-filter phase shifters. The performance degrades towards the lower frequency end, due to the shortened length of the couplers. Direct two-port S-parameter measurements on a separate test coupler indicated an equal power coupling of 3.5 dB at the frequency of maximum coupling of 12 GHz. Therefore, with the correct coupler length, the return loss would be 20 dB across the whole octave bandwidth. The group delay was confined within the range of 115 ±13 ps, at all bias levels, within the octave bandwidth.

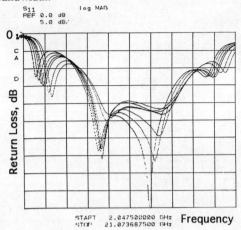

Fig. 7.12 Measured input return loss frequency responses

274 Phase shifters

7.2.4.2 Two octave bandwidth (4.4-16.1 GHz) design example

It can be seen from the frequency responses of the previous octave bandwidth design example that the potential bandwidth could far exceed that of one octave. With reference to Fig. 7.10(a), the relative phase shift responses have a positive gradient above 11.5 GHz. This gradient increases with an increasing reverse bias potential. This type of behaviour indicates that the resonant frequency tracking of the second stage reflection termination could be improved. With a positive gradient, the amount of coupling in the reflection termination of the second stage should be reduced with an appropriate reduction in the value of the second stage coupling capacitance. Also, it has been found that a multilayer coupler can have a similar octave bandwidth performance to that of a four-finger Lange coupler [24]. The multilayer coupler has the advantage of taking up less than half the chip space, while making bends much easier to implement, compared with the Lange coupler. As a result, a more compact two octave bandwidth CMRTPS was designed [25] employing this experimental coupler. After careful simulation, it was found that a 33% reduction in the second stage coupling capacitance and a 50% reduction in the values of all inductances were required in order to achieve a two octave bandwidth performance, when compared with the first MMIC design. The F20 foundry process was used to fabricate the MMIC. A microphotograph of the MMIC, which has dimensions of 2.75×2.0 mm^2, is shown in Fig. 7.13. It can be seen that there is ample scope for increasing the level of integration further.

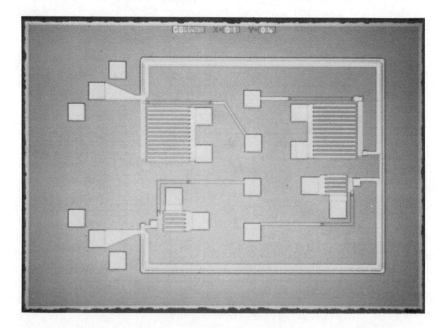

Fig. 7.13 Microphotograph of the two octave bandwidth monolithic CMRTPS

The measured performance of this second MMIC phase shifter design is shown in Fig. 7.14. A maximum relative phase shift of 98° was achieved with a capacitance ratio of C(0)/C(10 V)~4.6, using the F20 process. The maximum *rms* phase error is only ±2.8° across the 4.4 to 16.1 GHz frequency range and at all bias levels. The mean insertion loss is 7 dB, at the centre frequency of 10 GHz. This is more than double that found with the first MMIC design. The measured input return loss is better than 9 dB across the 4.8 to 16.1 GHz frequency range and at all bias levels. This modest input match performance is thought to be attributed to the poor experimental coupler designs. A very flat group delay response of 100 ±4 ps, in the 8.8 to 16.1 GHz frequency range, was also achieved.

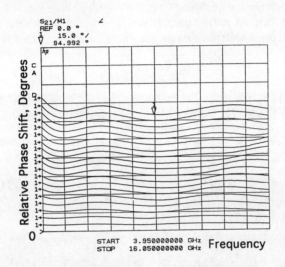

Fig. 7.14 Measured relative phase shift responses of the two octave bandwidth CMRTPS

7.2.5 Ultra-wide bandwidth CMRTPS implementations

7.2.5.1 Decade bandwidth (1-19 GHz) hybrid-MIC design example

The first realisation of a decade bandwidth 90° CMRTPS is described in this section [26]. Here, a 2 to 18 GHz 3 dB quadrature directional coupler was employed. As demonstrated with the previous two MMIC design examples, the potential bandwidth of the CMRTPS increases as the series inductance in the reflection terminations decreases. Therefore, a decade bandwidth CMRTPS should have negligible series inductance. It should be noted that a reduction in the series inductance values can also reduce the maximum relative phase shift level. However, if abrupt junction IPSVDs are employed, they will be the limiting factor on the maximum relative phase shift level that can be achieved.

276 Phase shifters

With a normal hybrid-MIC realisation, significant series inductance and stray parasitics are inherent. To effectively eliminate this inductance and minimise stray parasitic capacitances, monolithic technology was used to realise the reflection terminations. The MMICs sit on very low inductance chip mounts and are connected to the coupler's ports by a multitude of short bond wires. Monolithic technology also greatly simplifies the fabrication of the phase shifter.

A 6×150 μm IPSVD was previously characterised to a high degree of accuracy [13, 14]. The resulting equivalent circuit model was used in the design of the IPSVDs for the reflection terminations of both stages. The single IPSVD in the first stage reflection termination was implemented with a 18×300 μm MESFET. For the second stage reflection termination, two IPSVDs were connected in series, both being implemented with 6×150 μm MESFETs. The coupling capacitances for the first and second stage reflection terminations are 22 pF and 5 pF, respectively. A photograph of the complete phase shifter realisation, having dimensions of 2×1 in^2, is shown in Fig. 7.15.

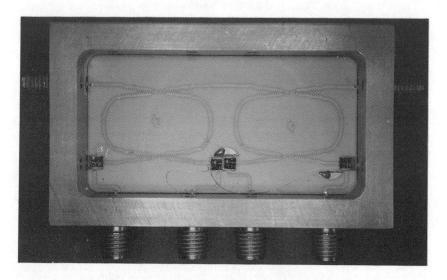

Fig. 7.15 Photograph of the decade bandwidth hybrid-MIC CMRTPS

The relative phase shift responses of the decade bandwidth design are shown in Fig 7.16(a). It can be seen that an overall flat response is achieved between 1 and 19 GHz. Since transmission phase is highly sensitive to impedance discontinuities, the ripples present in the relative phase shift measurements can be attributed to the coupler's 'wiggly lines' (used for phase velocity equalisation) and the production grade wedge shaped launchers (used for the coax-to-microstrip transitions)—no ripples were present in the simulations. The use of a 1-18 GHz stripline 3 dB quadrature directional coupler, such as the one manufactured by Krytar (Model 1831) [27], is ideal for ultra-wideband phase shifter applications–as they have no significant impedance discontinuities. The insertion loss performance, at zero bias, is shown in Fig. 7.16(b). The measured input return loss is better than 10 dB over 87% of the 1 to 19 GHz frequency range. An overall flat group delay response of 2.00 ±0.37 ns was also obtained across the decade bandwidth.

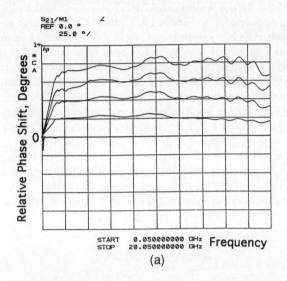

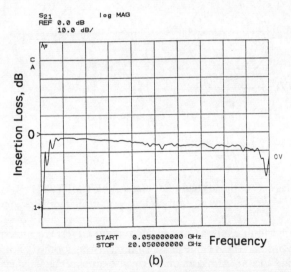

Fig. 7.16 Measured performance of the bandwidth hybrid-MIC CMRTPS: (a) Relative phase shift; (b) Insertion loss

7.2.5.2 Decade bandwidth (0.5-5.5 GHz) MMIC design example

For the past two decades, most RTPSs have employed directional couplers. Unfortunately, no practical decade bandwidth 3 dB quadrature coupler has yet been reported that is spatially compatible with monolithic technology. The design and simulated performance for a monolithic two-stage CMRTPS, which employs decade bandwidth active circulators, is described in Reference 28. A simplified representation of the decade bandwidth CMRTPS is illustrated in Fig. 7.17. The design of the active circulator is described in Chapter 12. It has been found that the potential bandwidth of a high performance CMRTPS increases as the simplicity of its reflection terminations increases. Also, it has been seen with the previous decade bandwidth CMRTPS design, that the series inductances must be omitted from the reflection terminations. The IPSVDs selected for the first and second stage reflection terminations are implemented with 48×150 μm and 6×150 μm MESFETs, respectively. The coupling capacitances for the first and second stage reflection terminations are 19.5 pF and 3.3 pF, respectively. From the initial mask layouts, the reflection termination of the first stage requires approximately half the area of that required by the active circulator. Therefore, one of the main advantages of using active circulators, in preference to directional couplers, is that only one reflection termination is required per stage, thus saving a considerable amount of expensive chip space. With careful design, the complete MMIC could fit within an active area of only 1.5×1.5 mm^2.

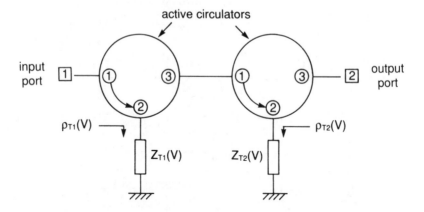

Fig. 7.17 Topology of the decade bandwidth monolithic CMRTPS

Analytical closed-form expressions for this type of CMRTPS implementation are given in Reference 28. The relative phase shift responses for the 0.5-5.5 GHz design are show in Fig. 7.18(a). A maximum *rms* phase error of ±1.5° is achieved for all relative phase shift levels up to 85°, over a full decade bandwidth. Fig. 7.18(b) shows the insertion loss responses of the phase shifter. The mean level of insertion loss is

15 dB, with an *rms* amplitude error of ±1.8 dB across the full decade bandwidth. It will be found that since the IPSVD of the second stage is physically much smaller than that of the first stage, the variations in the Q-factor of the second stage IPSVD will be much greater than that of the first stage. As a result, the high level of amplitude modulation is attributed to the second stage IPSVD. The input return loss responses of the phase shifter are shown in Fig. 7.18(c). It can be clearly seen that an excellent input impedance match is achieved across the whole decade bandwidth. This performance is attributed to the high return loss and high isolation characteristics of the active circulators [29].

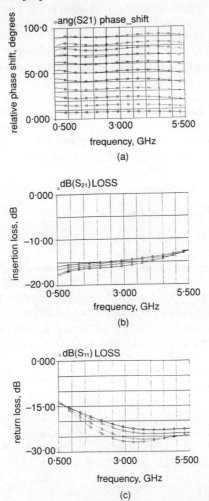

Fig. 7.18 Simulated performance of the decade bandwidth monolithic CMRTPS: (a) Relative phase shift; (b) Insertion loss; (c) Input return loss

280 Phase shifters

In conclusion, when compared with directional couplers, active circulators provide the following advantages:-

(1) they have non-dispersive characteristics
(2) ultra-wide bandwidths can be achieved
(3) only a single reflection termination is required per stage
(4) small chip size.

However, the complete phase shifter design using active circulators also has the following disadvantages (which can in nearly all cases be overcome):-

(1) it is non-reciprocal—which is inherent
(2) excessive total gate width of 9.9mm (reducing the yield)
(3) DC power limit on the common source bias resistors—ultra-wideband active loads can be employed [30]
(4) RF power limit—further investigation is required
(5) low upper frequency limit—consider using quasi-circulators [31] or unilateral circuits [32]
(6) poor noise figure—replace MESFETs with HEMTs
(7) excessive insertion loss—replace MESFETs with HEMTs and/or change topology.

7.2.6 Millimetre-wave implementations

At millimetric frequencies, much has been reported on monolithic mixers using Schottky barrier diodes and hybrid frequency multipliers using *p-n* junction varactor diodes. However, little has been reported on millimetre-wave monolithic analogue phase shifters. One exception to this is the work by Pao *et al.* [33] demonstrating a 60 GHz analogue phase shifter, realised using microstrip technology. Here, interdigitated *p-n* junction varactor diodes were employed in two single-bit branch-line phase shifters connected in cascade. The results demonstrated a worst-case *rms* phase error of ±21.0°, for a maximum relative phase shift of 160°, across a 59 to 61 GHz frequency range. Also, within this frequency range, the mean level of insertion loss was 6 dB with a worst-case *rms* amplitude error of ±1.8 dB. No results were given for the return loss performances.

Precision analogue phase shifters, for which the worst-case *rms* phase error is less than approximately ±1.0°, are very difficult to realise at millimetric frequencies, particularly with a monolithic implementation. This is because the insertion phase linearity is very sensitive to dispersion, unwanted modes of propagation and unwanted parasitic effects that can easily occur at these frequencies. A precision millimetre-wave monolithic analogue CMRTPS is described in Reference 34 which uses uniplanar technology to achieve the desired linear insertion phase responses.

The CPW 3 dB multi-layer directional coupler, described in Chapter 12, is employed for this MMIC phase shifter implementation. Microstrip IPSVDs have been accurately characterised well into the millimetric frequency range [13-15]. However, shunt IPSVDs are inherently more compatible with CPW technology, since the cathode (drain/source) electrodes make direct electrical contact with the upper ground plane. Moreover, because the cathode parasitic capacitance is greatly reduced in this CPW application, the total capacitance ratio of the varactor diode increases. Therefore, the maximum relative phase shift level increases from 90° to 120°, when compared with an equivalent microstrip implementation of a monolithic CMRTPS. The CPW varactor diodes were realised using GMMT's 0.5 μm MESFET technology. A photograph of the monolithic phase shifter, which has a total active area of 1.4 mm^2, is shown in Fig. 7.19.

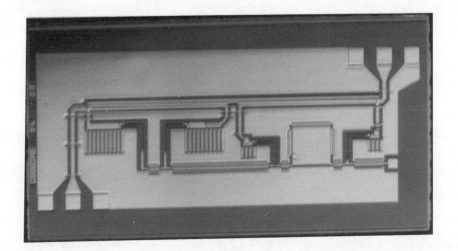

Fig. 7.19 Microphotograph of a millimetre-wave analogue phase shifter

The measured relative phase shift performance and corresponding tuning curve, for a centre frequency f_o=31.5 GHz, are shown in Figs. 7.20(a) and 7.20(b), respectively. A worst-case *rms* phase error of ±1.0° is achieved, for all levels of relative phase shift up to 120°, across the 31 to 32 GHz frequency range. It can be seen that linear tuning is almost achieved. The measured insertion loss performance and corresponding PM/AM conversion characteristic are shown in Figs. 7.20(c) and 7.20(d), respectively. The mean level of insertion loss is 16 dB and the worst-case *rms* amplitude error is ±1.0 dB. The worst-case input return loss for the phase shifter is 15 dB. These measured results demonstrate the potential of this novel uniplanar millimetre-wave analogue phase shifter. When compared with the results obtained by Pao *et al.*, a dramatic reduction was achieved in the levels of worst-case *rms* phase and amplitude errors, across the same fractional bandwidths.

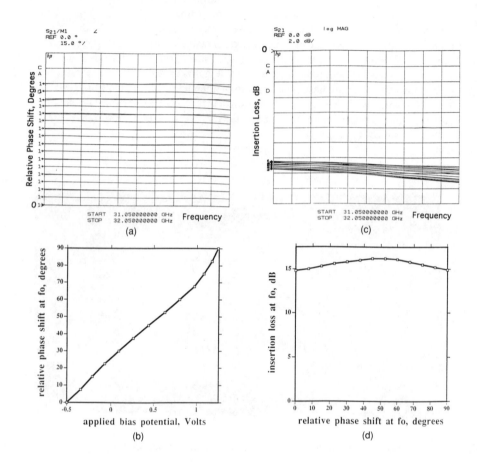

Fig. 7.20 Measured performance of the millimetre-wave CMRTPS: (a) Relative phase shift; (b) Tuning curve; (c) Insertion loss; (d) PM/AM conversion

7.2.7 Dual-gate MESFET

A dual-gate MESFET can be used to implement an analogue phase shifter [35-36]. The input RF signal is applied to one of the gate electrodes (G1) and a shunt inductance is connected to the other gate electrode (G2). A variable gate bias is then applied to G2. This will result in a voltage-controlled tuned circuit at G2, since the inductor is in parallel with the voltage-dependent gate-source capacitance associated with G2. The output RF signal is taken from the drain. This type of active phase shifter can achieve a maximum relative phase shift of approximately 120° and has the obvious advantage of being very compact. However, it has a relatively narrow bandwidth performance and can have poor PM/AM conversion characteristics with large values of relative phase shift.

7.2.8 Phase splitter-power combiner

7.2.8.1 Passive design

One method of implementing an analogue phase shifter is to combine a fixed phase-splitting network with a variable power-combining network. One common example of this type of phase shifter is the I-Q vector modulator, illustrated in Fig. 7.21. Here, the input signal power is equally split into two orthogonal channels. A balanced modulator (defined by Garver as a device with a voltage transmission coefficient that can vary between ±1, through zero [3]) in each channel controls the amplitudes of the two orthogonal signals. Therefore, any value of magnitude or phase angle can be achieved for the voltage transmission coefficient of the circuit.

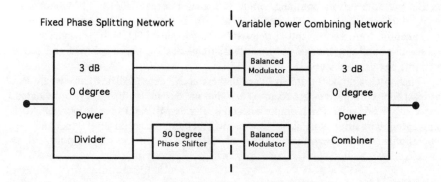

Fig. 7.21 Schematic diagram of an I-Q vector modulator

A noteworthy example of a monolithic I-Q vector modulator was reported by Devlin and Minnis [37]. A quadrature 3 dB power divider was implemented with a Lange coupler, at the input, and an in-phase 3 dB power combiner was implemented with a Wilkinson 0° coupler, at the output. The balanced modulators were implemented with a reflection-type attenuator [12].

A lumped element phase splitter-power combiner phase shifter was reported by Frost *et al.* [38]. Here, a lumped element Wilkinson 0° coupler was used for both the 3 dB power divider and combiner. A 120° phase difference between the two channels was achieved by inserting a lumped element high-pass filter in one channel (to increase its insertion phase by 60°) and an equivalent low-pass filter in the other channel (to decrease its insertion phase by 60°). A variable matched MESFET π-attenuator was used in each channel to control the amplitudes of the corresponding signals. Because variable attenuators were employed, instead of balanced modulators, the maximum theoretical level of relative phase shift is only 120°.

These passive designs have two drawbacks. The first is that at least two independent control lines are required. The second is that it has an inherent 3 dB insertion loss. However, they can operated at higher power levels than most other types of monolithic phase shifter.

7.2.8.2 Active design
Active phase splitting-power combining phase shifter designs have also been reported [39-41]. Mazumder *et al.* [39] employed an identical four-way power divider and combiner, consisting of three Wilkinson 0° couplers in each. The phase splitting network used four different lengths of transmission lines to produce the appropriate amount of phase offset in the four channels. A dual-gate MESFET was then assigned to each channel, in the variable power combiner, to control the amplitudes of the four orthogonal signals and to overcome the inherent 6 dB insertion loss. Salvage *et al.* [40] employed two all-pass networks and two differential amplifiers to realise a much smaller active four-way orthogonal phase splitter. As with the previous example, the active variable power combiner employed dual-gate MESFETs. The concept of weighting and combining the output signals from a phase splitting network in an architectural form was recently proposed by Magarshack [41]. Here, an arrangement of standard building blocks are used, consisting of active phase splitters and active variable power combiners.

There are a number of drawbacks with these active designs. The main two are that at least four control lines are required and that the circuit can be rather complicated, especially when differential amplifiers and dual-gate MESFETs are employed. Also, unwanted parasitics will limit the bandwidth and the maximum frequency of operation, and the noise figure and power handling performances may be poor.

7.3 Digital implementations

The beam-forming network of an adaptive phased-array antenna is one very important application of a variable phase shifter. Since phased arrays are often made up of many elements, it is advantageous to reduce the size, weight and cost of the electronics supporting each element. Size and weight constraints are increasingly challenging for airborne and space flight systems, especially when high performances are expected.

In the design of multi-bit digital phase shifters (where an n-bit phase shifter provides $M=2^n$ discrete phase states), the primary goal is to meet the required relative (or differential) phase shift for each discrete phase state, across the desired frequency range. Ideally, the insertion loss through the phase shifter should be invariant for all possible phase states. This invariant performance makes their application easier, as the insertion loss setting in other parts of the chain need not be adjusted. In addition, good input and output return losses are required in order to avoid any ripples in the insertion phase and insertion loss responses when they are cascaded with other components like amplifiers or attenuators. Moreover, good impedance matches are required for each individual phase shifter stage (or bit), in order to minimise interstage reflections. Other requirements imposed from overall system considerations are temperature stability and power handling. Not all of these requirements can be simultaneously fulfilled and a number of trade-offs are involved in the design of state-of-the-art MMIC phase shifters.

There are four main types of digital phase shifter: switched-line, reflection-type, loaded-line and switched-filter. In addition, there are a number of intrinsic one-bit digital phase shifters. This section reviews the various techniques for the design of digital phase shifters and their implementation in monolithic technology.

7.3.1 Switched-line

A simplified switched-line phase shifter is illustrated schematically in Fig. 7.22. In the switched-line approach, PIN diodes or switching-MESFETs [42, 43] are employed to route the input RF signal into the appropriate length of matched transmission line. Ideally, the amount of induced group delay is directly proportional to the difference in the physical lengths of the selected line and the reference line. This approach actually yields a time shifter response in the ideal case, as discussed in Section 7.1, and is more accurately described as a switched-line variable delay line.

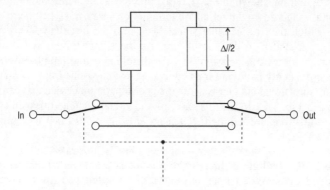

Fig. 7.22 Schematic diagram of a switched-line delay line

With this purely digital implementation, a wide bandwidth can be achieved, but the size of the switched-line implementation is dependant on the centre frequency of operation and the number of bits. As a result, it may only be cost effective to have a 1 or two-bit implementation for applications below a few gigahertz, whereas a 5 or 6-bit implementation is not uncommon for millimetre-wave applications. A number of

MMIC switched-line implementations have been reported over the past decade [44-49]. However, the most notable example of a state-of-the-art switched-line delay line was recently reported by Rogeaux et al. [50].

The satisfactory operation of a single-pole double-throw (SPDT) switch is essential for the performance of the variable delay line. A resonating inductor is usually placed between the drain and source terminals of the switching-FET, in order to tune-out the off-state drain-source capacitance. This technique was first demonstrated by Dunn [46]. Here, the on-state insertion loss of the FET switch was less than 2 dB and the off-state isolation was better than 18 dB, across a frequency range of 42 to 46 GHz. The four-bit phase shifter was designed and fabricated using a 0.6 μm MESFET process on a 90 μm thick GaAs substrate. The chip size was 2.5 × 1.3 mm². Measured performance of all the 16 phase states showed that, at centre frequency, the worst-case insertion phase deviation was 3.3° from the desired value. The insertion loss for all states was within 1.5 dB over the 42 to 46 GHz frequency range.

7.3.2 Reflection-type

7.3.2.1 Delay lines

Fig. 7.23 shows three variations of an ideal digital reflection-type delay line. The positions of the short circuits control the change of insertion phase through the device. The short circuit positions are controlled by at least one switch, as shown. The first configuration, shown in Fig. 7.23(a), employs a circulator; the input signal is circulated and routed down the reflection termination's transmission line to the short circuit end, which then reflects the signal back into the circulator. This reflected signal is then circulated to the output port. As the signal travels twice along the reflection termination's transmission line, the differential line length required for a given phase shift is only half that for an equivalent switched-line implementation. A 3 dB quadrature directional coupler is another signal separation device that can be used, as illustrated in Fig. 7.23(b). In this case, two identical switched transmission line reflection terminations are required. Nakahara et al. demonstrated this approach in monolithic form [51].

Another variation of the reflection-type delay line is illustrated in Fig. 7.23(c). A single-stage 3 dB quadrature directional coupler is employed. Identical lumped-element tapped transmission lines terminate the coupled and direct ports of the directional coupler. A number of equally spaced shunt switches can connect the transmission line to ground, but only one switch per reflection termination is in the on-state at any time. Ideally, the amount of induced group delay is directly proportional to the distance along the line from the on-state switch to the reference off-state switch. Although this provides a wide bandwidth and requires only minimal chip space, there is a practical limit on the number of possible phase states at higher microwave frequencies, because of the physical size of the switches. Also, where power is at a premium, as in the case of satellite or portable applications, considerable control power would be required for large aperture arrays. This type of reflection-type delay line has been demonstrated in monolithic form by Wilson et al. [18]. Here, a four-bit X-band phase shifter was designed, fabricated and tested on a single chip with dimensions of 3.7 × 2.3 mm². An important feature in their work was that Schottky diode structures were used as switching elements rather than MESFETs. These were fabricated using selective ion implantation techniques, compatible with MMIC processing.

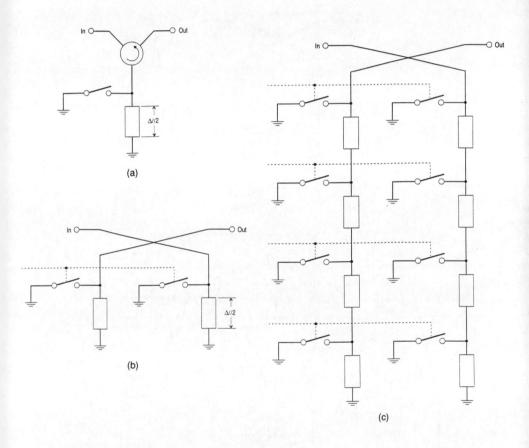

Fig. 7.23 Digital reflection-type delay lines: (a) One-bit employing an active circulator; (b) One-bit employing a 3 dB quadrature directional coupler; (c) M-state employing a 3 dB quadrature directional coupler

7.3.2.2 Phase shifters

Two examples of digital reflection-type phase shifters are illustrated in Fig. 7.24. With reference to Fig. 7.24(a), when the switch state is changed the device has an intrinsic relative phase shift of 180°, with ideal components. In practice, because of the non-idealities of the switches, an impedance-matching transformer is inserted between the switch and the directional coupler to minimise the amplitude modulation [52]. In addition, arbitrary values of relative phase shift are possible by suitable modification of these matching circuits. This principle was demonstrated in

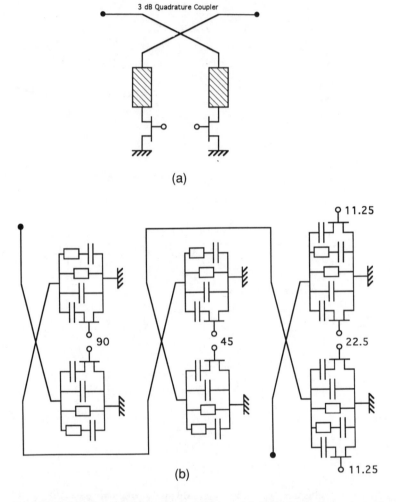

Fig. 7.24 Digital reflection-type phase shifters: (a) With impedance-transformed switches; (b) With switchable load impedances

monolithic form at C-band frequencies by Andricos et al. [53]. Here, a six-bit phase shifter was implemented with five digital bits and one analogue bit on a single chip. The 180° and 90° bits used the reflection-type arrangement, employing Lange couplers. Switching-FET sizes and circuit parameters were optimised for best input/output return losses and minimal insertion loss. The 90° bit had a single transmission line section between the Lange coupler and the switching-FET while the 180° bit required additional open-circuit stubs for impedance matching. Over a 5 to 6 GHz frequency range, the 180° and 90° bits had a maximum insertion loss of 2 dB, a maximum phase error of ±5°, and an input/output return loss of 14 dB. The disadvantage of using reflection-type phase shifters at these frequencies is the excessive length of the Lange coupler, which in this case is almost 5 mm long.

Another example of a digital reflection-type phase shifter is illustrated in Fig. 7.24(b). Here, the reflection terminations consist of a switchable load impedance. A single chip, 32-state implementation of this technique, covering the frequency range of 4 to 20 GHz, was demonstrated by Boire et al. [54].

7.3.3 Loaded-line

The loaded-line phase shifter was first introduced by Dawirs and Swarner [55]. A transmission line is periodically loaded with shunt capacitive reactance elements to ground. With this analogue implementation, the induced phase shift per section must be kept small (i.e. <60°) in order to achieve acceptable return losses. As a result, this type of phase shifter is considered inefficient, as considerable chip space would be required to realise a 360° design.

A digital version of this approach is illustrated in Fig. 7.25. A transmission line, with a characteristic impedance Z_c and an electrical length $\emptyset$, has two identical state-switchable load admittances, $Y(i) = G(i) + jB(i)$ (where $i = 1, 2$ and refers to the two bias states of the switching devices), connected in shunt on either side of the line. A change between one admittance state and the other produces a change in the insertion phase of the network.

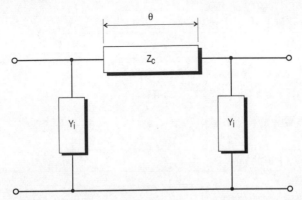

Fig. 7.25 Basic loaded-line phase shifter

The design of the loaded-line phase shifter requires the appropriate choice of the six unknown parameters:- $\emptyset$, Z_c, $G(1)$, $G(2)$, $B(1)$ and $B(2)$. For low-loss switching devices, the ohmic losses associated with the shunt admittances are small and $G(i)$ can be assumed to be zero, as a good approximation. A common design approach, assuming an arbitrary value of transmission line length, employs CAD techniques to select a configuration that results in the required relative phase shift over a wide bandwidth with low insertion loss. Atwater [56] has shown that the choice of the electrical length is not arbitrary but depends on the relative values of load admittance.

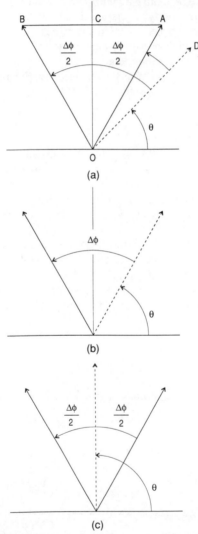

Fig. 7.26 Phasor diagrams for the different loaded-line modes of operation: (a) *Class I*; (b) *Class II* (unload/load); (c) *Class III* (complex conjugate)

A set of loading modes for the simple loaded-line phase shifter was defined by Opp and Hoffman [57] and analysed by Atwater [56]. *Class I* loading corresponds to the condition when $B(1)$ and $B(2)$ are non-zero and unequal. *Class II* corresponds to the case when $B(1)$ is zero and the insertion phase corresponds to the electrical length of the transmission line. The desired value of relative phase shift occurs when the other susceptance load, $B(2)$, is applied. This mode can also be called unload/load, as the line load is first removed and then applied. *Class III* loading corresponds to the condition when $B(1) = -B(2)$. In this condition, the loading admittances switch the insertion phase by half the maximum value of differential phase shift. The loading is switched between complex conjugate admittances and a uniform output amplitude is maintained at both phase states. Fig. 7.26 shows the phasor diagram for a loaded-line phase shifter operating in the different classes. Two approaches can be adopted for implementing the two states of the load admittances. In the first, ideal SPDT switches select the appropriate load susceptance, $B(1)$ or $B(2)$. In the second, a lossless impedance transformer is used to convert the terminal impedances of a non-ideal switch into the required values for $B(1)$ and $B(2)$.

The principal advantages of this loaded-line technique is that it is free from spurious resonances and it does not require a circulator or directional coupler. This technique is, however, only suitable for smaller values of relative phase shift, since it becomes more difficult to achieve a good impedance match with larger values of relative phase shift. Also, the return loss of a loaded-line phase shifter gets progressively worse as frequency deviates from band centre, making it a narrowband device.

Because of their simplicity, loaded-line phase shifters have received a great deal of attention and have been implemented in monolithic form up to X-band frequencies [47, 51-53, 58, 59] and at millimetre-wave frequencies by Slobodnik [60]. The C-band 6-bit phase shifter design reported by Andricos *et al.* utilised the loaded-line approach for the 11.25°, 22.5° and 45° bits. Here, a quarter wavelength 50 Ω through line with 100 Ω shunt stubs was used for the two least significant bits. The 45° bit shunt stub impedance was 56 Ω. The worst-case input/output return loss for the 11.25°, 22.5° and 45° bits were 20 dB, 15 dB and 12 dB, respectively. The maximum insertion loss was approximately 1 dB and the phase errors were ±2° over a 5 to 6 GHz frequency range. Ayasli *et al.* used the loaded-line approach for the design of the 22.5° and 45° bits in an X-band four-bit phase shifter. The switched-line approach was adopted for the 90° and 180° bits. Slobodnik *et al.* also employed the loaded-line approach for the design of the 22.5° and 45° bits of a four-bit phase shifter, operating over the 35 to 37 GHz frequency range [60].

7.3.4 Switched-filter

A conventional switched-filter phase shifter is illustrated in Fig. 7.27. It works by switching the RF signal path between a high-pass filter and a low-pass filter, by means of two SPDT switches. For this reason, this approach is also known as a high-pass/low-pass phase shifter. This technique is similar to a switched-line phase shifter but with the transmission lines replaced by filters. A low-pass filter consisting of series inductors and shunt capacitors provides phase lag (or delay) to signals passing through it. A high-pass filter consisting of series capacitors and shunt inductors provides phase lead (or advance) to signals passing through it. Both filters are matched for insertion loss at the operating frequency but have different insertion phases. Therefore, when the signal path is switched between the two filters the desired change in insertion phase can be achieved without any significant change in the insertion loss. For wider bandwidths, the number of filter elements needs to be increased. However, the bandwidth of the phase shifter depends not only on the characteristics of the two filters but also on the bandwidth and isolation of the two SPDT switches. The overall chip size of a switched-filter phase shifter is considerably smaller than an equivalent switched-line, reflection-type or loaded-line implementation, because the long distributed transmission lines are replaced with much smaller lumped elements. For this reason, it is the most popular type of digital MMIC phase shifter for low microwave frequency applications [58, 66-69]. With the switched-filter phase shifter, the amount of active area used is dominated by the size of the SPDT switches.

With the conventional switched-filter phase shifter, the design approach can be broken down into the individual designs for the two SPDT switches and the two filters. Unlike hybrid MIC digital phase shifters, which usually employ PIN diode-based SPDT switches, MMIC implementations traditionally employ MESFET-based SPDT switches. This is because additional and, therefore, very expensive foundry processing is required to realise PIN diodes. Two notable examples of a switched-

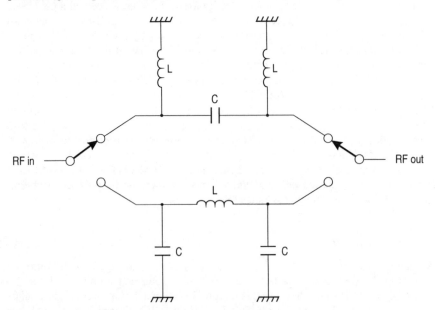

Fig. 7.27 A conventional switched-filter phase shifter

filter phase shifter that do employ PIN diode switches were reported by Coats et al. [62] and more recently by Teeter et al. [63]. With switching-FETs, when the control voltage is zero the FET is in the on-state and can be simply represented by a small resistance between its drain and source terminations. When the reverse bias control voltage is greater than the pinch-off voltage, V_p, the FET is in the off-state and can be simply represented by a capacitance between its drain and source terminations. This capacitance can be quite large. Therefore, if the FET is in a shunt configuration, the insertion loss of the switch can be unacceptably high. If the FET is in a series configuration, the RF isolation of the switch can be unacceptably low. In order to minimise both the levels of insertion loss and RF signal leakage through the switch, the FET can be resonated with a parallel inductor. However, with this approach the bandwidth of the discrete switches and, therefore, the phase shifter will be limited.

An alternative approach was proposed by Ayasli et al. [69] and others. Here, the MESFETs are embedded in the design of the filter, to absorb the unwanted parasitic capacitances, resulting in a significant improvement in the intrinsic bandwidth of the phase shifter. The bridge topology of an embedded switched-filter phase shifter is shown in Fig. 7.28(a). It can be seen that discrete SPDT switches no longer exist in the phase shifter design. Instead, the series and shunt MESFETs that make up the SPDT switches are dispersed throughout the phase shifter topology. The complementary control signal lines, C and $\overline{C}$ to the MESFETs, have been configured so that when C is at logic '0' (i.e. zero potential) the topology of the phase shifter is equivalent to a fifth-order high-pass filter, as illustrated in Fig. 7.28(b). Any capacitor in parallel with an inductor is assumed to have a capacitive reactance that is too large to shunt the associated inductive reactance. When C is at logic '1' (i.e. >V_p potential), the topology of the phase shifter is equivalent to a five section low-pass filter, as illustrated in Fig. 7.28(c). In both filter configurations, the MESFET's on-state resistance values are assumed to be negligible when compared with the reactances in series or parallel with them.

This technique was applied by Ayasli et al. in the design of a two-bit 2 to 8 GHz phase shifter [69]. The gate periphery and impedances of the individual MESFETs were treated as parameters for optimisation. The MESFET gate periphery used in the circuit varied from 83 μm to 1333 μm. The 180° bit was designed for a 50 Ω system while the 90° bit was designed for a 25 Ω system, inserted between impedance transformers. Similarly, Schindler and Miller [61] employed the bridge topology shown in Fig. 7.28(a) for the design of the 180° bit in a three-bit phase shifter operating from 18 to 40 GHz. Reber and Felde [64] designed a 6-bit phase shifter operating from 4 to 7 GHz. All except the 180° bit were implemented using the embedded MESFET technique. The 180° bit used the conventional switched-filter topology. This 6-bit phase shifter has an active GaAs area of only 3.8×3.0 mm^2. The insertion loss variation for all the phase states was measured to be ±2.6 dB over the 4 to 7 GHz frequency range and ±0.9 dB over the 5 to 6 GHz frequency range. The *rms* phase errors were only 3.8° and 2.4°, respectively, for the two frequency bands. However, modest return losses of 7 dB and 11 dB, respectively, were measured over the two frequency bands.

294 *Phase shifters*

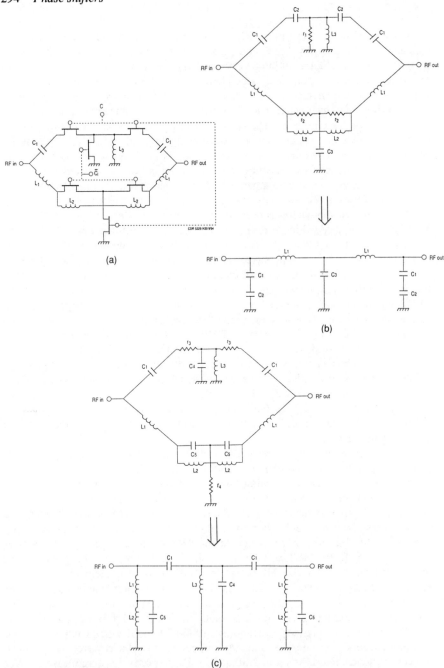

Fig. 7.28 Embedded MESFET switched-filter phase shifters: (a) Basic topology; (b) Equivalent high-pass filter configuration; (c) Equivalent low-pass filter configuration

7.3.5 Intrinsic phase shifters

The techniques described so far have been general, since they can be used to achieve most values of differential phase shift. Some techniques, however, provide an intrinsic value of differential phase shift. Phase shifters with intrinsic designs can have excellent performance characteristics.

7.3.5.1 Boire phase shifter

Boire *et al.* [70] demonstrated a wideband 180° bit phase shifter. The basic configuration of the phase shifter consists of two SPDT switches which change the RF signal path between a shorted 3 dB quadrature coupler and a π-network, as illustrated in Fig. 7.29. The coupler has its direct and coupled ports terminated with ideal short circuits, and the π-network is constructed with a series transmission line between two shunt transmission lines. The even- and odd-mode admittances of the directional coupler, with an electrical length $\emptyset$, are Y_{oe} and Y_{oo}, respectively. If the characteristic admittance of the shunt lines of the π-network are chosen to be Y_{oe}, with an electrical length $\emptyset$, and the characteristic impedance of the series line is $2(Z_{oe} \cdot Z_{oo})/(Z_{oe}-Z_{oo})$, with an electrical length $\emptyset$, then its transmission parameters are equivalent to those of the shorted directional coupler preceded by an ideal phase-reversing transformer. In other words, the two networks are similar in all aspects except that the transmission phase between the two is 180°. This relationship is independent of the electrical length of the two networks and, thus, independent of frequency.

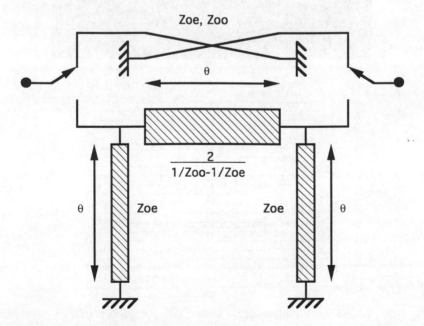

Fig. 7.29 The Boire 180° bit phase shifter

296 Phase shifters

Boire *et al.* first demonstrated this technique using hybrid MIC technology, operating across the 4.5 to 18 GHz frequency range [70]. They then went on to demonstrate the technique with monolithic technology, in a 5-bit phase shifter operating over the 1.5 to 7.0 GHz frequency range [54]. This technique has also been implemented in monolithic technology by Miller and Anderson [71] in a two-bit phase shifter operating over the 4 to 12 GHz frequency range. The 90° bit was implemented using an embedded MESFET switched-filter topology similar to that in Fig. 7.28(a). The microphotograph and measured performance of this MMIC phase shifter are shown in Fig. 7.30. More recently, Slobodnik *et al.* demonstrated the Boire 180° bit phase shifter in a four-bit monolithic design operating over the 35 to 37 GHz frequency range [60]. One drawback of the Boire 180° bit phase shifter is the large differential insertion loss between the two phase states.

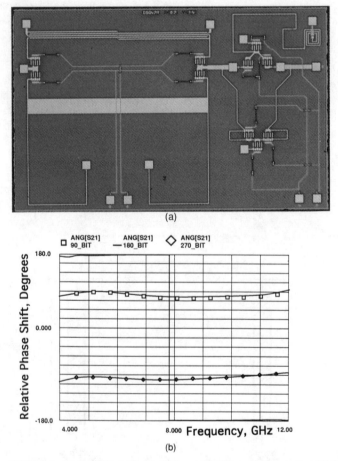

Fig. 7.30 X-band two-bit digital phase shifter: (a) Microphotograph; (b) Measured relative phase shift responses

7.3.5.2 Switched active balun

Another technique that has been adopted for implementing a 180° bit is to switch between the output ports of an active balun. This approach makes use of the inherent 180° phase difference between the source and drain terminals of a FET. Walters and Fikart used this technique in their 5-bit phase shifter design operating from 5 to 6 GHz [67]. The drawback of this technique is that unconditional stability needs to be maintained for the active balun, while simultaneously matching the gate, drain and source of the device. Moreover, the output port impedances and parasitics are unequal, thus, causing significant amplitude and phase imbalances. A technique for overcoming these problems, by cross-connecting identical active baluns, has been investigated and successfully implemented in a monolithic L-band bi-phase modulator by Goldfarb *et al.* [72].

7.3.5.3 Switched-coupler

A 90° bit can be implemented by switching between the direct and coupled ports of a 3 dB directional coupler. SPDT switches at the coupler's ports are used to ensure that the unused port is terminated in the coupler's impedance. This ensures that any mismatches at the ports do not degrade the performance of the phase shifter. Fig. 7.31 shows the topology of a switched-coupler 90° phase shifter. This technique was demonstrated using monolithic technology by Moore and Miller [73], and Fig. 7.32 shows a microphotograph of the chip. This type of phase shifter has a bandwidth that is comparable with that of the directional coupler used. Therefore, a one and a half octave bandwidth Lange coupler can, in principle, be employed to realise a one and a half octave bandwidth 90° bit phase shifter. The main disadvantage with this phase shifter is that it suffers from an inherent 3 dB insertion loss.

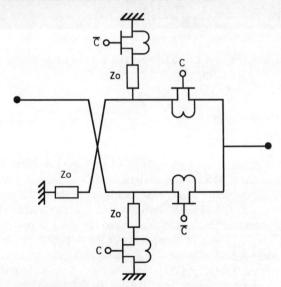

Fig. 7.31 Switched-coupler 90° bit phase shifter

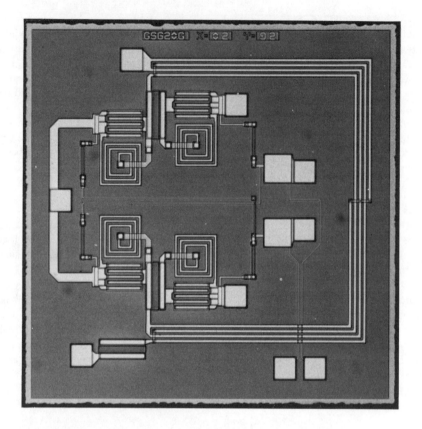

Fig. 7.32 Microphotograph of a switched-coupler X-band phase shifter

7.4 Summary

In this chapter the distinctive characteristics of a true phase shifter and a true delay line have been defined. The main techniques for the design of both analogue and digital MMIC implementations have been reviewed and their relative advantages and drawbacks discussed. It has been demonstrated that the traditional single-stage reflection-type phase shifter topology can indeed be used to implement a variable delay line. Conversely, the general topology of the switched-line delay line can be used to implement a digital switched-line phase shifter.

It can be seen that MMIC phase shifters and delay lines are capable of providing the high levels of performance required for future communication, radar and measurement applications.

7.5 References

1. HARDIN, R. N., DOWNEY, E. J., and MUNUSHIAN, J.: 'Electronically-variable phase shifters utilizing variable capacitance diodes', *Proc. IRE*, Vol. 48, May 1960, pp. 944-945
2. SEARING, R. M.: 'Variable capacitance diodes used as phase-shift devices', *Proc. IRE*, Vol. 49, Mar. 1961, pp. 640-641
3. GARVER, R. V.: 'Broadband binary 180° diode phase modulator', *IEEE Trans. Microwave Theory Tech.*, MTT-13, Jan. 1965, pp. 32-38
4. HENOCH, B. T., and TAMM, P.: 'A 360° reflection-type diode phase modulator', *IEEE Trans. Microwave Theory Tech.*, MTT-19, Jan. 1971, pp. 103-105
5. ULRIKSSON, B.: 'Continuous varactor-diode phase shifter with optimized frequency response', *IEEE Trans. Microwave Theory Tech.*, MTT-27, July 1979, pp. 650-654
6. DAWSON, D. E., CONTI, A. L., LEE, S. H., SHADE, G. F., and DICKENS, L. E.: 'An analog X-band phase shifter', IEEE Microwave and Millimeter-Wave Monolithic Circuits Symp. Dig., 1984, pp. 6-10
7. CHEN, C. L., COURTNEY, W. E., MAHONEY, L. J., MANFRA, L. J., CHU, A., and ATWATER, H. A.: 'A low-loss Ku-band monolithic analog phase shifter', *IEEE Trans. Microwave Theory Tech.*, MTT-35, Mar. 1987, pp. 315-320
8. KRAFCSIK, D. M., TMHOFF, S. A., DAWSON, D. E., and CONTI, A. L.: 'A dualvaractor, analog phase shifter operating 6 to 18 GHz', IEEE Microwave and Millimeter-Wave Monolithic Circuits Symp. Dig., 1988, pp. 83-86
9. BASTIDA, E. M., DONZELLI, G. P., and SCOPELLITI, L.: 'GaAs monolithic microwave integrated circuits using broadband tunable active inductors', 19th European Microwave Conf. Proc., Sept. 1989, pp. 1282-1287
10. BIANCHI, G., PINTO, G., and GIULIANI, C.: 'MMIC 12-13 GHz voltage controlled phase shifter', ESA Proc. of an International Workshop on Monolithic Microwave Integrated Circuits for Space Applications, ESTEC, Noordwijk, Mar. 1990
11. ALI, F., and MYSOOR, N.: 'An analog MMIC phase modulator for X-band satellite transponder applications', *IEEE Microwave and Guided Wave Lett.*, Vol. 2, no. 11, Nov. 1992, pp. 445-446
12. LUCYSZYN, S., and ROBERTSON, I. D.: 'Analog reflection topology building blocks for adaptive microwave signal processing applications', *IEEE Trans. Microwave Theory Tech.*, MTT-43, no. 3, Mar. 1995
13. LUCYSZYN, S., GREEN, G., and ROBERTSON, I. D.: 'Accurate millimeter-wave large signal modeling of planar Schottky varactor diodes', IEEE Int. Microwave Symp. Dig., Albuquerque, 1992, pp. 259-262
14. LUCYSZYN, S., LUCK, J., GREEN, G., and ROBERTSON, I. D.: 'Enhanced modelling of interdigitated planar Schottky varactor diodes', IEEE Asia-Pacific Microwave Conf. Dig., Adelaide, Aug. 1992, pp. 273-278

15. LUCYSZYN, S., GREEN, G., and ROBERTSON, I. D.: 'Interdigitated planar Schottky varactor diodes for tunable MMIC applications', Proc. of the IEEE/ESA European Gallium Arsenide Applications Symposium, ESTEC, Noordwijk, Apr. 1992
16. LANGE, J.: 'Interdigitated stripline quadrature hybrid', *IEEE Trans. Microwave Theory Tech.*, MTT-17, Dec. 1969, pp. 1150-1151
17. RUTZ, E. M.: 'A stripline frequency translator', *IRE Trans. Microwave Theory Tech.*, Mar. 1961, pp. 158-161
18. WILSON, K., NICHOLAS, J. M. C., McDERMOTT, G., and BURNS, J. W.: 'A novel MMIC X-band phase shifter', *IEEE Trans. Microwave Theory Tech.*, MTT-33, Dec. 1985, pp. 1572-1578
19. LUCYSZYN, S.: 'Ultra-wideband high performance reflection-type phase shifters for MMIC Applications', University of London PhD Thesis, Jan. 1993
20. LUCYSZYN, S., ROBERTSON, I. D., and AGHVAMI, A. H.: 'A high performance analogue time shifter for wideband phased-array antennas', IEE Coll. Dig. on Recent Advances in Microwave Sub-systems for Space and Satellite Applications, London, Mar. 1993, pp. 4/1-4
21. LUCYSZYN, S., ROBERTSON, I. D., and AGHVAMI, A. H.: 'High performance wideband analogue time shifter', *Electron. Lett.*, Vol. 29, no. 10, May 1993, pp. 885-887
22. LUCYSZYN, S., and ROBERTSON, I. D.: 'Synthesis techniques for high performance octave bandwidth 180° analog phase shifters', *IEEE Trans. on Microwave Theory Tech.*, MTT-40, no. 4, Apr. 1992, pp. 731-739
23. LUCYSZYN, S., and ROBERTSON, I. D.: 'High performance octave bandwidth MMIC analogue phase shifter', 22nd European Microwave Conf. Proc., Espoo, Aug. 1992, pp. 221-224
24. LUCYSZYN, S., and ROBERTSON, I. D.: 'An improved multi-layer quadrature coupler for MMICs', 21st European Microwave Conf. Proc., Stuttgart, Sept. 1991, pp. 1154-1158
25. LUCYSZYN, S., and ROBERTSON, I. D.: 'Two octave bandwidth monolithic analog phase shifter', *IEEE Microwave and Guided Wave Lett.*, Vol. 2., no. 8, Aug. 1992, pp. 343-345
26. LUCYSZYN, S., and ROBERTSON, I. D.: 'Decade bandwidth hybrid analogue phase shifter using MMIC reflection terminations', *Electron. Lett.*, Vol. 28, no. 11, May 1992, pp. 1064-1065
27. KRYTAR: 'Broadband, 3 dB 90° hybrids covering 1 to 18 GHz', *Microwave Journal*, Sept. 1985
28. LUCYSZYN, S., and ROBERTSON, I. D.: 'Decade bandwidth MMIC analogue phase shifter', IEE Coll. Dig. on Multi-octave Microwave Circuits, London, Nov. 1991, pp. 2/1-6
29. DOUGHERTY, R.: 'Circulate signals with active devices on monolithic chips', *Microwaves & RF*, June 1989, pp. 85-89
30. ROBERTSON, I. D., and AGHVAMI, A. H.: 'Ultrawideband biasing of MMIC distributed amplifiers using improved active load', *Electron. Lett.*, Vol. 27, no. 21, Oct. 1991, pp. 1907-1909

31. KOTHER, D., HOPF, B., SPORKMANN, T., and WOLFF, I.: 'Active CPW MMIC circulator for the 40 GHz band', 24th European Microwave Conf. Proc., Cannes, 1994, pp. 542-547
32. ROBERTSON, I. D., and AGHVAMI, A. H.: 'Novel monolithic ultra-wideband 4-port junction using distributed amplification techniques', IEEE Int. Microwave Symp. Dig., Albuquerque, June 1992, pp. 1051-1054
33. PAO, C. K., CHEN, J. C., LAN, G. L., WANG, D. C., WONG, W. S., and HERMAN, M. I.: 'V-band monolithic phase shifters', 10th Annual IEEE GaAs IC Symp. Tech. Dig., Nov. 1988, pp. 269-272
34. LUCYSZYN, S., ROBERTSON, I. D., and AGHVAMI, A. H.: 'Monolithic analogue phase shifter and cascode FET amplifier using uniplanar techniques', 24th European Microwave Conf. Proc., Cannes, Sept. 1994, pp. 554-559
35. TSIRONIS, C., and HARROP, P.: 'Dual gate GaAs M.E.S.F.E.T. phase shifter with gain at 12 GHz', *Electron. Lett.*, Vol. 16, no. 14, July 1980, pp. 553-554
36. PENGELLY, R. S.: 'GaAs monolithic microwave circuits for phased array applications', *IEE Proc. F*, Vol. 127, no. 4, Aug. 1980, pp. 301-311
37. DEVLIN, L. M., and MINNIS, B. J.: 'A versatile vector modulator design for MMIC', IEEE Int. Microwave Symp. Dig., 1990, pp. 519-522
38. FROST, R. D., FISHER, D. A, and PECK, D. E: 'A GaAs MMIC voltage-controlled phase shifter', *Microwave Journal*, Aug. 1991, pp. 87-94
39. MAZUMDER, S. R., TSAI, T. L., and TSAI, W. C.: 'A frequency translator using dual-gate GaAs FETs', IEEE Int. Microwave Symp. Dig., 1983, pp. 346-348
40. SALVAGE, S. T., HASH, R. J., and PETTED, B. E.: 'An octave band GaAs analog phase shifter', IEEE Int. Microwave Symp. Dig., 1989, pp. 1051-1054
41. MAGARSHACK, J.: 'A new digital phase shifter architecture suitable for MMIC's', *IEEE Trans. Microwave Theory Tech.*, MTT-42, no. 1, Jan. 1994, pp. 154-156
42. SHARMA, A. K.: 'Solid-state control devices: state of the art', *Microwave Journal*, 1989, pp. 95-112
43. CAVERLY, R. H.: 'Distortion in GaAs MESFET switch circuits', *Microwave Journal*, Sept. 1994, pp. 106-114
44. GUPTA, A., KAELIN, G., STEIN, R., HUSTON, S., LP, K., and PETERSEN, W.: 'A 20 GHz 5-bit phase shift transmit module with 16 dB gain', IEEE GaAs IC Symp. Dig., 1984, pp. 197-200
45. BAUHAHN, P., BUTTER, C., SOKOLOV, V., and CONTOLATIS, A.: '30 GHz multibit monolithic phase shifters', IEEE Microwave and Millimeter-Wave Monolithic Circuits Symp. Dig., 1985, pp. 4-7
46. DUNN, V. E., HODGES, N. E., SY, O. A., and ALYASSINI, W.: 'MMIC phase shifters and amplifiers for millimeter-wavelength active arrays', IEEE Int. Microwave Symp. Dig., 1989, pp. 127-130
47. OAKI, A., INOUE, M., TSUDA, Y., ORIME, N., and UCHINO, S.: 'A small X-band phase shifter with small phase increments', 3rd Asia-Pacific Microwave Conf. Proc., Tokyo, 1990, pp. 873-876

48. LAU, C. L., FENG, M., HWANG, T., LEPKOWSKI, T., ITO, C., DUNN, V., and HODGES, N.: 'Millimeter wave monolithic ICs using direct ion implantation into GaAs LEC substrates', IEEE GaAs IC Symp. Dig., 1990, pp. 73-76
49. AUST, M., WANG, H., CARANDANG, R., TAN, K., CHEN, C. H., TRINH, T., ESFANDIAR, R., and YEN, H. C.: 'GaAs monolithic components development for Q-band phased array application', IEEE Int. Microwave Symp. Dig., 1992, pp. 703-706
50. ROGEAUX, E., GROGNET, V., CAZAUX, J. L., COPPEL, F., GEEN, M., and GATTI, G.: 'Standard MMICs for space application: A broadband high resolution phase-shifter', 24th European Microwave Conf. Proc., Cannes, Sept. 1994, pp. 1483-1488
51. NAKAHARA, K., CHAKI, S., ANDOH, N., MATSUOKA, H., TANINO, N., MITSUI, Y., OTSUBO, M., and MITSUI, S.: 'A novel three phase-state phase shifter', IEEE Int. Microwave Symp. Dig., 1993, pp. 369-372
52. MILLER, P., and JOSHI, J. S.: 'MMIC phase shifters for space applications', ESA Proc. of an International Workshop on Monolithic Microwave Integrated Circuits for Space Applications, ESTEC, Noordwijk, Mar. 1990
53. ANDRICOS, C., BAHL, I., and GRIFFIN, E.: 'C-band 6-bit GaAs monolithic phase shifter', *IEEE Trans. Microwave Theory Tech.*, MTT-33, Dec. 1985, pp. 1591-1596
54. BOIRE, D. C., ONGE, G. ST., BARRATT, C., NORRIS, G. B., and MOYSENKO, A.: '4:1 bandwidth digital five bit MMIC phase shifters', Microwave and Millimeter-Wave Monolithic Circuits Symp. Dig., 1988, pp. 69-73
55. DAWIRS, H. N., and SWARNER, W. G.: 'A very fast, voltage-controlled, microwave phase shifter', *Microwave Journal*, June 1962, pp. 99-107
56. ATWATER, H. A.: 'Circuit design of the loaded-line phase shifter', *IEEE Trans. Microwave Theory Tech.*, MTT-33, no. 7, July 1985, pp. 626-634
57. OPP, F. L., and HOFFMAN, W. F.: 'Design of digital loaded-line phase shift networks for microwave thin-film applications', *IEEE Trans. Microwave Theory Tech.*, MTT-16, no. 7, July 1968, pp. 462-468
58. OZAKI, J., ASANO, T., WATANABE, S., TATEMATSU, M., and KAMIHASHI, S.: 'X-band MMIC phase shifter with small loss variation', 3rd Asia-Pacific Microwave Conf. Proc., Tokyo, 1990, pp. 869-872
59. AYASLI, Y., PLATZKER, A., VORHAUS, J., and REYNOLDS, L. D.: 'A monolithic single chip X-band four-bit phase shifter', *IEEE Trans. Microwave Theory Tech.*, MTT-30, no. 12, Dec. 1982, pp. 2201-2206
60. SLOBODNIK, A. J., WEBSTER, R. T., and ROBERTS, G. A.: 'A monolithic GaAs 36 GHz four-bit phase shifter', *Microwave Journal*, June 1993, pp. 106-111
61. SCHINDLER, M. J., and MILLER, M. E.: 'A 3 bit K/Ka band MMIC phase shifter', IEEE Int. Microwave Symp. Dig., 1988, pp. 95-98

62. COATS, R., KLEIN, J., PRITCHETT, S. D., and ZIMMERMANN, D.: 'A low loss monolithic five-bit PIN diode phase shifter', IEEE Int. Microwave Symp. Dig., 1990, pp. 915-918
63. TEETER, D., WOHLERT, R., COLE, B., JACKSON, G., TONG, E., SALEDAS, P., ADLERSTEIN, M., SCHINDLER, M., and SHANFIELD, S.: 'Ka-band GaAs HBT PIN diode switches and phase shifters', IEEE Int. Microwave Symp. Dig., 1994, pp. 451-454
64. REBER, R. H., and FELDE, H.-P.: 'A monolithic single-chip C-band 6-bit phase shifter', 21st European Microwave Conf. Proc., Stuttgart, 1991, pp. 479-482
65. KOMIAK, J. J., and AGRAWAL, A. K.: 'Design and performance of octave S/C-band MMIC T/R modules for multi-function phased arrays', *IEEE Trans. Microwave Theory Tech.*, MTT-39, no. 12, Dec. 1991, pp. 1955-1963
66. DEVLIN, L. M.: 'Digitally controlled, 6 bit, MMIC phase shifter for SAR applications', 22nd European Microwave Conf. Dig., Espoo, 1992, pp. 225-230
67. WALTERS, P. C., and FIKART, J. L.: 'A fully integrated 5-bit phase shifter for phased array applications', IEEE International Symp. Dig. on MMICs in Communications Systems, London, Sept. 1992
68. GOLDFARB, M. E., and PLATZKER, A.: 'A wide range analog MMIC attenuator with integral 180° phase shifter', *IEEE Trans. Microwave Theory Tech.*, MTT-42, no. 1, Jan. 1994, pp. 156-158
69. AYASLI, Y., MILLER, S. W., MOZZI, R., and HANES, L. K.: 'Wideband monolithic phase shifter', *IEEE Trans. Microwave Theory Tech.*, MTT-32, no. 12, Dec. 1984, pp. 1710-1714
70. BOIRE, D. C., DEGENFORD, J. E., and COHN, M.: 'A 4.5 to 18 GHz phase shifter', IEEE Int. Microwave Symp. Dig., 1985, pp. 601-604
71. MILLER, P., and ANDERSON, A. K.: 'Development of gain/phase control blocks for space applications', IEEE ESA Proc. of the European Gallium Arsenide Applications Symposium, ESTEC, Noordwijk, Apr. 1992
72. GOLDFARB, M. E., COLE, P. J. B., and PLATZKER, A.: 'A novel MMIC biphase modulator with variable gain using enhancement-mode FETs suitable for 3 V wireless applications', IEEE Microwave and Millimeter-Wave Monolithic Circuits Symp. Dig., 1994, pp. 99-102
73. MOORE, M. T., and MILLER, P.: 'Microwave IC control components for phased-array antennas', *IEE Electronics & Communication Engineering Journal*, June 1992, pp. 123-130

Chapter 8

Switches and attenuators

J. S. Joshi

8.1 Introduction

Microwave switches are essential elements for a number of widely varying applications ranging from highly sophisticated space communications systems to more common applications simply requiring the switching of an RF signal from one path to another–for the purpose of providing redundancy or antenna diversity, for example. Although the specific requirements for each application need to be tailored to individual needs, the essential characteristics of the switches remain the same. There is a continuing need for high performance, small and versatile microwave switches for use in a range of RF commercial and military systems applications. They allow routeing of microwave signals through desired paths in the system and sub-system and enable the realisation of different schemes for signal combinations necessary for system implementation.

The advent of phased-array antennas, smart antennas capable of providing shaping and electronic control of the main beam, and modern communications systems which require amplitude control and gain equalisation in different communication paths has meant that there are increasing requirements for gain setting and control circuits. These circuits are generally of digital architecture and provide binary weighted attenuation control to achieve the required performance objectives.

In this chapter we will first review the GaAs FET switch mechanism and the development of an equivalent circuit for switch operation. The advantages of monolithic switch operation will be described along with some of its limitations. The second part of this chapter deals with different schemes for the realisation of GaAs MMIC variable attenuators. A number of techniques will be examined and the performance characteristics offered by selected configurations compared.

8.2 GaAs FET MMIC switches

The implementation of solid state PIN diode switches several decades ago resulted in a significant improvement in mass, size and switching speed of microwave switches compared with their then available electromechanical counterparts. Such PIN diode circuits can be easily realised in hybrid integrated circuits in a compact outline. This enabled their widespread use in a range of different applications. PIN diodes can provide faster switching speeds and can handle medium to large RF power levels. The biasing requirements of PIN diodes are such that in the on-state they require high forward current and in the off-state they require large reverse bias across the semiconductor junction. Their switching characteristics are highly dependent on the individual characteristics of the PIN diode. In recent years these PIN diode switches have been increasingly replaced by GaAs FET based monolithic switches, especially for low power applications. Medium to high power GaAs FET MMIC switches have also been demonstrated.

The inherent advantages of GaAs FET-based switches over conventional PIN diodes are simplified bias networks, negligibly small DC power requirements of the FET switch itself, simplified design of driver circuits, faster switching speeds etc. Furthermore, monolithic implementation of such switches means that increased functionality can be designed within a single chip measuring only a few square millimetres. Monolithic implementation has resulted in some very high performance and extremely complex circuits being realised on very small chips.

8.2.1 GaAs FET switching mechanism

The FET switch is a three-terminal device in which the gate bias voltage V_g controls the states of the switch. The FET acts as voltage controlled resistor in which the gate bias controls the drain-to-source resistance in the channel. The intrinsic gate-to-source and drain-to-gate capacitances and device parasitics limit the performance of the FET switch at higher frequencies

In a typical switching mode a high impedance state is produced when a negative bias larger in magnitude than the pinch-off voltage ($|V_g|>|V_p|$) of the FET is applied across the gate terminal. When a zero bias gate voltage is applied across the gate terminal, the FET switch is in a low impedance state. These two linear operational regions of the FET are shown schematically in Fig. 8.1. It is important to note that in either state virtually no DC power is required by the FET switch itself. Thus, for all practical purposes, from a power consumption point of view, these FET switches can be considered as passive devices. This negligible DC power requirement of the FET switch results in a considerable simplification of its driver circuitry. Being a planar structure, with the gate stripe located between the source and drain terminals, the FET switch acts as a bi-directional switch.

Figure 8.2 shows the cross-section of a simple conventional FET structure. Under the zero gate bias condition the channel region is virtually open except for the zero field depletion layer thickness. Thus, for current levels less than the saturated channel current, I_{dss}, the FET can be modelled as a linear resistor. When a negative voltage V_g

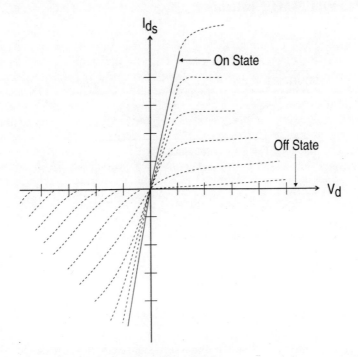

Fig. 8.1 Linear operating regions of a FET switch

is applied between gate and source terminals such that $|V_g|>|V_p|$, where V_p is pinch-off voltage, the channel region is completely depleted of free charge carriers. Under these conditions the FET can be modelled by series and parallel combinations of resistors and capacitors. The parts of the FET associated with resistive and capacitive regions are shown in Fig. 8.2. Assuming that the gate termination represents a high RF impedance at the frequency of operation, the off-state equivalent circuit can be expressed as a parallel combination of a resistor and a capacitor as shown in Fig. 8.2. For conditions when $1/\omega C_g \gg r_g$, the effective drain-to-source capacitance is simply $(C_{sd} + C_g/2)$ and the effective drain resistance is the parallel combination of r_d and $2/(\omega^2 C_g^2 r_g)$. The figure of merit for a switch FET can be expressed as a ratio of its effective off-state resistance to its on-state resistance [1, 2].

The physical parameters of the device strongly influence the equivalent circuit parameters of the switch FET. The important device physical parameters are the channel geometry, gate length, channel doping density and pinch-off voltage. The metallisation scheme, gate recess structure etc., influence the on-state resistance of the FET switch; the off-state impedance depends on the source-drain capacitance C_{sd}, its parallel resistance r_d, and the drain-to-gate and source-to-gate capacitance C_g and its associated series resistance r_g.

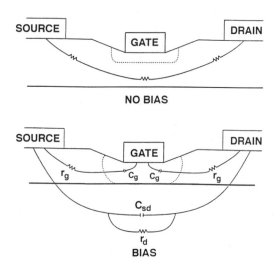

Fig. 8.2 Cross-section of a FET switch with resistive and capacitive regions

The capacitance C_{sd} represents the fringing capacitance between the source and drain electrodes. Its value mainly depends on the spacing between them and is virtually independent of device parameters such as channel doping and pinch-off voltage. Due to the symmetry of the device, the drain-to-gate and gate-to-source capacitances are equal and are represented by C_g. The gate pad capacitance to ground, if significant, needs to be added to the equivalent circuit. A simple but reasonable approximation of C_g is half the gate capacitance with the channel region fully depleted. C_g is thus dependent on both the channel doping and the pinch-off voltage of the device. The resistances which contribute to the off-state impedance are r_d and r_g (Fig. 8.2). Resistance r_d is in parallel with C_{sd} and represents the RF losses associated with the source-drain capacitance. Its value can be arrived at from a reasonable Q value for the capacitance. Resistance r_g represents the resistance of the channel outside the pinch-off region in the off condition.

The switching speed of the FET switch can also be calculated from the equivalent circuit parameters. Assuming that an internal gate resistance of 1.5 kΩ is included in series with the gate terminal and if 10 pF is used in the low-pass bias circuit filter, the charging time constant is determined mainly by the these two values and works out to be 15 ns. Reduction in switching times are quite possible by optimising the gate bias circuit design.

8.2.2 Switch FET equivalent circuit

It should be noted that the physical basis of these device parameters is applicable to discrete as well as MMIC FET devices. However, for monolithic circuit implementation special processing steps, such as ion-implantation etc., can be implemented by the GaAs foundries in order to achieve FET characteristics specially dedicated to switching applications.

The Philips GaAs foundry has developed equivalent circuit models for their D07M process [3] which have been extensively used by a number of different designers for switching and other related applications such as phase shifters and attenuators. The D07M process is available as a standard commercial foundry process. This process produces medium power ion-implanted FETs, which have 0.7 μm gate length with a guaranteed f_t of 17 GHz, on a 100 μm thick substrate with via-holes. Thickly plated tracks help to reduce resistance and air-bridges minimise parasitic capacitance at crossovers. Compared with the standard amplifying FET, the geometry of the switch FET device is modified so that the RF signal path is in-line between the source and drain terminals and the gate terminal, where the DC control voltage is applied, is at right angles to these. This separation of the gate terminal enables the use of larger gate width devices for switching applications. Also, for a given total gate-width, the footprint of the switch FET is nearly 25% smaller than a conventional FET. A typical switch FET layout is shown in Fig. 8.3.

To develop a model for switch FETs, a range of FET patterns from 50 to 300 μm unit gate width with devices having one, two, four and eight gate fingers, were defined on GaAs wafers. These switch FETs were RF-on-wafer probed in the on-state (V_{gs} = 0 V) and off-state (V_{gs} = -5 V) at frequencies ranging from 45 MHz to 26.5 GHz. A typical set of devices was selected and a twelve-element equivalent circuit model was generated. This model was then made scalable by introducing a physically meaningful dependency of all the parameters on the gate widths and the number of gate fingers. This equivalent circuit along with the resulting scaling rules is shown in Fig. 8.4.

The insertion loss of a four-finger gate geometry series FET switch in the on-state is shown in Fig. 8.5 over a 1 to 20 GHz frequency range as a function of total gate periphery. As expected, the insertion loss decreases as total gate periphery increases. Fig. 8.6 shows the performance characteristics of a series FET switch in the off-state. Due to increased capacitive coupling, the isolation in the off-state for larger gate periphery devices is lower than for smaller gate periphery devices. The switching characteristics are principally controlled by the total gate width of the transistor. The actual number of gate fingers has a marginal influence.

In the on-state the low channel resistance of the FET is the dominating parameter and this results in the insertion loss being virtually independent of frequency. In the off-state, however, the equivalent capacitance of the circuit determines the isolation characteristic which shows strong frequency dependence. Therefore, insertion loss for the series FET and isolation for the shunt FET configuration show frequency invariant behaviour. On the other hand, isolation for the series FET and insertion loss for the shunt FET show frequency dependent behaviour due to capacitive effects in

310 *Switches and attenuators*

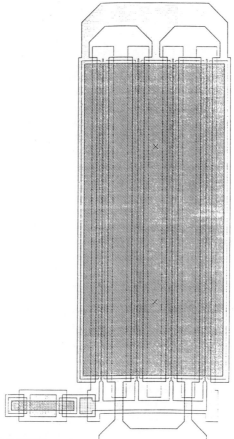

Fig. 8.3 Philips switch FET layout

the off-state. Increasing the size of the FET reduces its on-state resistance and thus reduces insertion loss. However, it also results in increased gate-source and gate-drain capacitance and increased source-drain fringing capacitance which limit the switch performance in the off-state.

8.2.3 Implementation of MMIC switches

FET switching circuits can be designed in the same way as conventional PIN diodes by using on- and off-state equivalent circuits as required in the overall circuit configuration. The FET can be in series or in shunt configuration with respect to the transmission lines to which these switches are connected. Before going into the details of these switch configurations, it is worthwhile to note some important differences between the conventional PIN diode-based switches and FET-based MMIC switches. Since the FET is a three terminal device the switching occurs only

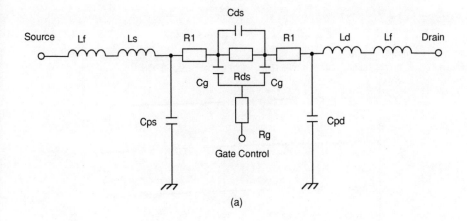

	ON	OFF
R1	0.52 + 750/W	1.21 + 1910/W
Rds	0.20 + 583/W	Infinite
Cds	Zero	0.11W + N + 5
Cg	W + 1.25N	0.17W + 1.25N

$Cpd = 1.2N + (0.057 + 0.015N)w$

$Cps = 1.23N + (0.057 + 0.009N)w$

$Lf = 80 \text{ pH}$

$Ld = [1.83N^2 - 18N + w(0.873 - 0.095N)] / [K + 1]$

$Ls = K \times Ld$

where
K = 1   for N=1
K = 2   for N=2
K = 1.5 for N=4
K = 1.25 for N=8

In general: $K = 1 + 2/N$ for $N > 1$
and $K = 1$ for $N = 1$

W is the total gate width in microns
w is the unit gate width in microns
N is the number of gate fingers

Capacitances ere in femtoFarads
Inductances are in picoHenries
Resistances are in Ohms

(b)

Fig. 8.4 D07M switch FET model: (a) Equivalent circuit; (b) Scaling rules

312 *Switches and attenuators*

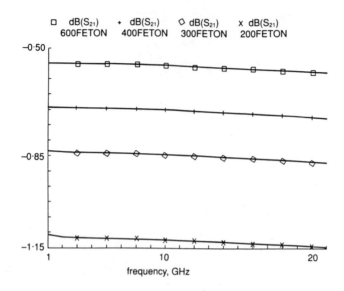

Fig. 8.5 Insertion loss against frequency for a series FET switch in the on-state

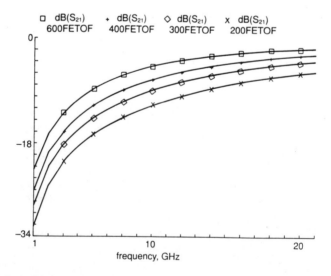

Fig. 8.6 Insertion loss against frequency for a series FET switch in the off-state

through the gate bias control voltage–no other voltage is necessary for switch operation. The RF transmission lines do not carry any DC voltage and therefore there is no necessity for using DC blocking capacitors between the various FET switching elements. This represents a very significant advantage in lowering the component count for MIC circuits and results in a simpler configuration for MMIC circuits.

In the two switch states the gate junction is either reverse biased or is biased at zero gate voltage. In both these states the gate draws negligible current which greatly simplifies the control circuit design requirements. In the off-state of the FET switch, the gate-drain and gate-source capacitances are equal as both the source and drain terminals are at ground potential. As a result, the gate and drain terminals are not isolated from each other and the RF impedance of the gate bias circuit affects the equivalent drain-source impedance. In the limit, when the gate terminating impedance is kept very high by including a built-in 1.5 kΩ resistor as in the Philips D07M process, for example, the equivalent drain-source capacitance can be approximated as $C_d + C_g/2$. The effect of this capacitive shunting should be included in the design of off-state circuits. At X-band frequencies or higher this capacitive shunting effect in the off-state may result in the off-state impedance of the FET being much smaller than the value of r_d itself. It may be necessary to resonate the capacitance with a shunt inductor across the source-drain terminals at the operating frequency of interest. This can be implemented in a number of ways ranging from a simple shunt inductor [4] to a distributed type structure as used by Ayasli *et al.* [5].

The performance characteristics of simple series or shunt configured monolithic FET switches shown in Figs. 8.5 and 8.6 may be adequate for certain limited applications, but generally a combination of series-shunt configurations is often necessary to provide the desired performance characteristics such as low insertion loss and high isolation.

Fig. 8.7 shows the layout of a reflective SPST switch for operation up to 6 GHz [3]. A π-type configuration is used with 4x75 µm switch FETs in the shunt arms and a 2x75 µm switch FET in the series arm. The insertion loss of the switch in the on-state is less than 1 dB up to 6 GHz and the isolation is better than 30 dB (Fig. 8.8). There is a good agreement between simulated and measured results. The insertion loss predicted by the Philips switch FET model is a bit pessimistic. Fig. 8.9 shows the layout of an absorptive SPDT switch for operation up to 6 GHz. 75 and 150 µm FETs are used in the main switching scheme. Additionally, two 2x50 µm FETs are used to switch the signal in the 'off' port to matched on-chip loads. The performance is shown in Fig. 8.10. Insertion loss is lower than 1 dB across the frequency range and the isolation is better than 20 dB. Again, the insertion loss predicted by the model is pessimistic.

Additional measurements of power handling, switching speed and video breakthrough showed that, depending on the FET size, the switches could handle up to 30 dBm before any significant loss in isolation. The switches could switch from on-to-off or off-to-on in less than 5 ns with 0 or -5 V switching on the gate control pads. A switching-frequency signal of less than 5 mV was measured in the RF ports [3].

MMIC switches capable of handling an RF power of more than 100 W at high switching speeds have also been reported by Katzin et al. [6]. A low-loss inductive gate bias network structure was developed which allowed high power RF switching. These were used to demonstrate S- and C- band SPDT switches capable of handling up to 300 W CW RF power.

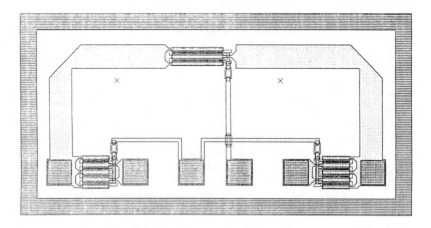

Fig. 8.7 Layout of a reflective SPST switch

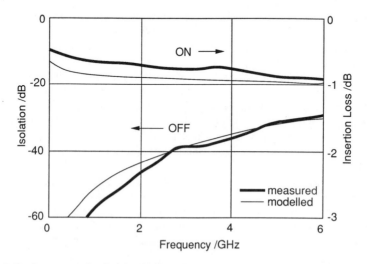

Fig. 8.8 Performance of reflective SPST switch

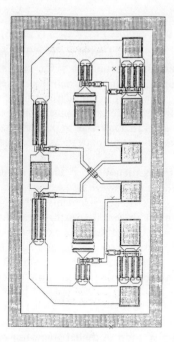

Fig. 8.9 Layout of an absorptive SPDT switch

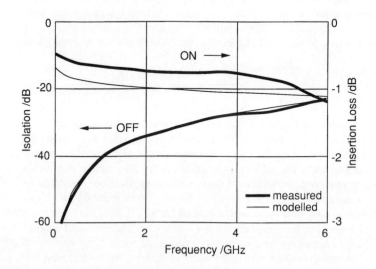

Fig. 8.10 Performance of an absorptive SPDT switch

8.3 Digital attenuators

Microwave communications systems are continually demanding high performance and sophisticated components and equipments for increasingly complex user requirements. Components for gain setting and control functions have applications in a number of areas. Such a component is useful for multichannel communication systems where the gains through different channels need equalisation. Digitally controlled attenuator circuits are also applicable to phased-array antennas and 'smart' antennas where a selective coverage pattern is required, and other similar applications.

Prior to the advent of GaAs monolithic microwave integrated circuits such complex circuit functions, even at microwave frequencies around 10 GHz, could only be realised using PIN diodes as switching elements in MIC configurations. These were bulky and required complex driver circuits for the PIN diodes. GaAs MMIC-based attenuator circuits can now be realised in an extremely small outline, have low mass, require negligible DC power consumption and have excellent performance characteristics. In fact, it can be safely stated that the realisation of such components in GaAs technology has made the implementation of complex new system architectures possible.

The attenuating function can be realised in analogue or digital form. Analogue implementations are briefly described in Section 8.5, but the majority of attenuators have been implemented in digital form. A digital attenuator consists of a number of cascaded units each containing an individual bit which can be switched in or out to achieve the desired value of attenuation. These cascaded bits are usually binary weighted with the least significant bit (LSB) level as the minimum resolvable level. The most significant bit (MSB) determines the upper level of attenuation. Minimum attenuation is obtained when all the individual bit states are turned off, while at the maximum attenuation level all the individual bit states are turned on. The minimum attenuation value is also called the zero- or reference-state insertion loss of the attenuator.

8.3.1 Design approaches

In the design of such digital attenuators the primary goal is to accurately meet the attenuation value requirements of each individual bit across the desired frequency range. System requirements for these attenuators are such that the insertion phase through the attenuator should ideally be invariant for all possible attenuation states. This phase invariant performance makes their application easier as no other parameters need to be adjusted to account for any finite insertion phase change for different states. This does away with any need for re-adjusting the insertion phase of the system every time the attenuator setting is changed. The input and output VSWRs of such digital attenuators have to be very good so as not to cause any insertion loss and insertion phase ripples across the band of interest when cascaded with other components like phase shifters, amplifiers etc. In fact, achieving excellent input and output matches for an individual attenuation bit itself is vitally important for the overall performance of the attenuator as any mismatch interactions within the chip itself will worsen the amplitude and phase ripple performance. The zero state attenuation value of the digital attenuator should be designed to be as low as possible. Other requirements imposed from overall system considerations are temperature performance, power handling requirements etc. Needless to say, all these requirements cannot be simultaneously fulfilled and considerable trade-off is involved in the design of such state-of-the-art GaAs MMIC digital attenuators. In the following sections we consider individual design approaches in detail. These are based on the use of dual-gate FETs. Alternative techniques for the realisation of MMIC digital attenuators based on the use of conventional single-gate FET switches are also considered.

8.3.2 Segmented dual-gate FET technique

An efficient way to achieve equi-phase but differential gain paths at microwave frequency is to use GaAs dual-gate FET devices. In this arrangement the gate 1 and drain ports of the dual-gate FET are used as input and output ports, and gate 2 is used as a control terminal. Gate 1 is biased for the required gain level with gate 2 biased at saturation (the on-state). When gate 2 bias is increased towards pinch-off, the gain of the dual-gate FET decreases while the transmission phase stays constant. In practice, however, gain variation is limited to a value of about 10 dB after which the change in transmission phase becomes excessive [7]. The transmission phase characteristics of the dual-gate FET are dependent on the device geometry, its material and processing parameters etc.

For GaAs MMIC implementation the dual-gate FET approach was modified by Hwang et al. [7] to produce segmented dual-gate FET (SDG FET) circuits. Fig. 8.11 shows the schematic diagram of a segmented dual-gate FET circuit. The technique was investigated further by Naster and Hwang [8], Snow, Komiak & Bates [9] and Siweris [10]. The implementation has been at L- and C/X-band frequencies. In SDG FETs, gate 1 remains as before but gate 2 is segmented into independently switched gates of different widths, allowing individual units of dual-gate FETs to be switched on or off for different gain settings. Since each dual-gate FET segment is only operated in either the on- or off-state, it is possible for the gain to be linearly scaled with the gate width. A variable gain amplifier with 5-bits of precision control results from the simple segmenting of the gate 2 sections into a binary form; that is 1:2:4:8:16. The DC power consumption, processing yield, chip size and dynamic range requirements determine the size of the most significant bit. The size of the least significant bit is thus automatically fixed from the overall bit requirements. In some cases the required gate 2 width for the LSB may be too small for practical realisation. Naster and Hwang [8] realised effective gate 2 widths of 5 and 10 µm by using a 4:1 capacitive voltage divider network with 20 and 40 µm gate 2 width segments. This segmented dual-gate technique has also been used by Devlin et al. [11] at Ku-band frequencies to realise a combined digital gain and phase control MMIC circuit. In this implementation, non-binary weighting was used for the 6-bit digital gain control function. A single chip unit containing both digital gain and phase control was designed and tested and had excellent performance characteristics.

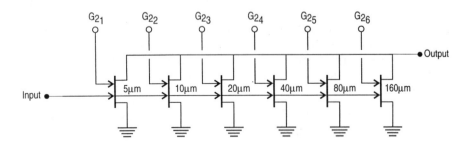

Fig. 8.11 Schematic diagram of a segmented dual-gate FET attenuator

The advantages of the segmented dual-gate FET approach are small chip size, the inherent gain, broadband performance etc. For the realisation of LSB units where smaller gate widths are required, the gate scaling model may need to be radically modified. Such techniques are, however, only available to designers with access to in-house foundries where individual optimisation is possible, and cannot be implemented by the majority of MMIC designers who only have access to commercial foundries with standard processes.

8.3.3 Switched attenuators

In the switched attenuator type, two single pole double throw (SPDT) switches are used to change the signal path between a through or reference line and an attenuator as shown in Fig. 8.12. For attenuation values in excess of 4 dB the attenuator can be formed by using a T- or π-type resistor network using either GaAs mesa resistors or by using the dedicated nichrome resistors available in the foundry process. The schematic diagram of a broadband SPDT switch as used by Gupta [12] is shown in Fig. 8.13. As the switch does not contain any capacitive elements there is no lower frequency limitation on the performance. The upper frequency limit on the useability of the configuration is determined by the isolation of the SPDT switch itself. The isolation of the switch in the off-state determines the leakage of the RF signal to the unwanted path and needs to be kept as high as possible to minimise any in-band ripple. The input and output VSWRs of individual bits are generally determined by the SPDT switch itself. Phase tracking between the two states can be easily achieved for larger bit values by equalising the reference path length to the resistive path length. The switched attenuator technique is attractive for the implementation of larger values of attenuation (>4 dB). Being a symmetrical structure its overall performance is less susceptible to the foundry process parameter variations and performance variations due to temperature are minimised. A similar SPDT switch configuration has been used by Anderson [13] for the design of a constant phase MMIC digital attenuator using the Philips D07M foundry. The insertion loss of the switch is less than 1.0 dB up to 5.5 GHz and increases with frequency, and the isolation is better than 28 dB for frequencies up to 10 GHz (Fig. 8.14).

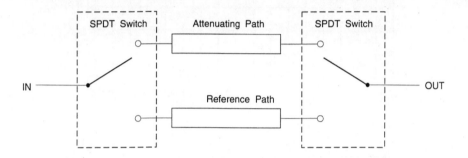

Fig. 8.12 Schematic diagram of a switched attenuator

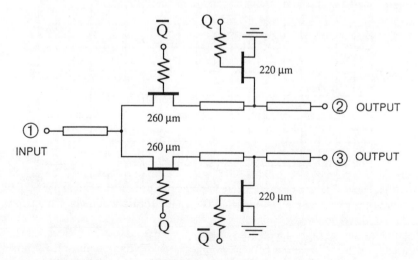

Fig. 8.13 Schematic diagram of a SPDT switch [11]

320 *Switches and attenuators*

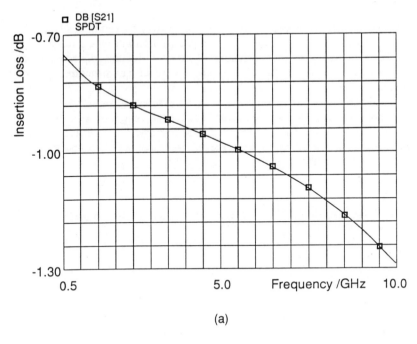

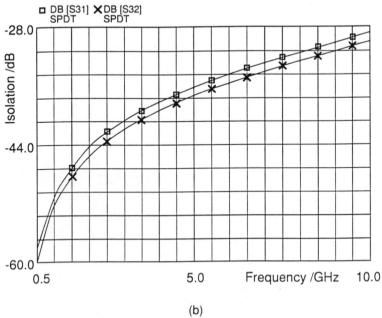

Fig. 8.14 SPDT switch performance [12]: (a) Insertion loss; (b) Isolation

8.3.4 Switched scaled FETs

The switched scaled FET technique uses different sizes of FETs in two paths connected between the input and output ports. For smaller attenuation bits of up to 2 dB, T-networks consisting of two passive FETs as switchable series resistive elements and a shunt resistor to ground are used. Wider FETs are used in the reference path to minimise the reference state insertion loss. The attenuation ratios for each bit can be changed by adjusting the device widths and shunt resistor values . For the switched scaled FET configuration, additional short transmission-line segments need to be included to equalise the attenuating path length and the reference path. This technique is usually confined to small values of attenuation (<4 dB, say) because for larger attenuation values the difference in FET sizes becomes excessive and makes it difficult to achieve phase matching between the two paths. Again, the symmetrical configuration makes the design more tolerant to process parameter variations, and the electrical performance is less sensitive to temperature variation. Fig. 8.15 shows a schematic diagram of the switched scaled FET configuration. The FET gate width in the reference path is kept to 600 μm to minimise the reference state insertion loss of the bit. Simpler topologies which only use a single shunt resistor to ground are also possible.

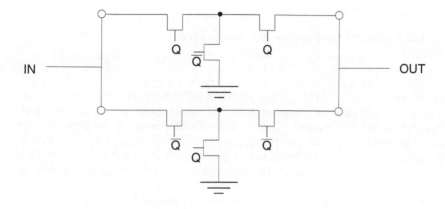

Fig. 8.15 Schematic diagram of the switched scaled FET configuration

8.3.5 Switched bridged-T attenuator

The switched bridged-T attenuator configuration consists of a classical bridged-T attenuator with a shunt and series FET as shown in Fig. 8.16. Switch FETs are placed across the bridge resistor R_1 and in series with the shunt resistor R_2. These two FETs are switched on or off to switch between the zero state and the attenuation state, whose value is determined by the bridged-T attenuator. The bridged-T attenuator inherently provides good input and output matches. The attenuation value in dB is given by:

Attenuation value = $20 \log(R_2/(R_2+Z_0)) = 20 \log(Z_0/(R_1+Z_0))$ (8.1)

where $R_1 R_2 = Z_0^2$.

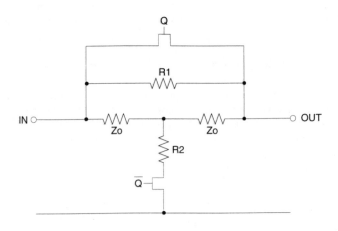

Fig 8.16 Switched bridged-T attenuator configuration

The performance of the individual bit is determined by the FET characteristics in the on- and off-states and by the realisability limit on the required resistance values and the resistors' parasitics. Bayruns *et al.* [14] used this technique to realise a 5-bit digital attenuator with integral TTL-compatible switch drivers for operation up to 1.6 GHz. Instead of using a classical arrangement for the shunt resistor with series FET in the bridged-T, a parallel combination of a fixed resistor to ground and a switch FET in series with a resistor was used. This helped to keep the shunt FET from saturating at large input power levels. The contribution of the FET on- and off-state resistances was taken into account in the design of individual attenuation bits. Compared with the switched attenuator configuration, the zero state insertion loss for the bridged-T configuration is lower. Also, the chip size for a multibit attenuator is smaller when the bridged-T type structure is used. For the 5-bit attenuator realised by Bayruns *et al.* the insertion loss was only 3 dB and the chip size was only 1.8 mm × 1.8 mm. The bridged-T resistor network provides excellent input and output matches for an individual bit, and therefore the effect of interaction between bits on in-band attenuation is minimal.

For lower attenuation bits the resistor in the bridged-T can be replaced by FETs with the correct on-state resistance. This configuration is called a switched FET bridged-T attenuator.

8.3.6 Switched T- or π-attenuators

The switched T- or π-attenuators are similar in principle to the above arrangement except the topology of the attenuator is in classical T- or π-form and the individual resistive elements can be switched in or out of the circuit. This technique can also be termed a switched resistor method. Fig. 8.17 shows a schematic representation of such an attenuator configuration. Bedard and Maoz [15] demonstrated a design for operation up to 10 GHz using a π-type configuration. Although similar to the bridged-T attenuator configuration, this technique offers lower insertion loss than the switched attenuator design as it uses much smaller FETs. As smaller FETs have lower parasitic capacitances this technique has good high frequency capability. However, unlike the bridged-T attenuator design which provides inherently good input and output matches for the individual bit, the designs which adopt this technique have to ensure that good port matches are achieved at the individual bits and that any mismatches do not affect the overall performance of the attenuator chip. Bedard and Maoz [15] realised a 5-bit design with a 1 dB LSB bit. The 4 and 8 dB bits were realised in the classical topology. However, for the 16 dB bit, computer simulations showed that the inductances associated with the shunt resistors and the capacitances of the FETs were limiting the attenuation flatness at the high frequency end. To resolve this problem the 16 dB bit was designed as a cascade of two 8 dB bits. For smaller attenuation values of 0.5 to 2 dB the classical T- or π-form can be replaced by either a series resistor shunted by a switch FET or a shunt resistor in series with a switch FET. The latter scheme was used in [15] to realise 1 and 2 dB bits. The measured results of the chip show that the reference or zero state insertion loss was less than 5 dB up to 10 GHz. Good accuracy for individual attenuation states was obtained but the relative phase shift for each attenuation state was as high as 60 degrees. For smaller attenuation bits, resistors can be replaced by FETs with the correct on-state resistance value. This arrangement can be termed a switched-FET T- (or π-) attenuator.

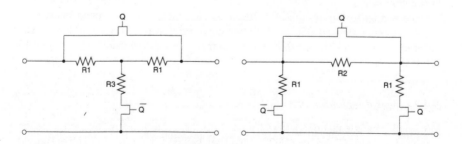

Fig. 8.17 Switched T- or π-attenuator configuration

Neither the bridged- nor the switched-T type attenuators have the degree of symmetry of the switched attenuator and switched scaled-FET attenuator. This could be a disadvantage as the performance of a symmetrical configuration will be less susceptible to variations in MMIC process parameter variation and the actual performance of the chip would be less sensitive to ambient temperature variations.

8.4 Attenuator design example

The selection of different attenuator topologies for the design and realisation of a multi-bit attenuator can be illustrated best by showing the step-by-step approach taken in the design of such an attenuator. The Philips D07M standard foundry process was used for this design [16]. The model for the switch FET developed by the foundry was described in Section 8.2.2 (Fig. 8.4). Initially, individual attenuation bits were designed and their performance evaluated for a cascaded 5-bit attenuator. A binary weighting arrangement with a 1 dB LSB was required. The chip was intended to be a generic functional block for applications up to 6.5 GHz.

8.4.1 Higher attenuation bits

For high attenuation states, i.e. the 4, 8 and 16 dB bits, the switched attenuator, bridged attenuator and switched resistor T-networks were investigated. For low attenuation bits, such as 1 and 2 dB, the switched scaled-FET, bridged FET T- and switched FET T- configurations were investigated.

The performance of the SPDT switches used in the switched attenuator type design is shown in Fig. 8.14. The optimised performance characteristics for an 8 dB bit are shown in Fig. 8.18. The reference state insertion loss is -1.9±0.2 dB/bit, the attenuation bit error is less than 0.02/-0.0 dB/bit, and the input and output matches are better than 22 dB. The path lengths in the two paths have been optimised to give an insertion phase error of less than 0.02 degrees between the reference state and the attenuation state.

The results for an optimised 8 dB attenuation bit using a switched resistor T-network design showed that the bit attenuation error was as high as 0.2 dB. The differential phase shift was ± 0.05 degrees in this case and port matches were only 15 dB or so.

8.4.2 Lower attenuation bits

For smaller bit attenuation values of up to 4 dB the techniques investigated were the switched scaled FET, bridged FET T- and switched FET T- networks. The optimised switched scaled FET circuit performance for a 2 dB attenuation bit is shown in Fig. 8.19. The attenuation setting error is less than +0.1/-0.05 dB and the port matches better than 25 dB over the frequency band. The phase error between the reference and

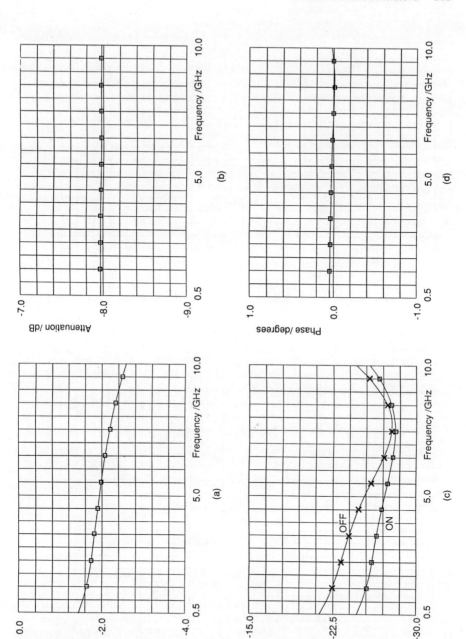

Fig. 8.18 Optimised switched attenuator response: (a) Insertion loss; (b) Attenuation; (c) Input and output return loss; (d) Differential insertion phase

326 Switches and attenuators

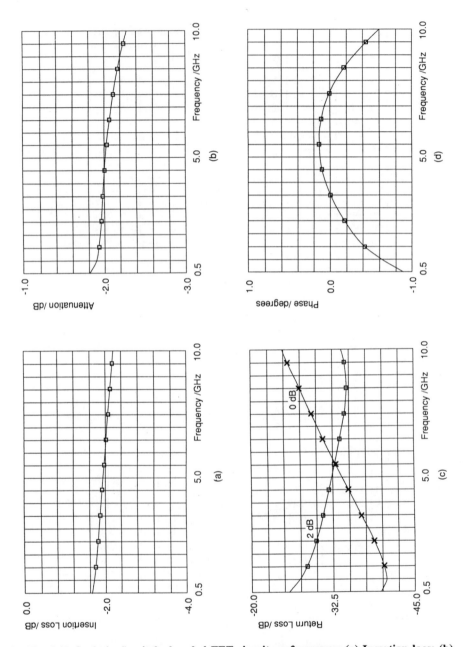

Fig. 8.19 Optimised switched scaled FET circuit performance: (a) Insertion loss; (b) Attenuation; (c) Input and output return loss; (d) Differential insertion phase

attenuation state is +0.15/-0.4 degrees. The techniques is not, however, suitable for the 4 dB bit as it is difficult to minimise the phase error between the two paths.

The optimised design for a 2 dB bridged-T attenuator bit showed that in the desired frequency range of 1.5 to 6.5 GHz the reference state insertion loss was 1.4 dB. The attenuation bit error was less than -0.02 dB and the input and output matches were better than 25 dB. The insertion phase error was optimised to give a phase error of less than ±1 degree between the reference state and the attenuation state.

The optimised results for a 2 dB bit using switched FET T-network technique showed that the attenuation error is ±0.1 dB and port matches are only 15 dB. The phase difference between the reference and attenuation path is +0.4/-2.0 degrees. The performance of the circuit was over a narrower frequency band and could not be implemented for a 1 dB bit.

Table 8.1 summarises the results of computer simulation and optimisation of the individual MMIC attenuator bits. It can be seen that for larger attenuation bits (> 4 dB) the switched attenuator or the switched resistor-Tee technique can be used. The bridged-T technique has poor port matches and limited operating frequency. On further examination it can be seen that the attenuation errors and port matches of the switched resistor-T configuration are poorer than the switched attenuator design. For smaller attenuation bits, the switched scaled FET or the bridged FET Tee techniques can be adopted. However, the bridged FET Tee design has larger phase error than the former. Moreover, the configuration is not symmetrical and the individual bit performance might be subject to process parameter variations. It can therefore be concluded that the switched attenuator and switched scaled FET techniques provide the best overall performance for the digital attenuator. The attenuation setting accuracy per bit is only ±0.07 dB and the phase error per bit is less than 0.4 degrees. Both the input and output matches are better than 22 dB, thus making it easier to cascade individual bits without giving rise to any amplitude or phase ripples due to mismatch.

8.4.3 Overall chip design and layout

Having selected the individual attenuation bits, a cascading arrangement was computer optimised. Due to the better matches of the switched scaled FET bits, it was decided to insert them between the switched attenuator type bits. The 4 dB switched attenuator was the first element in the unit followed by the 1, 8, 2 and 16 dB bits. This arrangement provides better power handling for the chip. The simulated performance of the cascaded chip including all modifications for layout is shown in Fig. 8.24 from 0.5 to 10 GHz. The desired frequency range was 1.5 to 6.5 GHz. Simulations show that the attenuation setting error for all possible states is less than 0.2 dB (less than one-quarter LSB). The insertion phase variation for all possible attenuator states is less than 1.5° over the frequency band of interest.

The chip was laid out using standard commercial layout software. Fig. 8.21 shows the chip photograph. Its overall dimensions are 5.0 mm x 2.5 mm. The RF input and output are at opposite ends with complimentary pairs of control lines along one edge.

Great care was taken to avoid coupling across the attenuators, to minimise insertion loss and to reduce the number of through GaAs via holes by the use of via sharing.

Table 8.1 Computer simulation results of individual attenuation bits

Attenuation value (dB)	Design approach	Attenuation error (dB)	Phase error (deg)	Frequency (GHz)	Match (dB)	Insertion loss per bit (dB]	Comments
4 8 16	Switched attenuator	+0.02/-0.0/bit	+0.04/ -0.0/bit	1.5 to 6.5	>22	1.9±0.2	Resistors limit attenuator values. Large area
4 8	Bridged-T attenuator	±0.1/bit	±1	< 2.0	>16	2±0.2	Resistors limit values
8 16	Switched resistor	0.2/-0.7/bit	0.05/ -0.3/bit	1.5 to 6.5	>15	1.9±0.15	Not symmetrical
1 2	Switched scaled FETs	±0.07/bit	0.2/ -0.4/bit	1.5 to 6.5	>28	1.95±0.15	FET geometries limit values
1 2	Bridged FET-T	±0.05/bit	0.65/ -0.5/bit	1.5 to 6.5	>24	1.15±0.15	Not symmetrical
2 4	Switched FET-T	0.3/bit	0.3/ -0.5/bit	1.5 to 5	>14	2.25±0.15	Not symmetrical

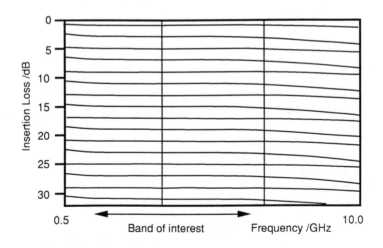

Fig. 8.20 Simulated performance of the cascaded chip

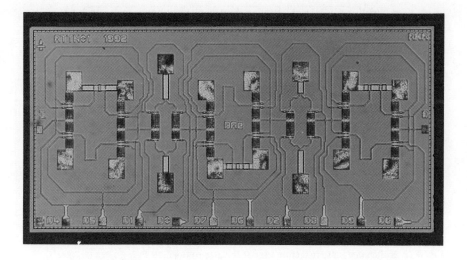

Fig. 8.21 Chip photograph

8.4.4 Measured results

The measured attenuation states are shown in Figs. 8.22 to 8.25. The zero state insertion loss was 8 dB up to a maximum operating input power of +21 dBm (Fig. 8.22). The attenuation settings relative to the zero state are shown in Fig. 8.23; peak setting errors are +0/-0.88 dB. These results show good correlation with predictions. The setting errors for all individual attenuation bits are negligible except for the 2 and 16 dB bits. For these the errors are 0.16 dB and 0.1 dB, respectively, in the band of interest. However, it is considered that the finite leakage in the test fixture itself may be a contributory factor for the 16 dB state. Normally, the cumulative errors for the 31 dB state (when all the individual bits are on) should be equal to the sum of individual bit errors. This has been observed at all the frequencies except at 3.5 GHz where the worst case attenuator setting errors occur in the higher attenuation states (>16 dB).

The input and output port matches are better than 13.5 dB for all states (Fig. 8.24). Differential insertion phase between states is shown in Fig. 8.25 which shows that the peak errors are +3°/-1.6°. There is a phase error associated with the 16 dB bit caused by coupling across the attenuator. This accounts for the two clusters of traces; the cluster about 0° is for states below 16 dB and the cluster about +2° is for states where the 16 dB bit is switched on.

330 *Switches and attenuators*

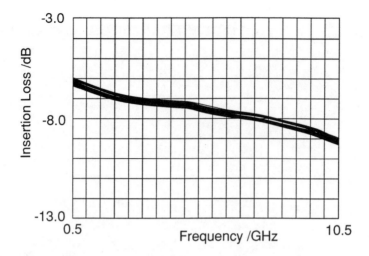

Fig. 8.22 Reference state insertion loss of the attenuator

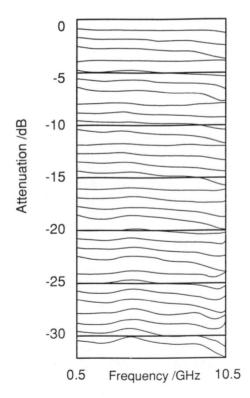

Fig. 8.23 Attenuation settings relative to the zero state

Switches and attenuators 331

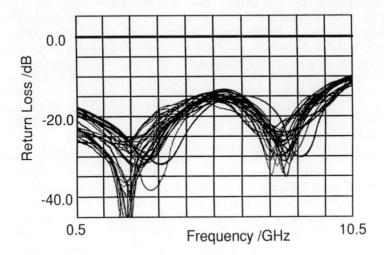

Fig. 8.24 Input and output port matches (all states)

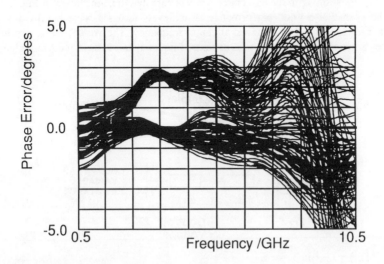

Fig. 8.25 Insertion phase tracking between states

8.5 Analogue attenuators

In contrast with a purely digital implementation, analogue attenuators can provide a continuous range of amplitude control. Hence, they have the advantage that they can correct for any degradation in performance attributed to fabrication process variations and enable calibration corrections to be performed when integrated into a subsystem. Analogue variable attenuators are often found in AGC control loops, vector modulators and adaptive beam-forming networks. In addition, they can be used as variable taps in direct implementation transversal and recursive filter architectures and a variety of other signal processing functions.

8.5.1 Analogue reflection-type attenuator

An analogue reflection-type attenuator topology was demonstrated by Devlin [17]. This technique, illustrated in Fig. 8.26, is similar to the reflection-type phase shifter described in Chapter 7. However, a variable resistance reflection termination is used so that the signal amplitude can be controlled, with a transmission phase which does not vary (in the ideal case). For MMIC implementations, this variable resistance is most easily realised with a cold-FET. Lucyszyn [18] reported the measured performance for an X-band analogue attenuator using this technique. The attenuator had a centre frequency of 10 GHz and a bandwidth covering 8 to 12 GHz. The chip used a folded four-finger Lange coupler with variable resistors realised using cold-FETs. The GaAs MMIC was fabricated at the GEC-Marconi (Caswell) foundry, using their standard F20 foundry process. A microphotograph of the attenuator is shown in Fig. 8.27. The chip has an active area of 2.4×1.0 mm^2. The measured levels of attenuation, with the associated tuning characteristic, are shown in Fig. 8.28. It can be seen that a good performance is achieved across a wide bandwidth. A dynamic range of 14 dB is achieved at centre frequency. The worst-case return loss decreases almost linearly from 24 dB at centre frequency to 19 dB at the band edges.

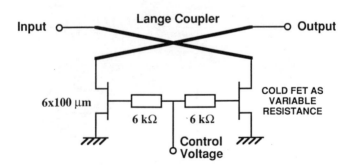

Fig. 8.26 Topology of the analogue reflection-type attenuator

Switches and attenuators 333

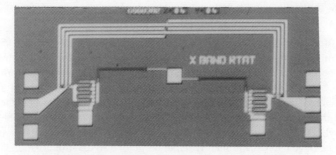

Fig. 8.27 Microphotograph of the analogue reflection-type attenuator

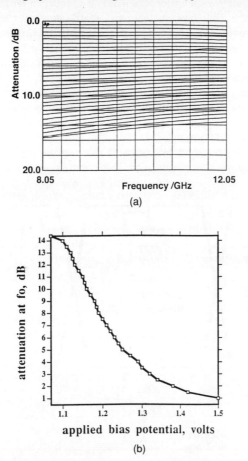

Fig. 8.28 Measured performance of the attenuator: (a) Attenuation against frequency; (b) Attenuation against control voltage

8.5.2 Other analogue attenuators

In addition to the reflection-type analogue attenuator, diodes can be used in a balanced attenuator configuration by employing the principle of the balanced amplifier. In this type of attenuator, illustrated in Fig. 8.29, the diodes are shunt-connected between a pair of Lange couplers: the variable impedance of the diode produces a continuously varying magnitude of transmission coefficient, S_{21}. The signals reflected from the diodes are absorbed in the terminations on the Lange couplers in the same way as in the balanced amplifier. This technique has been used to realise hybrid MIC attenuators with PIN diodes [19]. With special driving circuitry this design achieved an exceptional dynamic range of 120 dB in Ku-band. The same technique can be employed using Schottky barrier diodes on MMICs. The result is an octave bandwidth analogue attenuator which has an attenuation range of around 10 to 15 dB, with good input and output matches and a fairly constant transmission phase. Sweet [20] has provided a useful design procedure for this balanced type of attenuator.

Analogue attenuators can also be realised with π- and T-networks using FETs, PIN diodes or Schottky barrier diodes [21, 22]. However, it is very difficult to maintain a good input and output match over the whole attenuation range with this method. Finally, rather than employing an attenuator, a variable gain amplifier can be employed for continuous gain control. The most common method is to employ a dual-gate or cascode FET arrangement, and this is described in Chapter 5. It is important to note that the dynamic range of this type of amplifier can be very restricted when operating at the limits of the gain control range.

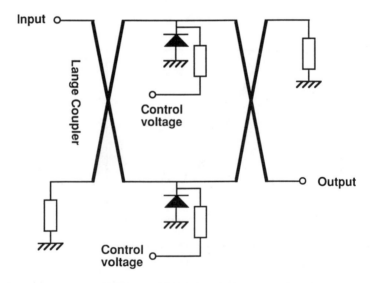

Fig. 8.29 Balanced analogue attenuator employing PIN or Schottky barrier diodes

8.6 Summary

This chapter has reviewed the switching mechanism of GaAs FET MMIC switches and has discussed the equivalent circuit of these MMIC switches. The advantages offered by these switches over conventional PIN diode switches have been described along with their limitations, such as lower power handling. Some design examples and applications of MMIC switches have been given. The second part of the chapter discussed the requirements for MMIC attenuators. A number of techniques available for the implementation of these attenuators have been reviewed and their performance trade-offs discussed. The design example of a single chip 5-bit digital attenuator operating from 1.5 to 6.5 GHz has been discussed. The measured performance of the chip showed a good correlation with computer simulations.

8.7 References

1. AYASLI, Y.: 'Microwave switching with GaAs FETs', *Microwave Journal*, Nov. 1992, pp.61-74
2. ERON, M.: 'Small and large signal analysis of MESFETs as switches', *Microwave Journal*, Jan. 1992, pp.128-140
3. BUCK, C., WILLIAMS, K., and LEBLANC, R.: 'The development of MMIC switched FET models for Philips D07M foundry process', IEE Colloquium on the Modelling, Design and Application of MMICs, June 1994
4. MALEVIGE, W., and SOKOLOV, V.: 'Resonated GaAs FET devices for microwave switching', *IEEE Trans.*, ED-28, Feb. 1981, pp. 196-204
5. AYASLI, Y., VOURHAUS, J., PUCEL, R., and REYNOLDS, L.: 'Monolithic GaAs distributed FET switch circuits', IEEE GaAs IC Symposium, San Diego, 1981
6. KATZIN, P., BEDARD, B., SHIFRIN, M., and AYASLI, Y.: 'High speed, 100+ W RF switches using GaAs MMICs', *IEEE Trans.*, MTT-40, no. 11, Nov. 1992, pp. 1989-1996
7. HWANG, Y., *et al.*: 'A microwave phase and gain controller with segmented dual gate MESFETs in GaAs MMICs', IEEE Microwave and Millmeter-Wave Monolithic Circuits Symposium, May 1984, pp. 1-5
8. NASTER, R, *et al.*: 'An L-band variable gain amplifier in GaAs MMIC with binary step control', IEEE GaAs IC Symposium, 1987, pp. 235-237
9. SNOW, K., KOMIAK, J., and BATES, D.: 'Wideband variable gain amplifiers in GaAs', IEEE Microwave and Millimeter-Wave Monolithic Circuits Symposium, 1988, pp. 133-137
10. SIWERIS, H.: 'A monolithic C-band variable gain amplifier with digital gain control', Proceedings of the 21st European Microwave Conference, 1991, pp. 1475-1480
11. DEVLIN, L.,*et al.*, ESTEC Conference on Active and Passive Beamforming networks, Noordwijk, 1992

12. GUPTA, R. K., et al.: 'Modelling and CAD of an ultra-broadband monolithic 5-bit digital attenuator', 1988, Institute of Monoelectronik
13. ANDERSON, A. K., and JOSHI, J. S.: 'Wideband constant phase digital attenuators for space applications', *Microwave Engineering Europe*, Feb. 1994
14. BAYRUNS, J., WALLACE, P., and SCHEINBERG, N.: 'A monolithic DC - 1.6 GHz digital attenuator', IEEE MTT-S Symposium Digest, 1989, pp. 1295-1298
15. BEDARD, B., and MAOZ, B.: 'Fast GaAs MMIC attenuator has 5-bit resolution', *Microwaves & RF*, Oct. 1991, pp. 71-76
16. ANDERSON, A. K., and JOSHI, J. S.: 'Generic constant phase digital attenuators', IEE Colloquium on the Modelling, Design and Application of MMICs, June 1994
17. DEVLIN, L. M., and MINNIS, B. J.: 'A versatile vector modulator design for MMIC', IEEE MTT-S Int. Symp. Dig., 1990, pp. 519-522
18. LUCYSZYN, S., and ROBERTSON, I. D.: 'Analog reflection topology building blocks for adaptive microwave signal processing applications', *IEEE Transactions on Microwave Theory and Tech.*, MTT-43, no. 3, March 1995
19. BLASER, R. J.: 'Control circuit produces linear attenuator response', *Microwaves & RF*, February 1994, pp. 59-64
20. SWEET, A.: *MIC and MMIC amplifier and oscillator circuit design*, Artech House, Dedham, Mass., 1990, pp. 168-174
21. FRANCO, D. P.: 'A miniaturised mounting for a 4-20 GHz PIN attenuator', *Microwave Engineering Europe*, June/July 1994, pp. 31-36
22. GOLDFARB, M. E., and PLATZKER, A.: 'A wide range analog MMIC attenuator with integral 180° phase shifter', *IEEE Trans. Microwave Theory Tech.*, MTT-42, no. 1, January 1994, pp. 156-158

Chapter 9

Oscillators

K. K. M. Cheng and U. Karacaoglu

9.1 Introduction

The oscillator is an essential component for microwave systems such as communications, radars and instrumentations. An ideal oscillator produces a pure sinusoidal carrier with fixed amplitude, frequency and phase. Practical oscillators, however, produce carrier waveforms with parameters (oscillation frequency, output power) that may vary in time due to temperature changes and component ageing. Oscillator frequency instabilities are characterised primarily by frequency drift (long term variations) and random noise (short term fluctuations). The latter appears as phase and amplitude fluctuations at the oscillator output and will always be of main concern in modulated systems. Phase noise generated by the local oscillator at the receiver can significantly affect the performance in digital communication systems.

The earlier development of microwave solid state oscillators was mainly based on Gunn and IMPATT diodes. Since the 1970s, the advent of GaAs MESFETs at microwave and millimetre-wave frequencies has given greater freedom to engineers in the design of microwave oscillators. GaAs MESFET devices, compared to Gunn and IMPATT diodes, offer several advantages such as suitability for monolithic integration, higher power efficiency, lower phase noise and lack of any threshold current requirements. Fixed frequency oscillators based on GaAs MESFETs or similar devices have been reported [13-15] for frequencies up to W-band. Broadband tuneable oscillators using YIG resonators and varactor diodes have also been described [2, 8, 9] in the literature. Recent investigations have been directed towards very low phase noise designs in MMIC form, especially those using HEMT and HBT active devices. Low phase noise oscillators based on high-temperature superconductive materials [14] operating at cryogenic temperature have also been demonstrated in the last few years. A summary of some research reports on the developments of solid-state microwave oscillators is depicted in Table 9.1. In recent years more and more oscillator circuits have been implemented using monolithic technology because the size and cost of the products can be significantly reduced.

Table 9.1 Summary of some scientific reports on microwave oscillator designs

Year	Active device	Types of resonators	Operating frequency (GHz)	Output power / efficiency	Phase noise level (dBc/Hz)	Remarks
1975 [1]	GaAs MESFET	Microstrip resonator	8.8	12.1 mW / 7.8 %		MIC
1977 [2]	GaAs MESFET	YIG	5.9 - 12.5	8 - 22 mW max. 8 %		MIC
1980 [3]	GaAs MESFET	Dielectric resonator	11.85	70 mW 20 %		MIC
1982 [4]	GaAs MESFET	On - chip varactors	11.1 - 14.4 16.0 - 18.7			MMIC
1984 [5]	GaAs MESFET	On-chip varactors	11.6 - 20.0	5.5 - 15.8 dBm		MMIC
1987 [6]	GaAs MESFET	Dielectric resonator	10.74	16 dBm	- 80 (10 kHz)	Quasi - MMIC
1988 [8]	GaAs MESFET	On - chip varactors	5.9 - 12.6	14-19 dBm	- 100 (1 MHz)	MMIC
1988 [9]	GaAs MESFET	External varactors	2.5 - 6.0	17.5 dBm 9.3 dBm	- 92 (100 kHz)	Quasi - MMIC
1988 [10]	Silicon Bipolar	DR - VCO	1.69 0.5% range	8-12 dBm / 3 %		MIC
1991 [11]	HBT	GaAs varactor	2.7 - 6.2	3.5 - 10 dBm	- 105 (100 kHz)	MIC
1991 [12]	HBT	Silicon varactors	6.9 - 15	9 - 12 dBm (buffered)	- 75 (100 kHz)	Quasi - MMIC
1991 [13]	InGaAs MESFET	Waveguide cavity	92.3	14 mW 11%	- 70 (15 kHz)	MIC + waveguide
1992 [14]	GaAs MESFET	HTSC resonator	6.5	4.9 dBm	- 90 (10 kHz)	MIC (77 K)
1993 [15]	PM-HFET	Dielectric resonator	81	1 mW 3.8%	- 90 (1 MHz)	MIC

In the quasi-MMIC construction the negative resistance circuit, along with all the necessary bias and decoupling components, is produced on a single chip and the only external element is the resonator. For this combination, the ultimate noise performance is retained by using high quality resonators, but the fabrication procedure is greatly simplified and the reliability of the circuits is also improved.

The design principles of oscillators are basically quite simple and straightforward, but there are many considerations when tight specifications must be met. Modern microwave systems require stable, wideband, low phase noise, tuneable oscillators with sufficient output power and efficiency. Although there are many different types of active devices, resonators and circuit topologies that can be used, these choices are usually bounded by other factors such as size, cost, operating frequency, reliability and frequency instabilities (both long and short term). A systematic study of the parameters which influence the oscillator performance is therefore important.

9.2 Design principles

There are two types of oscillator circuit, namely the feedback and negative resistance oscillators. They differ from each other mainly in the principles of operation, circuit topologies and noise performances.

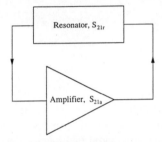

Fig. 9.1 Block diagram of a feedback oscillator

9.2.1 Feedback oscillators

A feedback oscillator consists of an amplifier and a resonant circuit (Fig. 9.1). The two elements are designed to fit the following oscillation condition:

$$S_{21_a} \cdot S_{21_r} = 1 \qquad (9.1)$$

where S_{21a} and S_{21r} are the forward transmission S-parameter for the amplifier and resonator circuits, respectively. Both the input and output impedances of the amplifier are assumed to be equal to Z_0, and practically this is achieved by employing bandpass matching networks to operate near the carrier frequency of the oscillator. The amplifier must also show unconditional stability over a wide range of frequencies. Eqn. 9.1 implies that the product of the two magnitudes at the oscillation frequency is unity and the sum of their angles is a multiple of 2π. In other words, the gain of the amplifier must be sufficiently large to compensate for the loss in the resonator, and the electrical delay round the loop must be an integral multiple of 360 degrees. The phasing condition can be achieved by adding a delay line or phase shifter circuit. In oscillators, the oscillation frequency and close-to-carrier noise performance are basically determined by the selectivity of the resonator. All oscillators are essentially Q multipliers which use positive feedback to enhance many times the natural Q-factor of the resonant element. The Q multiplier amplifies and filters the inherent circuit noise to give the output spectrum of the oscillator. Note that the amplifier gain has to be made higher than the minimum value to avoid start-up problems. The final gain will drop as the oscillation amplitude increases to its steady-state level, due to gain compression.

9.2.2 Negative resistance oscillators

Consider the circuit shown in Fig. 9.2, where Γ_L and Γ_R are the reflection coefficients of the left and right parts of the cut at an arbitrary plane (a popular choice is to use a plane between the resonator and the transistor) and Z_L and Z_R are the corresponding impedances. The oscillation condition of a negative resistance oscillator can be stated as:

$$\Gamma_L . \Gamma_R = 1 \qquad (9.2)$$

or

$$Z_L + Z_R = R_t + jX_t = 0 \qquad (9.3)$$

The first expression implies that, at the frequency of oscillation, the return loss of the resonator has to be equal to the return gain of the active block, and the sum of the arguments of Γ_L and Γ_R has to be an integral multiple of 360 degrees. A return gain implies that a negative resistance must be present between the input terminals of the active block. In other words, the negative resistance and reactance of the active block must equal in magnitude and opposite in sign from the resistance and reactance of the resonator. Note that the oscillation frequency does not necessarily have the same value as the resonant frequency of the resonator, provided that eqn. 9.2 is satisfied.

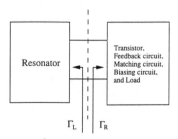

Fig. 9.2 Block diagram of a negative resistance oscillator

For oscillations to begin, the magnitude of the negative resistance has to be larger than the value determined from the small-signal condition. Owing to the excess negative resistance in the circuit, the oscillation will grow continually until the negative resistance is reduced in magnitude, by non-linear effects, to its steady-state value. In practice, the negative resistance is created by applying feedback to transistors. Two common feedback configurations, the common-gate inductive feedback and common-source capacitive feedback, are depicted in Fig. 9.3. The values of L_f and C_f are chosen such that, for a given transistor type, the negative resistance is maximised over the band of interest. In most circumstances, increased bandwidth and negative resistance can be achieved by adding a suitable matching circuit between the transistor's output and the load. Software tools such as Touchstone™ or SuperCompact™ may be used in this design procedure.

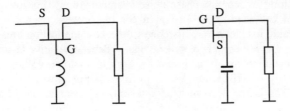

Fig. 9.3 Basic feedback topologies for negative resistance oscillators

9.3 Active device

The performance of oscillators is largely dependent on the choice of a suitable type of active device. Most microwave oscillators currently use either silicon bipolar transistors or GaAs devices. It is very important that the active device is capable of creating sufficient negative resistance, or forward gain, over the band of interest. Other factors such as flicker noise level, output power etc., must also be considered by the designer, subject to different applications and requirements.

9.3.1 GaAs MESFET versus silicon bipolar transistors

In most cases the choice of transistors is initially between a silicon bipolar transistor and a GaAs MESFET. It should be noted that the GaAs MESFET has higher frequency of oscillation, higher gain, higher output power and higher efficiency. The higher output power of the GaAs MESFET is a direct consequence of the greater critical field and higher saturated drift velocity. Higher gain is due to greater mobility of the electrons compared with silicon. However, silicon bipolar transistors usually have much lower flicker noise level than GaAs devices. A difference of 20 – 30 dB in close-to-carrier noise performance can easily be found in oscillators based on the two types of devices. In the past, bipolar transistors were the only practical devices to be used for a low phase-noise microwave oscillator. However, new types of devices such as HEMT and HBT are becoming potential candidates for low noise applications, especially at microwave and millimetre-wave frequencies. The effect of flicker noise can sometimes be reduced by low-frequency feedback circuitry and proper biasing of the active device [39].

9.3.2 Heterojunction bipolar transistors

Research into the combining of GaAs and other III-V compounds [7], has led to the recent demonstration of heterojunction bipolar transistors (HBTs). These new devices offer the prospect of obtaining performance features similar to those of Si bipolar transistors translated to substantially higher frequencies. They have shown excellent gain and noise performance in microwave and millimetre-wave circuits. The vertical structure of the HBT eliminates the surface-state problem associated with GaAs MESFETs, resulting in superior phase noise characteristics. This low phase noise characteristic makes them very attractive for microwave source applications. Broadband tuneable oscillators are also one of the key application areas for HBTs because these devices are capable of generating a wideband negative resistance due to their high transconductance. In addition, the high breakdown voltage and current handling capability of HBTs could lead to high output power in oscillators. A drawback of HBTs is the need to access several different layers of the vertical structure, which can lead to non-planar structures in some circumstances. This non-planarity complicates device processing, although it does not preclude monolithic integration of HBTs. A MMIC voltage controlled oscillator which uses an AlGaAs/GaAs HBT has been reported, having a very wide tuning range of 7 – 15 GHz and minimum output power of 9 dBm. This VCO circuit consists of a common base oscillator with a capacitively loaded emitter to generate the wideband negative resistance. The circuit requires two external varactors to form a complete VCO.

9.3.3 Biasing and spurious response suppression

The DC bias condition of transistors is usually established independently of the RF design. Power efficiency, stabilities and ease of use are the main concerns when selecting a biasing configuration. For a low phase noise oscillator design, the biasing circuit should be properly regulated and filtered to avoid any unwanted signal modulation or noise injection. Variation of the supply voltages or currents may also cause undesirable output power fluctuations and frequency drift. A few commonly used biasing configurations are given in Fig. 9.4. These circuits use either one or two power supplies.

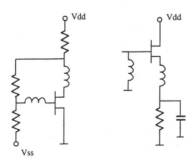

Fig. 9.4 Biasing circuits

All oscillators can produce spurious oscillations at one or more undesired frequencies. They can occur at any frequency where eqn. 9.1 is satisfied by the microwave and bias circuits. Prevention of these spurious oscillations requires great care. The rule is to avoid the magnitude and angle conditions required for oscillation occurring simultaneously. In practice, the amplifier gain or the negative resistance created by the transistor must be made as narrowband as possible but still operate over the designated frequency range, and, in reconciling the phasing between the active element and the resonator, the line length should be kept short. Furthermore, the bias lead inductance and DC blocking capacitors should be included in the design.

9.4 Resonators

When considering microwave oscillators it is often necessary to select an appropriate type of resonator to meet specific requirements including cost, size, noise performance and reliability. There are various kinds of resonators such as LC tuned circuit, cavity resonator, dielectric resonator, YIG resonator, and superconductive resonator. The LC tuned circuit is suitable for low frequency applications where noise performance is of minor importance. The cavity resonator is very bulky but allows substantially higher Q to be attained. The dielectric resonator (DR) is quite similar to the cavity resonator except that the DR has a very high dielectric constant medium instead of air, resulting in large reduction in size and weight. The quality factor of a DR is obviously lower compared with a cavity resonator due to dielectric and radiation loss. The recent advent of high temperature superconductors (HTS) has attracted much attention from microwave engineers for the design of filters, oscillators, delay lines and antennas. HTS resonators usually operate at 77 K, and have a very high Q factor even at microwave frequencies.

For broadband tuneable oscillator applications, YIG resonators or varactors are often used. YIG resonators are made from high Q ferrite which can be tuned over a wide bandwidth by changing the strength of a static magnetic field. The magnetic field is usually generated by an external coil with a high current running through it. Sometimes, this field is obtained by a combination of a magnetic coil and a permanent magnet. Multi-octave tuneable YIG oscillators have been demonstrated [2] with resonator Q of several thousands. These oscillators, however, have low tuning speed and high power dissipation due to the strong magnetic field required. An alternative solution is to use varactor diodes [8] for fast tuning rate, wideband oscillators. The varactor diode provides a variable reactance for tuning and is designed to have a very low series resistance. The varactor's Q available commercially is around 15 to 20 at X-band for some GaAs devices. The tuning range of the varactor is described by the capacitance ratio C_{max}/C_{min}, where this ratio can be 10-20 for hyperabrupt devices. The main drawback of the varactor-tuned oscillator is its poor phase noise performance resulting from the low diode Q, especially if tight coupling is used to achieve a wide tuning range.

9.4.1 Dielectric resonator

The dielectric resonator [3, 10] allows the design of compact, frequency-stable, low noise oscillators. Its resonant frequency is determined by the dimensions and dielectric constant of the DR and the dimensions of the conductor shield. Without the conductor shield, the Q of the dielectric resonator drops significantly due to radiation loss. Small resonant frequency adjustment is usually achieved mechanically by changing the distance between the resonator and a metal plate (Fig. 9.5). This distance should be at least half the resonator height to minimise the reduction in the resonator Q. The presence of the metal plate perturbs the field distributions in and around the DR, resulting in a shift in the resonant frequency. In practice, a tuning bandwidth of up to 10% can easily be achieved by this method.

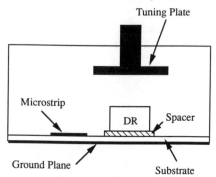

Fig. 9.5 Dielectric resonator oscillator assembly

Barium titanate compounds are normally used for dielectric resonators, with a dielectric constant value between 30 and 100. Choice of a dielectric constant is a compromise between high energy concentration within the resonator and high Q. A higher dielectric constant usually leads to increased loss. Dielectric resonators, nowadays, have Qs of several thousands up to tens of thousands and very small temperature coefficients. Temperature instability is caused primarily by the temperature dependency of the dielectric constant and the thermal expansion of the material. Temperature-compensation may be achieved by using a circuit configuration which gives a negative temperature coefficient, or by housing the oscillator in a temperature-controlled environment. In practice, a low-loss spacer is usually used to keep the DR at a certain distance from the ground plane, thus minimising conductor loss. Some varactor-tuned DR configurations are shown in Fig. 9.6. Varactor tuning of dielectric resonators is important and enables them to be used where a high quality resonator and fast tuning are required, such as in radio base stations where dynamic channel allocation is employed. However, dielectric resonators are generally difficult to tune electronically over a wide bandwidth. Other disadvantages of dielectric resonators are that they are not amenable to MMIC implementation, they are sensitive to mechanical vibration and for frequencies above 20 GHz they are not easy to handle.

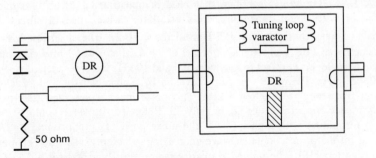

Fig. 9.6 Varactor tuning of dielectric resonators

9.4.2 Transmission-line resonators

Recently, a new type of resonator has been proposed for use in low noise feedback oscillators. These resonators have been realised in several different forms, including helical [17], microstrip [16] and coplanar waveguide structures [18]. The basic structure consists of a low-loss transmission line and two shunt or series elements. Note that different types of components including inductors or capacitors can be used as the shunt/series elements (Fig. 9.7).

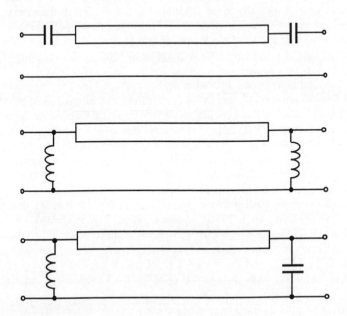

Fig. 9.7 Possible configurations of transmission-line resonators

A 1.49 GHz hybrid oscillator has been demonstrated [16] using a microstrip resonator with shunt capacitors. This oscillator, which uses a silicon bipolar transistor, was fabricated on RT Duroid ($\varepsilon_r = 10.2$). The insertion loss of the resonator was 7 dB, and the output power of the oscillator was 1 mW. The measured phase noise levels at offset frequencies of 5 and 10 kHz were, respectively, -95.3 and -100.9 dBc/Hz. The noise spectrum showed a $1/f_m^2$ characteristic and a flicker noise corner frequency well below 5 kHz. This noise level represented state-of-the-art performance for a resonator Q_o of only 83. Better performance could be obtained using resonators with higher Q_o, as the noise power is proportional to $1/Q_o^2$. A similar resonator on a coplanar waveguide using inductors as the shunt elements was also reported [18]. The unloaded quality factor was over 500 for a 5 GHz CPW resonator fabricated on Duroid. This coplanar resonator was unshielded, but demonstrated very low radiation loss.

A 1.5 GHz oscillator which uses a GaAs MESFET (similar resonator) demonstrated considerably higher sideband noise due to the effect of transposed flicker noise. However, the efficiency of this oscillator was over 14% whereas that of the previous one was only 1%. The measured noise levels of the GaAs oscillator at 10 and 20 kHz offset were -77 dBc/Hz and -85.5 dBc/Hz, respectively, thereby showing a $1/f_m^3$ characteristic. Hence, transposed flicker noise is dominant at these offset frequencies. Note that the noise level (10 kHz offset) of the silicon oscillator is about 24 dB lower than the GaAs oscillator. It can therefore be seen how important it is to reduce the effect of flicker noise both in the device, the bias circuit and the resonator.

At frequencies of 1 GHz or below, however, the transmission line can take up considerable space especially if the substrate has a low dielectric constant. A new form of helical transmission-line resonator [17] has been developed which is compact and, as the screening is part of the resonator, the Q is not degraded. It differs from conventional helical resonators both in length and form of the coupling. The new resonator is directly coupled and resonant for lengths approximately equal to $\lambda/2$, allowing easy fabrication and operation up to several GHz. This resonator can be used both for low noise oscillators as well as for front end filtering in communication systems. Two bipolar feedback oscillators (900 MHz & 1.6 GHz) were demonstrated using helical resonators. Both helical resonators had an insertion loss of between 7 and 8dB. The phase noise level of the 900 MHz oscillator was -127 dBc/Hz at 25 kHz offset, with 0 dBm output power. The 900 MHz helical resonator had an unloaded Q of 582. The 1.6 GHz oscillator had a phase noise level of -120 dBc/Hz at 25 kHz offset, unloaded resonator Q of 382 and output power of 0 dBm.

Obviously, the microstrip and coplanar waveguide versions of this form of resonator are very amenable to MMIC implementation. Two monolithic circuits have recently been developed at King's College London. Both resonators used inductors as the shunt elements to reduce radiation loss. A microstrip version showed a resonant frequency at 21.8 GHz and an unloaded Q of 65. A coplanar waveguide design resonated at 22.5 GHz, with an unloaded Q of 75. Both chips measured 1 mm by 3 mm.

The above mentioned fixed resonant frequency resonators were further developed [21] in an attempt to obtain a broadband tuneable resonator. The proposed structure, as shown in Fig. 9.8, consists of a transmission line and two varactor diodes. Biasing of the diodes is achieved by inserting a resistor near the centre of the resonator.

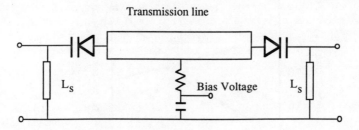

Fig. 9.8 Varactor-tuned transmission-line resonator

A prototype was demonstrated [21] on alumina (ε_r=9.8) using microstrip circuit techniques. The varactor diode used was the Alpha CVE7900D GaAs tuning diode chip with C_{j0} = 1.5 pF, Q(-4 V, 50 MHz) = 7000, breakdown voltage = 45 V and k = 6. The two shunt inductors were realised by gold bond wires of 0.4 mm long. Fig. 9.9 shows the theoretical resonant frequency and experimental results against the bias voltage of the diodes. The discrepancies at high resonant frequencies are caused by the parasitic capacitances. The measured insertion loss and Q_L of the resonator are given in Fig. 9.10 . The results show a fairly constant insertion loss (5 ± 1dB) over the tuning range from 3.5 – 6 GHz (53% tuning bandwidth). It was predicted that a tuning bandwidth of over 100% could be possible if diodes of higher capacitance tuning ratio were used.

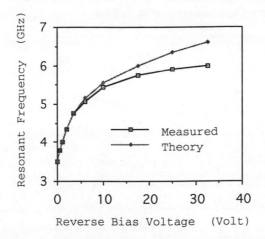

Fig. 9.9 Comparison between the predicted and measured resonant frequency

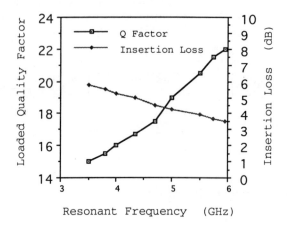

Fig. 9.10 Measured insertion loss and loaded Q against resonant frequency

An MMIC version of this tuneable resonator was fabricated [22] using the Plessey F14 foundry process. The chip area measured 1 mm by 2 mm. The varactors used were planar interdigitated Schottky-barrier diodes. The diode had a punch-through voltage of 1.5 V, and a breakdown voltage of 16 V. These diodes were chosen to operate in the region beyond punch-through, and so the capacitance tuning ratio was very limited. This MMIC resonator had a tuning range of 8.6 – 9.75 GHz when the bias voltage was varied from 2 to 14 V. The Q factor of the resonator had been found to drop significantly [4, 22] when the diodes were operated very close to punch-through. The insertion loss and loaded quality factor of the resonator are shown in Fig. 9.11. The insertion loss flatness is 4.6 ± 0.25 dB.

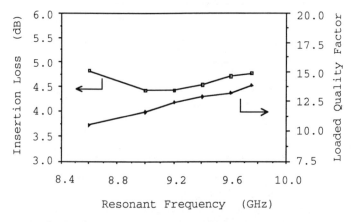

Fig. 9.11 Variation of insertion loss and loaded Q factor against resonant frequency

In VCOs where it is necessary to have a broad tuning range, the loss resistance of the varactor can become the dominant noise source in the oscillator. Under this circumstance the design strategy for minimum phase noise must be to maximise the carrier signal effectively appearing across the noise resistance. Hence, the phase noise performance is fundamentally limited by the maximum RF voltage that can appear across the varactor. However, planar diodes produced to date have low punch-through voltages (usually between 1 and 2V) and hence the maximum allowable voltage swing across the diode is small.

9.4.3 Superconductive resonators

The discovery of high temperature superconductors (HTS) has sparked a world-wide research and development effort. HTS films are usually made from cuprate-based materials which show critical temperatures of below 77 K; the boiling point of liquid nitrogen. HTS resonators are usually based on coplanar waveguides since these have a particularly convenient thin-film structure, requiring the coating of only one side of a substrate with superconductor and thereby allowing the other side to be thermally anchored during deposition and *in situ* annealing of the superconductor. Although superconductors do dissipate energy at microwave frequencies, these losses are far less than that of copper or gold. For example, CPW resonators made of YBCO [14] have shown losses superior to those of copper at 77 K. The PEM model of Lee and Itoh [19] for films of thickness comparable with the skin depth has been used with microwave CAD software for design and analysis of superconductive resonators. It has been reported that the Q of superconductive resonators is ten to one-hundred times better than gold or copper resonators, all operating at 77 K. The loss mechanisms contributing to the intrinsic Q of a resonator include: conductor loss, dielectric loss and radiation loss. The conductor loss is proportional to the surface resistance of the superconductor. It has been observed that the Q of superconductive resonators often degrades with increasing input power due to the current density dependency of the surface resistance [20]. Therefore, optimising the current density distribution on these resonators is a crucial step in maximising their Qs. However, the Q of superconductive resonators is not usually limited by the conductor loss but rather by the loss tangent of the substrate and radiation losses. Radiation loss is minimised by enclosing the resonator in a specially-designed assembly. HTS-based ring resonators, edge-coupled microstrip resonators, and centre-coupled half-wave CPW resonators, and their realisation in microwave oscillators, have been published in the literature. Klieber reported a CPW resonator [14] with an unloaded Q of 3850, which is 43 times larger than the Q value obtained with a resonator out of copper on $LaAlO_3$ substrate with the same geometry but a considerably larger conductor thickness. Both the active device (MESFET) and the resonator were operated at the same temperature of 77 K. This oscillator had a fundamental output power of 4.9 dBm at 6.5 GHz. The measured single-sideband noise power of the oscillator at 10 kHz offset from carrier was -90 dBc/Hz. Oscillators based on HTS material may be suitable for application in space, provided that cooling may be accomplished reliably.

9.5 CAD techniques for non-linear oscillators

Although an oscillator is in reality a non-linear component, small-signal considerations are usually sufficient to ensure that oscillation conditions are met and to establish the desired frequency. However, the linear theory is incapable of predicting output power precisely and determining the voltage waveforms across critical components in the circuit, such as a tuning varactor. The latter may be very valuable in maximising power and efficiency or minimising noise in VCOs. The traditional methods of designing oscillators are mainly based on empirical methods, quasi-linear models [23] or large-signal S-parameters. The accuracy of these methods, however, is questionable and they do not give data on harmonic content. Today, high performance PCs and workstations are so popular that the application of full numerical methods to these non-linear problems is becoming increasingly attractive. Current research activities have been concentrating on developing accurate device models (both linear and non-linear) and efficient analysis methods. Analysis techniques generally fall into three categories: time-domain methods, harmonic balance methods and Volterra series analysis.

9.5.1 Time-domain methods

Time-domain methods enjoy a wide range of applications [24], including the steady state and transient solutions of both analogue and digital circuits. Circuit differential equations are solved in the time domain by numerical integration, requiring the solution of a set of non-linear algebraic equations at each time step. The starting point is usually chosen as the result of a DC analysis. Methods that work entirely in the time domain represent the most natural and straightforward approach to the simulation problem. However, there are several factors which limit the applicability of these methods to microwave circuit analysis. Firstly, the actual circuits may contain dispersive transmission lines that are difficult to analyse in the time domain. Secondly, the circuit may have time constants that are large compared to the inverse of the fundamental excitation frequency. If the steady-state response is of major interest, a large number of time steps may be required, thereby consuming much CPU time. Considerable effort [25] has been spent on devising time-domain methods which allow the transient calculation at least to be partially bypassed and the steady state solution to be reached more quickly. Long computer time and dynamic range limitations are the major burdens of this method.

9.5.2 Harmonic balance method

In this method, variables are represented by their Fourier coefficients rather than as functions of time. The transients are naturally avoided and therefore the computational complexity depends only on the size of the circuit and not on the actual frequencies or the time constants present in the circuit. In this method, the linear elements are analysed in the frequency domain whereas the non-linear elements are simulated in the time domain. The interfacing between the two domains is accomplished by fast Fourier transformation (FFT). The Fourier coefficients of the unknown variables are then optimised to give the approximated solution. Harmonic balance techniques have been used extensively to analyse virtually all kinds of non-linear microwave subsystems [26-28]. However, some problems arise when applying this technique to analyse oscillators. Firstly, in an oscillator the frequency can change as the oscillation develops and, therefore, can only be known approximately. If the frequency used in the harmonic balance analysis is not equal to the large-signal oscillation frequency, the results may be erroneous. A way to overcome this is by treating the unknown oscillation frequency as an extra optimising variable. The second problem is due to the fact that non-physical, spurious oscillations might exist. An algorithm is therefore required to find the physical solution and separate it from the spurious ones. Newton's method is known to offer fast convergence but it is generally difficult to find suitable initial values especially for an oscillator. Hence, the chance of obtaining spurious solutions is high. Furthermore, the harmonic balance method also requires FFTs and an optimisation algorithm, which could use up considerable computer time.

9.5.3 Volterra series analysis

Volterra series theory was first introduced in 1930 and was further developed by Weiner in the 1950s for the expansion of functionals in terms of orthogonal polynomial series. Weiner's functional expansions, now known as Volterra non-linear transfer functions, can handle frequency dependent systems with single valued input-output characteristics. For a linear system, the output $y(t)$ can be related to its input $x(t)$ by

$$y(t) = \int_{-\infty}^{\infty} h_1(u_1) x(t - u_1) du_1 \qquad (9.4)$$

where $h_1(u_1)$ is the impulse response of the system. When the system becomes non-linear, the output can be related to its input by the Volterra functional series [27],

$$y(t) = \int_{-\infty}^{\infty} h_1(u_1) x(t - u_1) du_1 + \int_{-\infty}^{\infty}\int_{-\infty}^{\infty} h_2(u_1, u_2) x(t - u_1) x(t - u_2) du_1 du_2 + \ldots (9.5)$$

where $h_n(u_1, \ldots, u_n)$ is known as the 'nth order impulse response' of the system. Volterra series analysis in a non-linear system can be viewed as the generalisation of convolution integrals used in linear systems. In the frequency domain, the output spectrum $Y(f)$ can be expressed as a function of the input spectrum $X(f)$ given by

$$Y(f) = \sum_{n=1}^{\infty} \int_{-\infty}^{\infty} \ldots \int_{-\infty}^{\infty} H_n(f_1, \ldots, f_n) \cdot \delta(f - f_1 - \ldots - f_n) \prod_{1}^{n} X(f_i) df_i \qquad (9.6)$$

where $H_n(f_1, ..., f_n)$ is known as the non-linear transfer function which is the Fourier transform of the impulse response $h_n(u_1, ..., u_n)$. For a system with mild non-linearity, it has been suggested that the first few terms of the expansion may be sufficient to represent $y(t)$. An efficient method of solving the response of a non-linear system with power series type non-linearity is the current method described by Bussgang et al. [29]. Volterra series analysis is very attractive because computations are performed in the frequency domain and neither Fourier transformation nor iteration is required. Volterra series analysis is, in general, restricted to weakly non-linear systems because of the algebraic complexity of determining Volterra non-linear transfer functions of high order (as required with more strongly non-linear systems).

Volterra series analysis has recently been used by Cheng and Everard [30] in designing microwave oscillators. In their method they converted an oscillator into a one-port network by breaking a short-circuit point somewhere in the circuit. This leads to the oscillation criterion whereby the impedance looking into this port is zero at the fundamental and at all harmonics. As shown in Fig. 9.6, the idealised impedance Z_k is considered to present to the circuit an open-circuit at the fundamental, and a short-circuit at DC and higher harmonic frequencies.

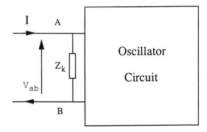

Fig 9.12 Modified oscillator circuit

The input impedance looking into the circuit is then derived by applying Volterra series analysis. This impedance is therefore a function of I and ω, which can be expressed as:

$$Z_{in}(\omega, I) = Z_1(\omega) + Z_3(\omega, \omega, -\omega)I^2 + Z_5(\omega, \omega, \omega, -\omega, -\omega)I^4 + ... \quad (9.7)$$

where we denote $Z_3(\omega, \omega, -\omega)$ and $Z_5(\omega, \omega, \omega, -\omega, -\omega)$ as the third- and fifth-order non-linear impedances, respectively. $Z_1(\omega)$ is simply the input impedance of the linear circuit. The essence of this approach is therefore to adjust the variables, I and ω, until Z_{in} becomes zero. Since there are only two unknowns in this equation, the chance of finding the solution is high. Numerical techniques for solving this non-linear equation are available in many textbooks. For illustration, this method of analysis is now applied to the van der Pol oscillator (Fig. 9.13) which consists of a shunt resonant circuit in parallel with a non-linear conductance.

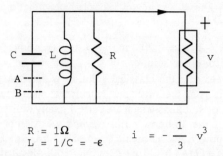

$$R = 1\Omega$$
$$L = 1/C = \varepsilon$$
$$i = -\frac{1}{3}v^3$$

Fig 9.13 Van der Pol oscillator

The behaviour of the circuit is well-described by the van der Pol equation given by

$$\frac{d^2v}{dt^2} - \varepsilon(1-v^2)\frac{dv}{dt} + v = 0 \qquad (9.8)$$

This equation has been extensively studied in the literature and its solution for $\varepsilon = 0.2$ is $V = 2$ and $\omega_0 = 0.9975$, where V and ω_0 are the voltage amplitude across the resonant circuit and frequency of oscillation, respectively. It can be shown that the linear impedance between A & B terminals is given by

$$Z_1(\omega) = \frac{\omega^2\varepsilon^2}{1+\omega^2\varepsilon^2} + j\left(\frac{\varepsilon}{\omega} - \frac{\omega\varepsilon}{1+\omega^2\varepsilon^2}\right) \qquad (9.9)$$

By equating the imaginary part of $Z_1(\omega)$ to zero, the resonant frequency ω_r is simply

$$\omega_o = \frac{1}{\sqrt{1-\varepsilon^2}} \qquad (9.10)$$

Note that the real part of $Z_1(\omega)$ is positive at all frequencies. However, the third-order non-linear impedance of this example, which is given by

$$Z_3(\omega,\omega,-\omega) = \frac{-\omega^4\varepsilon^4}{4(1+\omega^2\varepsilon^2)^3}(1-\omega^2\varepsilon^2 + 2j\omega\varepsilon) \qquad (9.11)$$

gives a negative real part if $\omega < 1/\varepsilon$. The 'third-order' solutions of eqn. 9.7 are found to be

$$\omega_o = \frac{1}{\sqrt{1+\varepsilon^2}} \qquad (9.12)$$

$$V = 2(1+2\varepsilon^2)\sqrt{1+\varepsilon^2} \qquad (9.13)$$

Consequently for $\varepsilon = 0.2$, $V = 2.203$ and $\omega_o = 0.9806$. Eqn. 9.7 has been solved numerically giving 'fifth-order' solutions of $V = 1.976$ and $\omega_o = 0.9977$. The values thus found are very close to the classical solution with a maximum error of 1%. Repeating this analysis using the impedance equation of a sufficiently high order, one can in principle generate solutions which are correct to any desired accuracy.

A practical use of the Volterra series method was demonstrated [30] on the prediction of output power and harmonic level of a 5.5 GHz oscillator (Fig. 9.14). The circuit consists of a GaAs MESFET chip (NE71000) located at the centre of the alumina substrate, wherein the FET gate is grounded via a bond wire. The source is connected to a stripline via another bond wire, and the other end of the stripline is terminated by a 100 pF chip capacitor which presents a short-circuit at RF. The DC power supply is fed to the drain of the FET through an inductor and a bias tee at the drain terminal. The biasing resistor at the source provides a DC return and current stabilisation. The RF output power is coupled directly from the source terminal via a DC blocking capacitor.

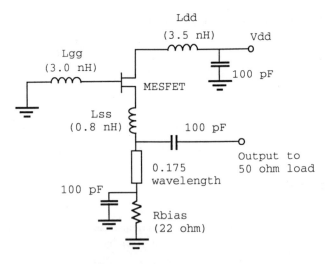

Fig. 9.14 5.5 GHz GaAs MESFET oscillator

An optimum design has been realised by adjusting the length of the stripline to reduce the power levels of the second and third harmonics. The analysis and optimisation of the design is performed by a computer program based on the Volterra series method and a non-linear model of the MESFET [31]. Computer simulations show that the stripline should be about 0.7 of a quarter wavelength long for optimum suppression of the harmonics. A comparison between the predicted and measured results of the constructed oscillator is shown in Fig. 9.15. In fact, when higher accuracy is required, it is possible to resort to a harmonic balance technique, using the results obtained by this method as the starting values.

	Predicted	Measured
Resonant Frequency	5.48 GHz	5.41 GHz
IDS	26.5 mA	24.0 mA
VGS	-0.58 V	-0.53 V
Fundamental	8.3 dBm	8.6 dBm
1 st harmonic	-15.5 dBm	-17 dBm
2 nd harmonic	-22.4 dBm	-24 dBm

Fig. 9.15 Comparison between predicted and experimental results

9.6 Oscillator noise analysis

There are two main sources of phase noise in an oscillator, namely thermal and flicker noise. As shown in Fig. 9.16, thermal noise of the active device has a flat spectrum and is responsible for the $1/f_m^2$ spectral characteristic of an oscillator. Flicker or $1/f$ noise is primarily a baseband noise, being upconverted by non-linearities in the active devices to the carrier frequency. The transposition gain is dependent upon the device and the operating conditions. The loop transfer characteristic then filters the output voltage, producing a $1/f_m^3$ characteristic [32]. The value of the offset frequency at which the transition between $1/f_m^3$ and $1/f_m^2$ laws occurs is the transition frequency f_c and often is different from the baseband flicker noise corner frequency. The transition frequency is usually less than a few kHz for a silicon bipolar transistor and over hundreds of kHz for a GaAs transistor. Although the $1/f$ baseband characteristic will still be converted to a $1/f_m^3$ characteristic in an oscillator, modulation noise actually behaves very differently compared with thermal noise. For example, increasing the oscillator power level does not result in a direct reduction in oscillator phase noise as it does for thermal noise. Another consequence of direct modulation noise is that, if the amplifier is the dominant source of noise, increasing Q_L will decrease the oscillator flicker noise level. A theoretical analysis which is capable of predicting the noise performance of oscillators as well as VCOs is obviously an important subject.

356 *Oscillators*

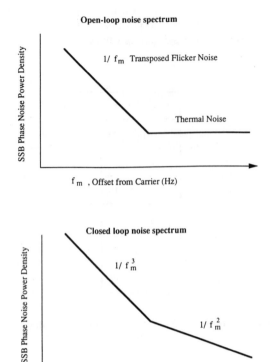

Fig. 9.16 Open and closed loop noise spectrum in oscillators

9.6.1 Noise analysis of negative resistance oscillator

A negative resistance oscillator is considered to be stable if any perturbation (such as noise) in the instantaneous RF voltage or current of the circuit decays, bringing the oscillator back to its point of equilibrium. The general 1st-order treatment was due to Kurokawa [33] who analysed the oscillator quasi-statically. The condition for stable oscillation is realised for:

$$\frac{\partial R_t}{\partial I_o}\frac{\partial X_t}{\partial \omega} - \frac{\partial X_t}{\partial I_o}\frac{\partial R_t}{\partial \omega} > 0 \qquad (9.14)$$

The graphical interpretation of the above expression is shown in Fig. 9.17. This work indicates that the angle of interaction between the circuit locus and the device line is a measure of the oscillator's stability and noise performance at that particular operating point. Low noise is associated with an intersection angle of 90 degrees,

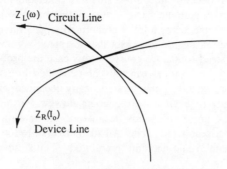

Fig. 9.17 Graphical representation of a negative resistance oscillator

which means that no frequency fluctuations will amplitude modulate the oscillation signal and, more importantly, that no amplitude fluctuations will modulate the phase of the resonator and hence the frequency of oscillation. Therefore, for phase-noise minimisation, the resonator phase shift with frequency must be maximised, which implies maximising Q. Kurokawa's method was later generalised by Eddison and Howes [34] to account for the frequency-dependency of the device conductance. As a result, the graphical interpretation of the device admittance should be considered as a surface instead of a line. Esdale and Howes [35] extended Kurokawa's method by converting it to a reflection coefficient analysis, which uses readily available device and circuit reflection coefficient information to predict oscillation conditions. In theory it should be possible, for any given device, to design a passive circuit to give an oscillator structure with a given stability and noise performance. However, it has been pointed out [34] that the orthogonality condition is not easily satisfied with practical component values. Consequently, the intersection angle is likely to be small, a fact which is reflected by the large noise output of microwave negative-resistance oscillators, particularly the phase noise which is by far the largest noise component.

A number of investigations into the upconversion of baseband noise in fixed-frequency oscillators have been published in the literature. Debney and Joshi [36] presented the first detailed version of Kurokawa's analysis, and involved the transformation of a baseband noise source through a non-linear, two port device model to an equivalent RF noise source in the one port representation of the oscillator. The interaction of this source with the device and the load impedances was analysed to produce the amplitude and phase noise spectra. A more general model was considered by Siweris and Schiek [37] in their analysis of a fixed frequency oscillator using the quasi-static approach. They included the gate-source capacitance in the intrinsic, non-linear device model, an element neglected by Debney and Joshi, and showed this to be the important element in the upconversion of baseband noise to close-to-carrier phase noise. Siweris and Schiek have also considered the inclusion of a non-linear, baseband noise source to give improved agreement between measured and simulated results for fixed frequency oscillators. A more general solution of the frequency conversion process by injecting a small RF deterministic signal into a non-linear circuit supporting a periodic steady-state solution has also been reported [38]. When the RF source is replaced by a set of random noise generators as the perturbing mechanism, it is quite reasonable to expect that the same arguments will lead to a noise analysis of the steady state. Of course in this case the problem is much more complicated, since the free sources can only be described in a statistical sense. If several noise generators exist they may be statistically dependent and their correlation must be accounted for in evaluating the noise power. Further correlations are established among the noise sidebands because of the intermodulation of noise waveforms with the periodic steady state. All such effects need to be included in the noise analysis if a general and accurate prediction of noise level in oscillators is desired.

9.6.2 Linear noise analysis of feedback oscillators

Excellent noise performance is easily found in feedback oscillators. This is because, unlike negative resistance oscillators, the conditions for minimum sideband noise in feedback oscillators are much more easily satisfied. This is certainly true for thermal noise when mixing is not a problem, for example, in a highly linear amplifier. The oscillator output spectrum can be considered as noise, amplified and filtered by the effective high Q circuit provided by positive feedback. For flicker noise, however, the situation is different since the non-linearities of the active element also play a major role in determining the transposition coefficient of the upconverted $1/f$ noise. A linear theory has been used quite satisfactorily in predicting the phase-noise level in feedback oscillators caused by flat thermal noise. This model is quite accurate for a good quality oscillator except in predicting carrier/noise ratios for small offset frequencies such as transposed flicker noise. The noise equation has been derived [16] as

$$L(f_m) = \frac{FkTG}{8Q_L^2 P}\left(\frac{f_r}{f_m}\right)^2 \qquad (9.15)$$

where G is the amplifier gain, F is the amplifier noise figure, Q_L is the loaded quality factor of the resonator, f_r is the resonant frequency, f_m is the offset frequency from carrier and P is the amplifier output power and in general tends to be close to the saturated output power of the amplifier. Note that the occurrence of saturation in amplifiers not only limits the output amplitude but also removes any AM noise. This assumes that the limiting does not cause extra components due to mixing. The above expression shows that for small frequency offset from the carrier the noise level decays ultimately at 6 dB/octave. Noise level far from the carrier is the same in both the open- and closed-loop conditions and is independent of frequency. However, as the noise frequency gets closer to the carrier, the closed-loop noise level begins to deviate from the open-loop level. This occurs when the noise frequency falls within the 3dB bandwidth of the resonator.

9.6.3 Optimum Q_L for minimum phase noise

On examining the noise equation in eqn. 9.15, it appears that phase noise may be minimised by reducing G, F or kT and by increasing Q_L. However, the parameters G and Q_L are interrelated, such that

$$S_{21} = \frac{1}{G} = 1 - \frac{Q_L}{Q_o} \tag{9.16}$$

where S_{21} is the magnitude of the transmission coefficient of the resonator at resonance. Therefore, one cannot arbitrarily increase Q_L without increasing G. If one takes the power as the power available at the output of the amplifier, and assuming that F and P are constants, a minimum value for eqn. 9.15 would occur at $Q_L = Q_o/2$. It should be noted that in principle the noise factor is dependent on the source impedance seen by the amplifier and the optimum noise impedance of the active device. The source impedance seen by the amplifier is the combination of the resonator output impedance and the matching network at the amplifier input. Furthermore, the available power P is dependent on the output impedance seen by the amplifier. Therefore, in practise, F and P are both dependent upon the ratio Q_L/Q_o.

9.6.4 Effects of gain compression

Noise in oscillators may be modified by gain compression present in the amplifier. This is a non-linear process and it is not possible to use a simple theory to describe it accurately. Even for thermal noise, where mixing is unlikely to cause any problems at low power levels, the noise figure may still be affected by gain compression. In the case of flicker noise, gain compression is certainly an influential factor on the oscillator's upconversion coefficient. Parker [40] measured flicker noise levels versus gain compression in both silicon and GaAs devices. However, these experiments were conducted for amplifiers only, and not for oscillators. Cheng and Everard [41] studied this effect on oscillators experimentally, based on the set-up shown in Fig. 9.18. The transfer power characteristic of the amplifiers was measured firstly, under open loop conditions. This allowed the amount of gain compression versus output power to be determined. The loop was then closed and the noise level of the oscillators recorded. Different levels of gain compression were adjusted by the variable attenuator provided. An isolator was inserted in the experimental system to ensure that a constant source impedance was seen by the amplifier, thereby ensuring a constant noise factor F. A constant load impedance for the amplifier was provided by a second isolator. Both bipolar and GaAs oscillators were tested in these experiments. Fig. 9.19 shows the measured noise level of the oscillator as a function of gain compression of the amplifier. These noise measurements were performed at 25 kHz offset frequency from carrier. For the bipolar oscillator, this was within its thermal noise region. For the GaAs oscillator, it was well inside its flicker noise region. Only a slight increase in noise level was observed in both oscillators even for as much as 3 dB gain compression. However, these observations do not necessarily lead to a general conclusion, and a new set of experiments must be conducted when the type of device, circuit configuration or operating conditions are changed.

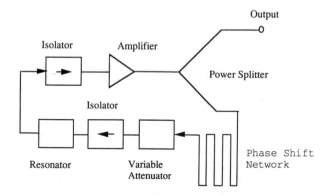

Fig. 9.18 Experimental setup for the investigation of the effect of gain compression on oscillator output noise

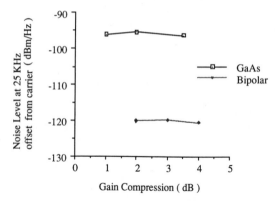

Fig. 9.19 Output noise against gain compression for bipolar and GaAs MESFET oscillators

9.6.5 Effects of non-zero open loop phase error

When the loop of a feedback oscillator is closed, oscillation will occur if two conditions are met. Firstly, the amplifier must give sufficient signal gain to overcome the loss in the other components and, secondly, the phase shift round the loop must be an integral multiple of 360 degrees. It is therefore possible to vary the frequency of oscillation by changing the phase shift in some of the components. The oscillation frequency will move to a value at which the extra phase shift added by these components is compensated by an equal and opposite phase shift in the resonator, provided that the first condition is satisfied. Hence, the oscillator does not necessarily oscillate at the resonant peak f_r of the resonator. If ϕ is defined as the phase error in the open loop, then the shift in oscillation frequency $\Delta f = f_o - f_r$ can be evaluated by

$$\Delta f = \frac{f_r}{2Q_L} \tan \phi \tag{9.17}$$

To examine the effect of a non-zero value of ϕ on the noise performance of oscillators, it is necessary to look at the thermal and flicker noise separately. If thermal noise is considered, the output noise spectrum of the oscillator $L(f_m)$ is then given by [41]

$$L(f_m) = \frac{F k T G_o^2}{8 P Q_L^2 \cos^4 \phi} \cdot \left(\frac{f_r}{f_m} \right)^2 \tag{9.18}$$

where G_o is numerically equal to $1/S_{21}$. In the case of modulation noise (e.g. flicker noise), which is due to direct phase fluctuations in the electronic components of an oscillator such as the amplifier, if the phase perturbation is denoted as ϕ_m, then the transmission coefficient of the amplifier can be represented by

$$G_v \exp\{j(\phi + \phi_m)\} \tag{9.19}$$

where G_v is the voltage gain of the amplifier. Assuming that $f_m \ll \Delta f$, the transmission coefficient of the resonator at $f = f_r + \Delta f + f_m$ can be approximated as

$$S_{21_r}(f) \approx S_{21} \cos \phi \, \exp(-j\psi)$$
$$\psi = \phi + 2Q_L \cos^2 \phi \, \frac{f_m}{f_r} \tag{9.20}$$

For oscillation to occur, the following conditions have to be satisfied:

$$G_v = S_{21} \cos \phi$$
$$\phi + \phi_m - \psi = 0 \tag{9.21}$$

In fact, the second requirement in the above expression may be rewritten as

$$\frac{f_m}{f_r} = \frac{\phi_m}{2Q_L \cos^2 \phi} \tag{9.22}$$

Since $f_m/f_r \ll 1$, the phase noise level is therefore proportional to $(f_m/f_r)^2$. It is assumed that the upconversion coefficient remains unchanged. The analysis shows that the phase noise level (both thermal and modulation noise) will degrade for an oscillator if it is pulled off its natural resonant frequency by the factor $\cos^4 \phi$. The effect of non-zero phase error on the noise performances of both bipolar and GaAs FET oscillators has been demonstrated experimentally (Fig. 9.20). The value of φ was controlled by a variable delay line. The gain compression level of the amplifier was kept the same throughout the experiments by an adjustable attenuator. Noise levels were measured at an offset frequency of 25 kHz. The measured noise spectrum of the GaAs FET oscillator showed a flicker noise characteristic well above 100 kHz. The flicker noise corner of the bipolar oscillator was found to be less than 5 kHz.

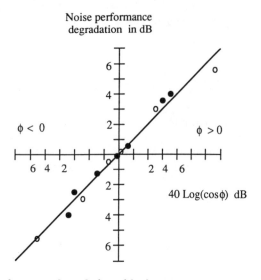

Fig. 9.20 Noise performance degradation with phase error

Both the theoretical and experimental results show that the noise level degrades very rapidly at a large value of φ, and the consequence of this observation is significant. For example, for a 10 GHz feedback oscillator having a resonator Q of 4000, a noise degradation of 6 dB is produced when the oscillation frequency is pulled off by only 1.25 MHz.

9.7 MMIC VCO design example

Oscillators form an essential part of microwave and millimetre-wave transceivers and for many applications a voltage-controlled oscillator (VCO) is desirable; either for frequency tuning or for use as a phase-locked oscillator. The design of varactor-tuned FET VCOs has received widespread attention in the literature and convenient design methods have been identified. This section describes the modelling and design of an X-band MMIC VCO [42] which uses MESFET devices and interdigitated planar Schottky diodes based on the FET geometry.

9.7.1 Varactor diode modelling

Varactor diodes are usually realised on MMICs by using the standard MESFET structure, with source and drain fingers connected together to form the cathode, and the gate forming the anode. In this way a varactor with acceptable performance can be realised without additional processing steps. An accurate model has been developed [43] for these interdigitated planar Schottky varactor diodes (IPSVDs). In the model, the significant extrinsic elements are combined with the reverse and forward bias intrinsic elements of the low frequency equivalent circuit. The extrinsic parameters are based on the physical structure of the device. In addition, frequency dispersion and delay models are included. The result is a bias-dependant model which is accurate well into the millimetre-wave frequency range, with either reverse or forward bias applied. Small-signal modelling using this bias-dependant varactor diode model and FET S-parameters, or an FET model, can be used for most of the design procedure. A bias-dependant varactor diode model enables the tuning range of the VCO to be simulated and optimised. However, a full large-signal model is required to optimise the output power of the oscillator, best carried out using a harmonic-balance simulator, and to check the oscillator's start-up performance with a time-domain simulator.

9.7.2 VCO design

The circuit diagram of the VCO is shown in Fig. 9.21. The four major parts of the circuit are:

(1) the FET with a feedback capacitance, C_f, in the source
(2) the resonator on the gate terminal, consisting of L_1 and the varactor
(3) the output matching on the drain
(4) the DC bias networks.

The feedback capacitor on the source is required in order that the FET presents a negative resistance at the gate terminal. In the ideal case this negative resistance is given by

$$R = -\frac{g_m}{\omega^2 C_{gs} C_f} \quad (9.23)$$

where C_{gs} is the gate-source capacitance of the FET, g_m is its transconductance and ω is the angular frequency. The resonator at the gate terminal consists of L_1 in series with the tuning varactor. These create a series resonance, effectively short-circuiting the gate and allowing positive feedback to occur. The output matching is accomplished by L_5 along with L_3 which is also the drain bias choke. The FET is operated in the self-biased single supply mode; hence the gate is shorted at DC through L_2, whereas the source is connected to ground through a bias choke (L_4) and a resistor R_s which ensures that the gate-source voltage is negative.

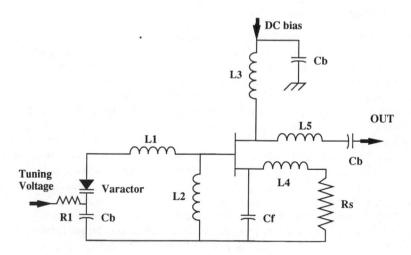

Fig. 9.21 VCO circuit diagram

The major part of the design is to use small-signal analysis to design the feedback and the resonator elements in order to satisfy the oscillation conditions over the required bandwidth. The circuit is analysed as a two-port network by breaking the connection between the tuning varactor and ground. The components are optimised to meet the oscillation conditions; at resonance the S_{21} response should show a large gain peak, and the input impedance response should show a negative resistance of more than approximately 50 Ω with a reactance of zero. The tuning range of the oscillator can be ascertained by varying the bias in the bias-dependant varactor model and observing the frequency range over which the required level of negative resistance is maintained. For this X-band design the chosen FET was the 4 × 50 μm device, and the 6 × 150 μm varactor was found to be optimum. The tuning range of the oscillator can be increased considerably by the use of a varactor diode as the feedback capacitance C_f. This enables the negative resistance value to be varied, and if the two varactors are controlled with independent tuning voltages then a four-fold increase in tuning range can easily be achieved. Having carried out the initial small-signal optimisation, Microwave Spice was used with the foundry large-signal FET model in order to verify the correct operation of the VCO from start-up. Some manual adjustment of the circuit was also carried out using Microwave Spice™ in order to maximise the output power. Fig. 9.22 shows the output signal from the VCO, simulated with Microwave Spice™. The start-up performance of the oscillator must be checked with time-domain simulations of this type, but the long CPU time required for these simulations means that most of the design would be completed using a linear or harmonic balance simulator first.

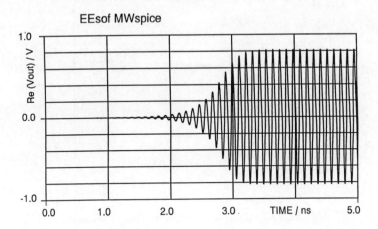

Fig. 9.22 VCO output signal simulated using Microwave Spice™

For comparison, the VCO design was fabricated both with a fixed feedback capacitance and with the second varactor diode. The chips were fabricated on a Eurochip multi-project wafer, using the standard GEC Marconi F20 process. The chips were measured using a Cascade prober and HP8562A spectrum analyser. The drain voltage and current were nominally 6 V and 15 mA, respectively, and the varactor bias voltage varied from 0 to 8 V. The frequency against voltage tuning characteristic is plotted in Fig. 9.23 for the design with a fixed feedback capacitance. The output power, corrected for probe and cable losses, is 7 – 8 dBm over a tuning range of 420 MHz.

A photograph of the second chip, which employs a varactor as the source feedback capacitor, is shown in Fig. 9.24. This additional flexibility enables the second oscillator chip to tune over the range 8.6 – 11.4 GHz, whilst the increase in circuit complexity is quite modest. The chip size is less than 1.5×1.5 mm^2. Note that the measured output power (Fig. 9.25) and tuning range are considerably improved compared with the first design. The output power is over 10 dBm for the range 9.4 – 11.4 GHz, and the peak output power is 12.5 dBm. These results are very satisfactory considering that no buffer amplifier is used. For very demanding applications it would be advantageous to consider alternative devices such as the HBT. In the case of a VCO using a bipolar transistor, the device can be used in a common-base configuration with an inductor as the feedback element. Apart from this, and minor changes to the bias network, the VCO design principle is the same.

366 *Oscillators*

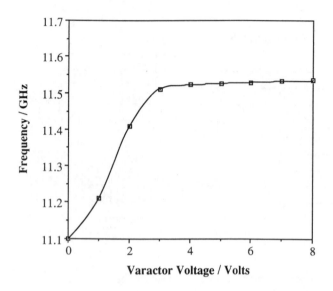

Fig. 9.23 Measured VCO tuning characteristic with fixed C_f

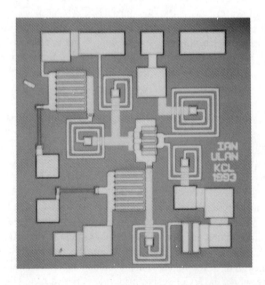

Fig. 9.24 Photograph of the MMIC VCO with varactor as source feedback

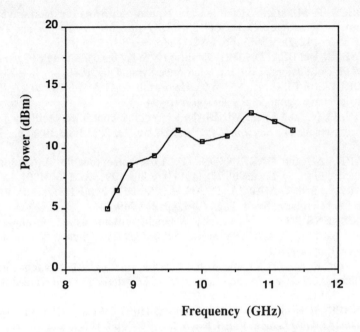

Fig. 9.25 Output power against frequency for the VCO chip with a source varactor

9.8 References

1. PUCEL, R. A., *et al.*: 'Experiments on integrated gallium-arsenide FET oscillators at X band', *Electron. Lett.*, Vol. 11, no. 10, 15th May 1975, pp. 219-220
2. TREW, R. J.: 'Octave-band GaAs FET YIG-tuned oscillators', *Electron. Lett.*, Vol. 13, no. 21, 13th Oct. 1977, pp. 629-630
3. ISHIHARA, O., *et al.*: 'A highly stabilized GaAs FET oscillator using a dielectric resonator feedback circuit in 9-14 GHz', *IEEE Trans. Microwave Theory Tech.*, MTT-28, no. 8, August 1980, pp. 817-824
4. SCOTT, B. N., and BREHM, G. E.: 'Monolithic voltage controlled oscillator for X- and Ku-bands', *IEEE Trans. Microwave Theory Tech.*, MTT-30, no. 12, Dec. 1982, pp. 2172-2177
5. SCOTT, B. N., *et al.*: 'A family of four monolithic VCO MIC's covering 2-18 GHz', IEEE Microwave and Millimeter-wave Monolithic Circuits Symposium Digest, May 1984, pp. 58-61
6. MOGHE, S. B., and HOLDEN, T. J.: 'High-performance GaAs MMIC oscillators', *IEEE Trans. Microwave Theory Tech.*, MTT-35, no. 12, Dec. 1987, pp. 1283-1287

7. ASBECK, P. M., et al.: 'Heterojunction bipolar transistors for microwave and millimeter-wave integrated circuits', *IEEE Trans. Microwave Theory Tech.*, MTT-35, no. 12, Dec. 1987, pp. 1462-1468
8. REESE, E., and BEALL, J. M.: 'Optimized X & Ku band GaAs MMIC varactor tuned FET oscillators', IEEE Int. Microwave Symp. Dig., 1988, pp. 487-490
9. ANDREWS, J. E., et al.: '2.5 - 6.0 GHz broadband GaAs MMIC VCO', IEEE Int. Microwave Symp. Dig., 1988, pp. 491-494
10. CHAN, O. Y., and KAZEMINEJAD, S.: 'Voltage-controlled oscillator using dielectric resonator', *Electron. Lett.*, Vol. 24, no. 13, 23rd June 1988, pp. 776-777
11. DEARN, A., and PARKINSON, G.: 'Low noise tunable GaAs bipolar oscillator', *Electron. Lett.*, Vol. 27, no. 14, 4th July 1991, pp. 1300-1301
12. ADAR, A., and RAMACHANDRAN, R.: 'An HBT MMIC wideband VCO', IEEE Int. Microwave Symp. Dig., 1991, pp. 247-249
13. SCHELLENBERG, J. M., et al.: 'W-band oscillator using ion-implanted InGaAs MESFET's', *IEEE Microwave and Guided Wave Letters*, Vol. 1, no. 5, May 1991, pp. 100-101
14. KLIEBER, R., et al.: 'A coplanar transmission line high-Tc superconductive oscillator at 6.5 GHz on a single substrate', *IEEE Microwave and Guided Wave Letters*, Vol. 2, no. 1, Jan. 1992, pp. 22-24
15. WENGER, J., et al.: ' Ka- and W-band PM-HFET DRO's', *IEEE Microwave and Guided Wave Letters*, Vol. 3, No. 6, June 1993, pp. 191-193
16. EVERARD, J. K. A., and CHENG, K. K. M.: 'Novel low-noise 'Fabry-Perot' transmission line oscillator', *Electron. Lett.*, Vol. 25, no. 17, 17th August 1989, pp. 1106-1108
17. EVERARD, J. K. A., CHENG, K. K. M., and DALLAS, P. A.: 'High Q helical resonator for oscillators and filters in mobile communications systems', *Electron. Lett.*, Vol. 25, no. 24, 23rd Nov. 1989, pp. 1648-1649
18. EVERARD, J. K. A., and CHENG, K. K. M.: 'High performance direct coupled bandpass filters on coplanar waveguide', *IEEE Trans. Microwave Theory Tech.*, MTT-41, no. 9, Sept. 1993, pp. 1568-1573
19. LEE, H. Y., and ITOH, T.: 'Phenomenological loss equivalence method for quasi-TEM transmission lines with a thin normal conductor or superconductor', *IEEE Trans. Microwave Theory Tech.*, MTT-37, Dec. 1989, pp. 1904-1909
20. KOBAYASHI, Y., et al.: 'Microwave measurement of temperature and current dependences of surface impedance for high-Tc superconductors', *IEEE Trans. Microwave Theory Tech.*, MTT-39, no. 9, Sept. 1991, pp. 1530-1538
21. CHENG, K. K. M., and EVERARD, J. K. A.: 'Novel varactor tuned transmission line resonator', *Electron. Lett.*, Vol. 25, no. 17, 17th August 1989, pp. 1164-1165
22. CHENG, K. K. M., and EVERARD, J. K. A.: 'X band monolithic tunable resonator/filter', *Electron. Lett.*, Vol. 25, no. 23, 9th Nov. 1989, pp. 1587-1588
23. ABE, H.: 'A GaAs MESFET oscillator quasi-linear design method', *IEEE Trans. on Microwave Theory Tech.*, MTT-34, no. 1, Jan. 1986, pp. 19-25

24. ROBERTSON, I. D., CHENG, K. K. M., and AGHVAMI, A. H.: 'L-band 20 Watts GaAs FET amplifier for on-board digital mobile satellite systems', IEEE International Conference on Communications Digest, June 1988, pp. 532-536
25. APRILLE, T. J., and TRICK, T. N.: 'A computer algorithm to determine the steady-state response of nonlinear oscillators', *IEEE Trans. on Circuit Theory*, CT-19, no. 4, July 1972, pp. 354-359
26. RIZZOLI, V., and LIPPARINI, A.: 'A computer-aided approach to the nonlinear design of microwave transistor oscillators', IEEE Int. Microwave Symp. Dig., 1982, pp. 453-455
27. CHENG, K. K. M., and EVERARD, J. K. A. 'Nonlinear circuit analysis using the Newton-SOR continuation method', *Electron. Lett.*, Vol. 26, no. 25, 6th Dec. 1990, pp. 2120-2121
28. CURTICE, W.: 'Nonlinear analysis of GaAs MESFET amplifiers, mixers, and distributed amplifiers using the harmonic balance technique', *IEEE Trans. Microwave Theory Tech.*, MTT-35, no. 4, April 1987, pp. 441-447
29. BUSSGANG, J. J., et al.: 'Analysis of nonlinear systems with multple inputs', *Proc. IEEE*, Vol. 62, August 1974, pp. 1088-1119
30. CHENG, K. K. M., and EVERARD, J. K. A.: 'A new and efficient approach to the analysis and design of GaAs MESFET microwave oscillators', IEEE Int. Microwave Symp. Dig., May 1990, pp. 1283-1286
31. CURTICE, W. R.: 'GaAs MESFET modeling and nonlinear CAD', *IEEE Trans. on Microwave Theory Tech.*, MTT-36, no. 2, Feb. 1988, pp. 220-230
32. LEESON, D. B.: 'A simple model of feedback oscillator noise spectrum', *Proc. IEEE*, Vol. 54, no. 2, 1966, pp. 329-330
33. KUROKAWA, K.: 'Some basic characteristics of broadband negative resistance oscillator circuits', *The Bell System Technical Journal*, July-August 1969, pp. 1937-1955
34. EDDISON, I., and HOWES, M. J.: 'Circuit aspects of the noise performance of microwave oscillator modules', *IEE Microwaves, Optics and Acoustics*, Vol. 1, no. 3, April 1977, pp. 103-109
35. ESDALE, D. J., and HOWES, M. J.: 'A reflection coefficient approach to the design of one-port negative impedance oscillators', *IEEE Trans. Microwave Theory Tech.*, MTT-29, no. 8, August 1981, pp. 770-776
36. DEBNEY, B. T., and JOSHI, J. S.: 'A theory of noise in GaAs FET microwave oscillators and its experimental verification', *IEEE Trans. on Electron Devices*, Vol. ED-30, no. 7, July 1983, pp. 769-775
37. SIWERIS, H. J., and SCHIEK, B.: 'Analysis of noise upconversion in microwave FET oscillators', *IEEE Trans. on Microwave Theory Tech.*, MTT-33, no. 3, March 1985, pp. 233-241
38. ANZILL, W., and RUSSER, P.: 'A general method to simulate noise in oscillators based on frequency domain techniques', *IEEE Trans. Microwave Theory Tech.*, MTT-41, no. 12, Dec. 1993, pp. 2256-2263
39. PRIGENT, M., and OBREGON, J.: 'Phase noise reduction in FET oscillators by low-frequency loading and fedack circuitry optimization', *IEEE Trans. Microwave Theory Tech.*, MTT-35, no. 3, March 1987, pp. 349-352

40. PARKER, T. E.: 'Characteristics and sources of phase noise in stable oscillators', ieee 41st annual freq. Control Symp. Dig., 1987, pp. 99-110
41. CHENG, K. K. M., and EVERARD, J. K. A.: 'Noise performance degradation in feedback oscillator with non-zero phase error', *Microwave and Optical Technology Letters*, Vol. 4, no. 2, Jan. 1991, pp. 64-66
42. KARACAOGLU, U., LUCYSZYN, S., and ROBERTSON, I. D.: 'Modelling and design of X-band MMIC varactor-tuned VCOs', IEE Colloquium on Modelling, Design, and Application of MMICs, London, June 1994, pp. 13/1-5
43. LUCYSZYN, S., LUCK, J., GREEN, G., and ROBERTSON, I. D.: 'Enhanced modelling of interdigitated planar Schottky varactor diodes', The IEEE Asia-Pacific Microwave Conference Digest, Adelaide, August 1992, pp. 273-278.

Chapter 10

Silicon millimetre-wave ICs

J.- F. Luy

10.1 Millimetre-wave circuits on high resistivity silicon

The development of MMICs is still mainly associated with the substrate material GaAs. This is due to its semi-insulating properties and the high electron mobility. The latter enables the fabrication of MESFETs for operation up into the millimetre-wave range. The lattice matched GaAs/AlGaAs material system provides the basis for the realisation of HEMTs. However, there are some drawbacks: III/Vs are compound semiconductor materials, which makes them difficult to produce and precludes high temperature processing. There is no natural oxide, and there is some risk from the toxicity of arsenic. Wafer sizes used in most foundries and research labs are 4 inches or less in diameter. The thermal conductivity is only 0.5 W/cmK at room temperature. The recombination rate is high, and the carrier lifetime in bulk material is in the ns-region.

If silicon can be used as a substrate material many of these drawbacks are overcome. The primary question is about the dielectric losses of the non semi-insulating material. The highest resistivities are in the range of several 10000 Ωcm for float zone material. Fig. 10.1 shows the calculated attenuation of microstrip lines on high resistivity silicon substrates [1]. The dominant contribution of conductor losses in the microstrip line and in the backside metallisation is calculated using an approach of Wiesbeck [2]. The dielectric losses are calculated following Hammerstad and Jensen [3]. It can be seen that the conductor losses dominate for microstrip line widths below 0.3 mm and that the most significant contribution comes from the microstrip conductor strip losses.

Measurements by Hyltin [4] yielded attenuation values of between 3.6 dB/cm on 100 Ωcm material and 0.4 dB/cm on 1400 Ωcm material at 10 GHz. However, it was observed that the high resistivity characteristic of the material was changed by high temperature processes. Later on, attenuation values of between 0.28 dB/cm and 0.6 dB/cm were measured using 5000 Ωcm material in the frequency range of 2 – 18 GHz [5]. A low temperature (< 800°C) technology based on ion-implantation and laser annealing was developed to realise silicon millimetre-wave integrated circuits

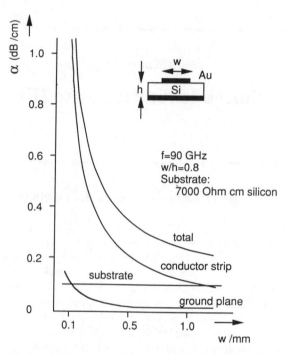

Fig. 10.1 Attenuation of microstrip lines on 7 000 Ωcm silicon substrate as a function of the microstrip line width w. The ratio $w/h=0.8$ is kept constant in the calculation to maintain 50 Ω lines [1]

(SIMMWICs) [6]. At 90 GHz an attenuation of 1.4 dB/cm on 650 Ωcm and 0.6 dB/cm on 10 000 Ωcm material was measured [7]. Simultaneously, it was shown [8] that even thermal oxidation at 1100°C for several hours may now be tolerated. No change in the resistivity was observed using high purity silicon. Compared with earlier investigations [9] this advance in silicon technology now opens the way for a millimetre-wave integration technique which is in principle compatible with CMOS technology.

In this chapter the key devices of this technology are treated: Silicon IMPATT diodes, which are still the most powerful active semiconductor devices at frequencies beyond 60 GHz, PIN diodes, Schottky diodes and finally SiGe bipolar devices. In order to understand the advantages of the SiGe heterojunction bipolar transistor (HBT) compared with bipolar (homo)junction transistors (BJTs), first some physics of the material system silicon/germanium is discussed.

10.2 The Si/SiGe material system

Silicon and germanium are group IV elements which crystallise in the cubic diamond lattice. SiGe alloys are completely miscible between Si and Ge whereas other possible group IV heterostructures (SiSn or GeSn) are only stable within a very small range. The Si/SiGe heterosystem is the most device-relevant group IV combination. Between the indirect semiconductor materials Si and Ge a lattice mismatch of 4.17 % at room temperature is observed. The lattice constant of silicon is 5.431 Å and the lattice constant of germanium is 5.657 Å. This lattice mismatch causes the strain in the silicon/germanium system, if a critical layer thickness is not exceeded [10]. Elastic accommodation by strain is observed for thin films, whereas above the critical thickness plastic accommodation occurs by misfit dislocations relaxing the built-in strain. The strain and the critical thickness depend strongly on the lattice mismatch.

Pseudomorphic growth of $Si_{1-x}Ge_x$ on Si is possible in a stable region limited by thermodynamic equilibrium [11]. Experimentally observed critical thicknesses of molecular beam epitaxy (MBE) grown layers exceed the calculated values significantly. It is assumed that kinetic effects (nucleation and motion of

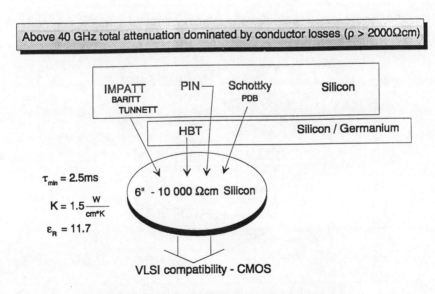

Fig. 10.2 Basic concept of SIMMWIC technology

dislocations) are responsible for this deviation from equilibrium. The density of misfit dislocations in this metastable growth regime depends critically on the growth temperature and the substrate quality. Electronic properties are influenced by strain. Compared with an unstrained SiGe alloy the fundamental bandgap of a strained SiGe alloy is lowered [12]. The actual band alignment is also dependent on the strain situation. People [12] discussed three distinct cases.

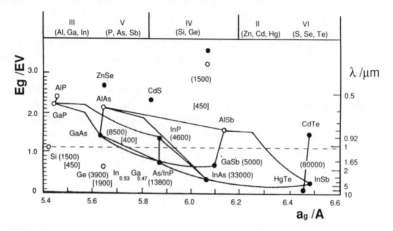

Fig. 10.3 Energy gap as a function of the lattice constant for several semiconductor materials

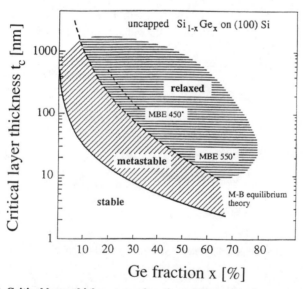

Fig. 10.4 Critical layer thickness as a function of the germanium content [10]

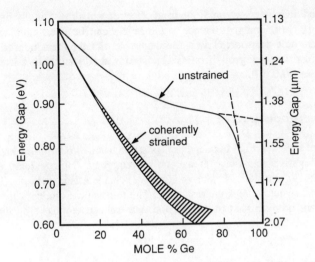

Fig. 10.5 Fundamental indirect bandgap of SiGe alloys on Si <001> substrates after Reference 12

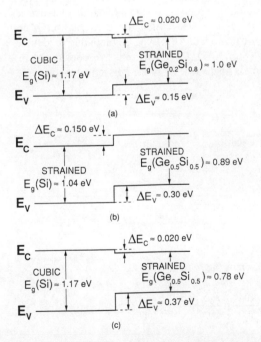

Fig. 10.6 Band alignments for (a) Si/Si$_{0.8}$Ge$_{0.2}$ heterostructures on <001> Si substrate; (b) Si/Si$_{0.5}$Ge$_{0.5}$ heterostructures on <001> Si$_{0.75}$Ge$_{0.25}$ substrate; (c) Si/Si$_{0.5}$Ge$_{0.5}$ heterostructures on <001> Si substrate after [12]

If strained SiGe is grown on Si the band offset occurs almost totally in the valence band (Fig. 10.6 (a) and (c)). Hence, SiGe layers can be used advantageously in devices where hole transport plays a dominant role. For the formation of a barrier in the conduction band a growth on Ge substrates or on suitable buffer layers is necessary (Fig. 10.6 (b)).

10.3 Transit-time devices

Generation of power at high frequencies is possible using two terminal devices, which provide a negative resistance at a certain frequency. This resistance does not necessarily occur in a static current-voltage (I-V) characteristic as in the Gunn device but it can be a dynamic negative resistance. The transit-time diodes which show this behaviour will be discussed in this section with special emphasis on the IMPATT diode.

10.3.1 A unified transit-time diode model

Transit-time diodes for high frequency operation provide an injection mechanism and a transit-time retardation in order to obtain a dynamic negative resistance. The effect of the transit-time region, also called the drift region, is investigated using a generalised analytical model. An n-type semiconductor device is operated in the reverse direction (Fig. 10.7). Only electrons are considered in a one-dimensional model and diffusion is neglected. From the Poisson equation,

$$\frac{dE}{dx} = \frac{q}{\varepsilon}(N_D - n) \tag{10.1}$$

and the continuity equation,

$$\frac{dn}{dt} = -\frac{1}{q}\frac{dJ}{dx} \tag{10.2}$$

we obtain with,

$$J = -qnv \tag{10.3}$$

a differential equation of the second order:

$$\frac{d^2E}{dx^2} = \left[-j\frac{\omega}{v_0} - \mu\frac{n_0 q}{\varepsilon v_0} \right]\frac{dE}{dx} \tag{10.4}$$

The usual small signal assumptions have been made, i.e. higher order products have been neglected and the time dependence is $\propto e^{j\omega t}$. With $v = \mu E$ a solution of eqn. 10.4 is:

Silicon millimetre-wave ICs 377

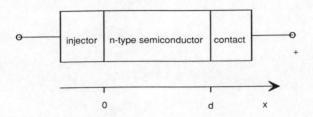

Fig. 10.7 Scheme of a generalised transit-time device with an n-doped drift layer, an injector at the left side and a collector contact at the other side of the device

$$E(x) = a.e^{px} + b \qquad (10.5)$$

with

$$-p = \frac{\mu n_0 q}{\varepsilon v_0} + j\frac{\omega}{v_0} \qquad (10.6)$$

To determine the constants a and b appropriate boundary conditions are necessary. The total current J_t consists of a drift current $J_1(x)$ and a displacement current $j\omega \varepsilon E(x)$:

$$J_t = J_1(x) + j\omega \varepsilon E(x) \qquad (10.7)$$

The comparison with eqn. 10.5 yields

$$a = \frac{-J_1(0)}{j\omega \varepsilon} \qquad (10.8)$$

and

$$b = \frac{J_t}{j\omega \varepsilon} \qquad (10.9)$$

and the result for the electric field in the drift region is:

$$E(x) = \frac{J_t}{j\omega \varepsilon}\left\{1 - \frac{J_1(0)}{J_t}e^{px}\right\} \qquad (10.10)$$

378 Silicon millimetre-wave ICs

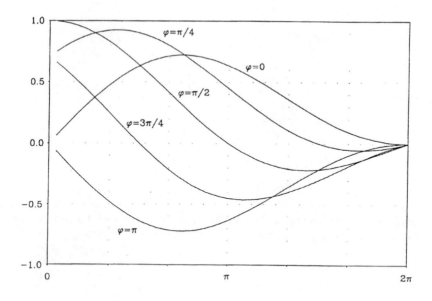

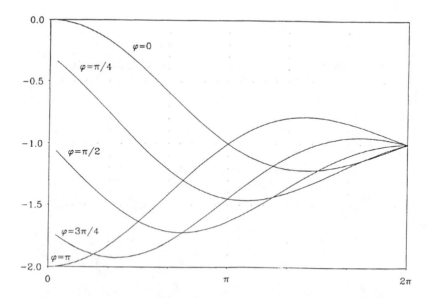

Fig. 10.8 Real part (upper diagram) and imaginary part (lower diagram) of the drift region impedance of a generalised transit-time device as a function of the transit angle $\theta = \omega d / v$. Parameter is the injection angle φ

The drift region impedance is calculated by integration over the drift region length:

$$Z_d = \frac{\int_0^d E(x)dx}{J_t} = -j\frac{d}{\omega\varepsilon}\left\{1 + \frac{J_1(0)}{J_t}\left(\frac{1-e^{pd}}{pd}\right)\right\} \quad (10.11)$$

with the complex injection ratio:

$$\frac{J_1(0)}{J_t} = A(\omega)e^{-j\phi(\omega)} \quad (10.12)$$

The real part R and the imaginary part X of the complex drift region impedance are given by

$$R = \frac{A(\omega)}{\omega C}\left\{\frac{\cos\phi - \cos(\theta+\phi)}{\theta}\right\} \quad (10.13)$$

$$X = -\frac{1}{\omega C}\left\{1 + A(\omega)\frac{\sin\phi - \sin(\phi+\theta)}{\theta}\right\} \quad (10.14)$$

with the transit angle $\theta = \frac{\omega d}{v}$ and $C = \frac{\varepsilon_r\varepsilon_0 A}{d}$.

Fig. 10.8 shows the real part and the imaginary part of the drift region impedance as functions of the transit angle for different injection angles. $A(\omega)/\omega C = 1$ is assumed in the calculation. It can be seen that the injection of carriers with an injection angle $\varphi > \pi/5$ can lead to a negative resistance if the transit angle is chosen accordingly. The largest amount of negative resistance is obtained at an injection angle of π. These conditions correspond to the IMPATT mode $(\phi = \pi, \theta = 3\pi/4)$. In-phase injection transit-time diodes $(\phi = \pi/2, \theta = 3\pi/2)$ generally provide a lower impedance level. The imaginary part is always capacitive.

10.3.2 The IMPATT diode

Carriers can be generated by impact ionisation and subsequent avalanche multiplication. This mechanism is extremely non-linear; the number of generated holes per unit distance (cm) increases from 1 at an electric field of 200 kV/cm up to 7000 at an electric field of 400 kV/cm.

This is the basis for a phenomenological explanation of the operation principles of the IMPATT (impact avalanche transit-time) diode: a p^+n^+ i n^+ structure is biased in the reverse direction as first suggested by Read [13]. The maximum electric field is at the p-n junction. The field in the intrinsic (i-) region is still high enough to ensure drift of the injected carriers with saturated velocity. An RF voltage is superimposed on the DC voltage. The number of carriers generated by impact ionisation increases if the

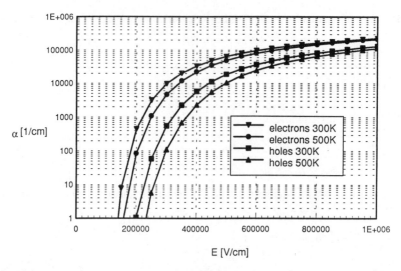

Fig. 10.9 Impact ionisation rate as a function of the electric field for holes and electrons at 300 and 500 K

threshold field for impact ionisation is exceeded – the maximum number of carriers which can be injected into the transit region is reached when the driving force (RF + DC voltage) is just equal to the threshold value for impact ionisation. The strong non-linear behaviour of the impact ionisation causes an injection current which has the form of a sharp pulse. An inductive phase shift of $\pi/2$ and an injection angle $\varphi = \pi$ is observed. The latter can also be deduced solving the Read equation to obtain an analytical expression for the IMPATT diode impedance [14]. This injection delay (referring also to Fig. 10.8) and the drift mechanism lead to an induction current which is positive when the voltage is negative; a negative resistance results and oscillations can grow if a resonator is matched to the diode.

This single drift diode structure is capable of producing an output power of up to 500 mW at 94 GHz [15]. However, impedance matching becomes increasingly difficult at higher frequencies due to the low level of the diode impedance. Double drift diodes are more promising device structures at millimetre-wave frequencies because they can be considered as a series connection of two single drift diodes and so the impedance level is doubled. In waveguide technology these diodes enable output powers of up to 1 Watt in CW operation at 100 GHz [14]. The large modulation depths required (up to 0.5) lead to correspondingly low impedance levels. These impedance requirements can be met in waveguide technology. In planar integrated circuits it is difficult, however, to achieve impedence values below 5 Ω. The imaginary part of a transit-time diode's impedance is usually a factor of 10 to 20 times larger than the real part. It is therefore essential to maximise the real part of the diode impedance via optimisation of the doping profile and the geometry. This is performed by applying a large signal drift-diffusion model [16].

Silicon millimetre-wave ICs 381

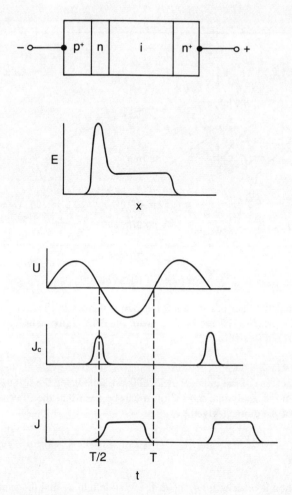

Fig. 10.10 Doping profile, electric field distribution, voltage, injection current and induced current of an idealised Read-type IMPATT diode

Fig. 10.11 shows the impedance of flat profile double drift IMPATT diodes in the impedance plane. Lines of constant input power density for the design frequencies 70 and 80 GHz are displayed with the drift region length as a parameter. The drift regions are assumed to be symmetrical. The optimum drift region length for 70 GHz operation is 300 nm, where the optimum is more pronounced at the lower input power density of 100 kW/cm^2. At this operation point a real part of -1×10^{-5} Ωcm^2 and an imaginary part of -12.5 Ωcm^2 is expected. Due to the large bandwidth negative resistance of IMPATT diodes, the optimum drift region length is almost unaffected by a shift to an operating frequency of 80 GHz.

382 Silicon millimetre-wave ICs

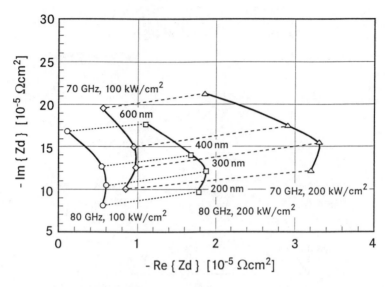

Fig. 10.11 Calculated lines of constant input power density in the impedance plane for two different frequencies (70 and 80 GHz) with the drift region length as a parameter. Double drift *flat profile* structure

A further increase of the impedance level is possible using the original Read diode approach: From the analytical modelling it can be seen that the internal conversion efficiency, η_i, of the diode is given by

$$\eta_i \propto \frac{l_d}{l_{dl}} = 1 - \frac{l_a}{l_{dl}} \tag{10.15}$$

with the depletion layer length l_{dl}, the drift region length l_d and the avalanche region length l_a [13, 14]. A confinement of the avalanche region length can be achieved by the introduction of doping spikes on both sides of the p-n junction in a double drift diode. In a first order design the lower limit of the intrinsic avalanche region length in this double low high low (DLHL) diode is given by the breakdown condition

$$\alpha_{eff} \cdot l_a \approx 0.85 \tag{10.16}$$

where the effective ionisation rate is given by [17]

$$\alpha_{eff} = \frac{\frac{\alpha_p}{\alpha_n} - 1}{\ln\left\{\frac{\alpha_p}{\alpha_n}\right\}} \cdot \alpha_n \tag{10.17}$$

with the ionisation rate α_n for electrons and α_p for holes. The factor 0.85 is empirically determined and attributed to the fact that carriers drift already in the avalanche region.

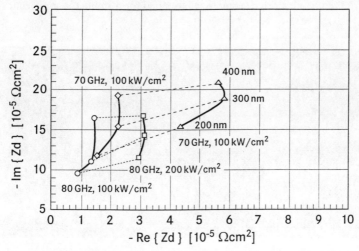

Fig. 10.12 Calculated lines of constant input power density in the impedance plane for two different frequencies (70 and 80 GHz) with the drift region length as a parameter. Double drift *low-high-low* structure

Using the ionisation rates from Fig. 10.9 an effective ionisation rate of 65000 cm^{-1} results at an electric field of 7×10^5 V/cm, which leads to a minimum required avalanche region length of 130 nm. The sheet carrier density in the doping spikes is designed to guarantee a sufficient electric field in the drift region to reach the saturation velocity.

Fig. 10.12 shows the impedance of a DLHL diode again in the impedance plane, with lines of constant input power density with the drift region length as a parameter, for 70 and 80 GHz. We can see that the influence of the drift region length on the real part is even weaker than in flat profile structures. However, the absolute value of the real part is more than a factor of two larger in DLHL diodes: 2.2×10^{-5} Ωcm^2 at 70 GHz with a drift region length of 300 nm at an input power density of 100 kW/cm^2 !

Double low high low IMPATT diodes provide a negative resistance which is high enough to be matched in planar circuits. In a fabrication process of monolithic IMPATT diode circuits a self stopping etchant, self-aligned contacts, silicon-nitride passivation and air-bridge technology are used [18]. First, the high resistivity silicon substrates are thermally oxidised and highly doped p$^+$-layers are formed by B-diffusion. A surface concentration greater than 10^{20} cm^{-3} is achieved. Next, the active layers of IMPATT profiles are grown by Si-MBE . The growth temperature is 550°C. Then the top contact of the IMPATT diode is defined and the diode is mesa etched in a self stopping etchant (aqueous KOH solution). A slight undercut is performed. This undercut enables self-aligned technology to be used for defining the lower contact and reduces series resistance. To reduce surface leakage currents the mesa edges are passivated by a 150 nm thick, low temperature (300°C) plasma-enhanced deposited Si_3N_4 film. Finally, air-bridge technology is applied for forming the top contact of the diode.

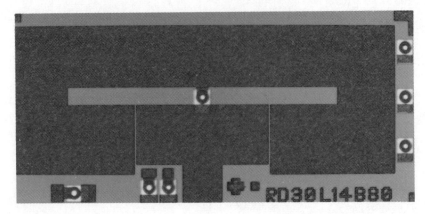

Fig. 10.13 Photograph of an active slot line antenna with monolithically integrated IMPATT diode

Monolithic single chip transmitter circuits have been realised [18], and a photograph of such a transmitter chip is shown in Fig. 10.13. A coplanar slot resonator also serves as the antenna and an IMPATT diode is monolithically integrated in the centre of the slot. Due to the high permittivity of silicon, more radiation is emitted through the substrate than directly into the air. The CW power is radiated from the backside of the chip into a W-band waveguide measurement system. No additional heat sink is applied. The maximum radiated power is 4.4 mW at 89 GHz with a S/N of -82 dBc at 100 kHz off carrier. Due to their higher impedance levels, the DLHL structures have lower oscillation threshold currents than the double drift flat profile devices. This operation mode is also extracted by mounting the chip upside down on a metal carrier with an integrated cavity to improve frequency and temperature stability [19]. In addition, the slot transmitters can be operated front-side up with a grounded substrate. The maximum power radiated directly into the air is 1.3 mW at 75 GHz.

10.3.3 Heterojunction transit-time diodes

The main advantage of the IMPATT diode over other transit-time diodes is the inductive phase shift of $\pi/2$ between injection current and voltage caused by the impact ionisation mechanism. Together with the additional transit-time delay of $\pi/2$ a total phase shift of π results between the voltage and the external current: the

IMPATT diode can provide the highest conversion efficiency of any millimetre-wave two-terminal device due to these favourable phase relationships. However, from eqn. 10.15 it is evident that the efficiency could be increased further by a reduction of the generation region length and of the voltage across the generation region, respectively. The use of different semiconductor materials for the generation and the transit-time regions is an efficient way to reduce the losses in the generation region [20]. In this concept, the avalanche region of an IMPATT diode is made from a low band-gap material where the ionisation rate is high and the drift region consists of a larger band-gap material where the ionisation rate is low. With this combination the avalanche generation is spatially limited and the efficiency can be increased.

Fig. 10.14 shows the application of this concept to silicon based IMPATT diodes. The injection region is made from a $Si_{1-y}Ge_y$ alloy (Fig. 10.14 (a)) or can be configured as a superlattice. In a Si/SiGe superlattice with a suitable period duration, the Ge percentage y in the $Si_{1-y}Ge_y$ layers of the superlattice can be selected to be greater, if the layer thickness is the same as for an individual SiGe layer, than in a corresponding individual $Si_{1-y}Ge_y$ layer. By increasing the Ge percentage y of the $Si_{1-y}Ge_y$ layers and optimally configuring the period duration of the Si/SiGe superlattice, it is possible to produce a miniband in the energy band scheme of the Si/SiGe superlattice whose band gap towards the conduction band Ec is smaller than the band gap between the valence and the conduction bands of an individual SiGe layer having the same layer thickness. The critical layer thickness of the $Si_{1-y}Ge_y$ superlattice is greater, with the identical Ge percentage y, than that of an individual $Si_{1-y}Ge_y$ layer.

Experimental work on double drift IMPATT diodes with a 10 nm thick $Si_{0.6}Ge_{0.4}$ alloy injection region has shown that interband tunnelling has to be considered as an additional injection mechanism [21]. 25 mW of output power has been achieved at 103 GHz. The first samples of superlattice avalanche transit-time diodes (SLATT), with an injection superlattice which has a Ge content of 40 % and 5 periods each 4 nm thick, showed oscillations in a pulsed mode of operation at 100 GHz.

10.3.4 A SiGe Gunn device ?

A negative differential resistance in the I-V characteristic may be observed in some compound semiconductor materials (GaAs, InP). This effect may be used for the generation of microwave oscillations [22]. The reason for this negative resistance is the scattering of electrons from the conduction band minima in the centre of the Brillouin zone at <000> into a satellite valley in the <100> direction [23]. Of importance for the application of this transferred-electron mechanism is that the energy difference between the two conduction band minima is larger than the thermal energy and smaller than the bandgap in order to avoid impact ionisation. If the electrons have a large effective mass in the satellite valley compared to that in the lowest valley, a velocity-field characteristic with a negative differential mobility section results. These conditions may be fulfilled by an artificial semiconductor formed by an ultra thin SiGe superlattice.

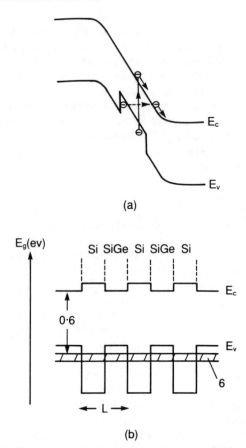

Fig. 10.14 (a) Energy band diagram and germanium content of a Si/SiGe heterostructure IMPATT diode in reverse direction; (b) Energy band diagram of a superlattice injection region

The idea of a silicon-based Gunn device relies on the zone folding effect in ultra thin SiGe superlattices: the electronic band structure of the superlattice is modified compared with a pure silicon crystal. The properties of this artificial semiconductor can be adjusted by a proper selection of the number of monolayers of silicon (n) and germanium (m) and the period length, $L=n+m$. The strain situation in a Si_nGe_m superlattice grown on a $Si_{1-y}Ge_y$ buffer leads to a twofold degenerated minimum perpendicular to the layers and a fourfold degenerated minimum parallel to the layers [10]. With a suitable superlattice period the twofold degenerated minimum is folded into the centre of the Brillouin zone, with the fourfold degenerated minimum being unaffected. Calculations of the energy dispersions in the two conduction band minima show that the effective masses are indeed significantly different ($m^*_1=0.194\ m_0$; $m^*_2=1.004\ m_0$ [24]).

These preconditions may be met by a structure as proposed in Fig. 10.15. On a high resistivity silicon substrate a slightly doped buffer layer is grown with a thickness of 50 nm. The superlattice is then grown with approximately 100 periods; each period with 10 monolayers silicon and 10 monolayers germanium. With this period length the lowest conduction band minimum will be folded into the centre of the Brillouin zone. The superlattice is n-doped to obtain a population of the lowest minimum. Highly doped contacts are diffused into the superlattice to guarantee low contact resistances and an electric field component parallel to the superlattice layers.

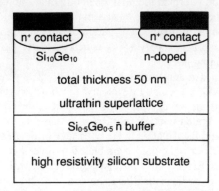

Fig. 10.15 Layer sequence of the proposed Si/SiGe Gunn device

10.4 PIN diodes

The 'ON'-resistance of diodes with an intrinsic (I) region between a p-doped (P) and an n-doped (N) contact layer may be described by [25]

$$R_{on} = \frac{w_i^2}{(\mu_n + \mu_p) I_f \tau_b} \tag{10.18}$$

which shows the favourable influence of a large effective carrier lifetime τ_b in the intrinsic region. In bulk silicon we expect mobilities μ_n=1500 cm²/Vs for electrons and μ_p=450 cm²/Vs for holes. With an effective carrier lifetime of 2×10^{-4} s the 'ON'-resistance of silicon PIN diodes is below 1 Ω at a forward current of 20 mA. If biased in reverse direction the intrinsic region is fully depleted and a constant capacitance of

$$C = \varepsilon \frac{A}{w_i} \tag{10.19}$$

is obtained. We expect a capacitance of 62.5 fF for an effective i-region width of 0.4 µm and an area A=5 µm × 50 µm.

Planar PIN diodes may be realised without any epitaxial processes. The high resistivity silicon substrate provides the i-region, the p⁺ and n⁺ contact layers are formed by implantation. The i-region width and the capacitance of the diode can be adjusted by changing the separation of the implantations. Monolithically integrated single-pole double-throw (SPDT) switches have been reported [26]. For operation at E-band frequencies, PIN diodes with a nominal 5 µm i-region width are used. The SPDT switch comprises two symmetrical cells for the ON and OFF arm connected to a common input port. Each arm consists of two diodes in a parallel or shunt configuration, with one side connected to the ground. A radial stub is used as an RF short circuit and the DC is shorted with a via-hole and a small transmission line, $\lambda/4$ long. The via-hole is transformed to an RF open circuit at the input port of the radial stub. In one arm the anodes, and in the other arm the cathodes, of the two shunt diodes are connected to ground. With this anti-parallel combination of the diodes in the different arms, only one bias supply $+/-V_d$ is necessary for biasing both arms. The bias is injected at the input line.

Millimetre-wave measurements of the SPDT PIN switch were made in the E-band from 60 to 90 GHz with a specially developed four-port waveguide test fixture. A broadband transition from waveguide to microstrip is achieved with E-field probes. The test fixture is calibrated with 50 Ω microstrip transmission lines. The insertion loss for the ON-state and the isolation in the OFF-state of the SPDT-switch are shown in Fig. 10.16. In the ON-arm the diodes are biased in reverse direction by $V_d=-1.2$ V and simultaneously in the OFF-arm in forward direction with Id = 20 mA. In this frequency range of 67 to 80 GHz the SPDT shows an insertion loss of 2.0 to 2.5 dB in the ON-state. In the OFF-state the isolation was measured to be better than 25 dB.

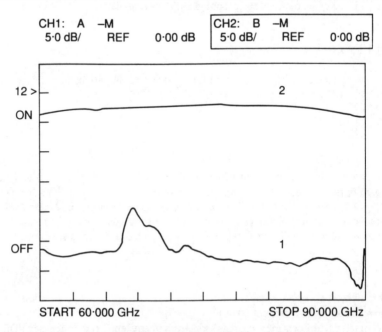

Fig. 10.16 Insertion loss and isolation of the PIN-switch against frequency

The switching time is another important measure of the diode performance. When a reverse voltage is applied, the carriers take some time to leave the intrinsic layer. This time constant is determined by the semiconductor bulk lifetime τ_b and a time constant given by the diffusion of the carriers to the surface and their decay at the surface with lifetime τ_s. The resultant carrier lifetime is then [27]:

$$\frac{1}{\tau} = \frac{1}{\tau_b} + \frac{1}{\tau_s} \tag{10.20}$$

The time T_s denotes the time at which the intrinsic layer is totally depleted of mobile charges and, assuming rectangular current pulses, is given by:

$$T_s = \tau \ln\left\{1 + \frac{I_F}{I_R}\right\} \tag{10.21}$$

where I_F and I_R are the forward and reverse currents. It can be seen that, besides the resultant carrier lifetime, the measured switching time depends on the ratio of the forward to reverse current. This requires an optimisation of the driver circuit for high speed switches. Additionally, the long carrier lifetime in silicon may be reduced by gold implantation, for example, with some increase of the ON-resistance.

10.5 Schottky diodes

Schottky barrier diodes are efficient detector elements for millimetre waves. In order to use them in a receiver or mixer chip they may be realised in coplanar form [28]. For proper working of these diodes in the millimetre-wave region, the series resistance R_s and junction capacitance C_j should be as small as possible. To achieve the best mixer characteristics the cut-off frequency

$$f_{c0} = \frac{1}{2\pi R_s C_j} \tag{10.22}$$

is aimed to be 10 times the detecting frequency. The junction capacitance is related to the diode area, doping of the epitaxial layer, and diffusion voltage (barrier height). The series resistance is also related to the diode area and the doping of the epitaxial layer, but in the opposite sense. Therefore, an optimisation of geometrical parameters (small area and minimal parasitic components) and doping parameters has to be performed to achieve a high cut-off frequency [29].

For the fabrication of the monolithically integrated planar Schottky barrier diodes, high resistivity silicon substrates ($\rho > 4000\Omega$cm) are used. These substrates are thermally oxidised and windows are etched in the oxide. Highly doped n$^+$ buried layers are formed by ion implantation and a subsequent diffusion process. Doping concentrations up to 10^{20} cm^{-3} and a sheet resistance of 4.3 Ω/square have been achieved. A lightly doped n-epitaxial layer is grown by silicon molecular beam epitaxy (Si-MBE). The doping concentration is 1×10^{16} cm^{-3}, and the layer thickness is 80 nm. Next, the Schottky anode contact is formed by photolithographic patterning and a lift-off process with 50 nm Ti, 50 nm Pt and 300 nm Au. Then, the epitaxial layer is plasma etched using the Schottky contact as an etch mask. The ohmic contact is formed using a second lift-off process with Ti, Pt, and Au metallisation. The diodes are then passivated with a low temperature silicon nitride. Finally, the gold pads are electroplated to a thickness of several μm [30]. DC measurements show an ideality factor of less than 1.08. The barrier height as determined from the saturation current is 0.5 V. C-V measurements on mesa-diodes yield a depletion width of 70 nm. This corresponds to a junction capacitance at zero bias of 1.5 fF/μm².

Diodes with a Schottky anode area of 12 μm² were mounted upside down in a single ended mixer and tested at 94 GHz. With an LO power of 1 mW a conversion loss of 10 dB is measured, and with an LO power of 10 mW a conversion loss of 6.5 dB is achieved. No bias is applied to the diodes (zero bias operation). The integration of these diodes in slot line detectors [30] or their combination with microstrip antennas on high resistivity silicon yields complete single chip receivers [8].

10.6 SiGe heterojunction bipolar transistor

If we consider a conventional npn bipolar junction transistor (BJT) it is well known [31] that the collector current is given by:

$$J_C = \frac{qD_n n_i^2}{Q_{Beff}}\left(e^{V_{BE}/V_T} - 1\right) = J_n\left(e^{V_{BE}/V_T} - 1\right) \tag{10.23}$$

where Q_{Beff} is the effective base Gummel number, D_n is the diffusion constant of electrons in the base, the intrinsic carrier concentration of silicon $n_i = 1.5\times10^{10}$ cm^{-3}, and the base-emitter voltage is V_{BE}. It is assumed that transport of minority carriers across the base is the dominant mechanism and that recombination can be neglected. In a similar manner, the expression for the base current is

$$J_B = \frac{qD_p n_i^2}{Q_{Eeff}}\left(e^{V_{BE}/V_T} - 1\right) \tag{10.24}$$

where D_p is the diffusion constant of holes and Q_{Eeff} is the effective emitter Gummel number. The physical mechanism described is the injection of holes into the emitter. The current gain β can now be calculated by

$$\beta = \frac{J_c}{J_B} = \frac{D_n Q_{Eeff}}{D_p Q_{Beff}} \approx \frac{D_n N_E w_E}{D_p N_B w_B} \tag{10.25}$$

As the diffusion constants and the emitter and base width w_E and w_B are in the same order of magnitude, the base doping N_B has to be lower than the emitter doping N_E.

To evaluate the RF potential of the BJT we find the frequency at which the current gain approaches zero: the current gain cut-off frequency, f_T, which is inversely proportional to the emitter-collector transit time is given by

$$f_T = \frac{1}{2\pi \tau_{EC}} \tag{10.26}$$

In a rough approximation, τ_{EC} is governed by the base transit time

$$\tau_B = \frac{w_B^2}{2D_n} + \frac{w_B}{v_s} \tag{10.27}$$

and an additional time constant which is estimated to be 0.8 ps [32]. Eqn. 10.27 shows the contributions originating from carrier transport by drift and diffusion. Clearly, in order to obtain a high f_T we have to choose a very thin base width w_B. However, this has a severe effect on the base sheet resistance, which is given by

$$R_b = \frac{1}{qN_B \mu w_B} \tag{10.28}$$

where μ is the mobility. The base sheet resistance is inversely proportional to the base width w_B. Additionally, the maximum base doping is limited in order to maintain a sufficient current gain. If we also look at the expression for the maximum oscillation frequency, f_{max}, (the key figure of merit for analogue RF applications) which is given by

$$f_{max} = \sqrt{\frac{f_T}{8\pi R_B C_{BC}}} \tag{10.29}$$

where C_{BC} is the base-collector capacitance, we now recognise the limitations of the RF potential of the conventional bipolar transistor: the thin base required for high f_T results in a large R_b, such that the f_{max} is reduced.

The way out is the silicon/germanium heterojunction bipolar transistor (SiGe HBT). If we introduce a silicon/germanium alloy in the base, the band-gap decreases and the number of injected electrons increases according to

$$J_C = \frac{qDn_{i(SiGe)}^2}{Q_{Beff}} \left(e^{V_{BE}/V_T} - 1 \right) \tag{10.30}$$

which describes the collector current in a small base band-gap transistor. However, in contrast to eqn. 10.23, the intrinsic carrier concentration of the alloy $n_{i(SiGe)}^2$ may be approximated by

$$n_{i(SiGe)}^2 = n_{i(Si)}^2 \cdot e^{\Delta E_g/(kT)} \tag{10.31}$$

and this results in an expression for the current gain of the HBT:

$$\beta = \frac{J_C}{J_B} \approx \frac{D_n N_E w_E}{D_p N_B w_B} e^{\Delta E_g / kT} \tag{10.32}$$

The exponential dependence of the intrinsic SiGe-carrier concentration on the bandgap difference ΔE_g between Si and Ge leads already at a Ge content of only 20 % to a value of the exponential expression of more than 1000 ! Therefore, the base doping is no longer limited to values below the emitter doping but can now be increased beyond the latter! Theoretically, the increase of the base sheet resistance at very thin base widths can be compensated for by an increased base doping.

10.6.1 Design considerations

In order to achieve a high f_{max} it is not necessary to reduce w_B below certain limits [33]. The dependence of the base sheet resistance on the base width and of the transit time on the base width lead to an optimum base width and current gain cut-off frequency, respectively. The advantage of SiGe HBTs over standard bipolar transistors results from the possibility of doping inversion of the emitter and the base; i.e. the base can be more highly doped than the emitter. Hence, low base sheet resistances are also attainable with bases as thin as 10 nm. For example, typical high frequency BJTs with base widths of less than 100 nm reach base sheet resistances as high as 10kΩ/square, whereas SiGe HBTs exhibit sheet resistances of approximately 1kΩ/square for a 20 nm wide base. The base width is the essential parameter for obtaining high f_T values. In reality, the trade-off between low base resistance and high cut-off frequency limits the f_{max} values: analytical calculations can predict f_T and f_{max} values and show that a maximum exists for f_{max} as a function of f_T, as shown in Fig. 10.17. This is due to the increase of f_T with decreasing base width and the simultaneously rising base sheet resistance as discussed above.

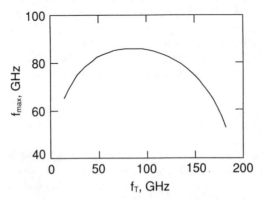

Fig. 10.17 Maximum oscillation frequency as a function of the transition frequency for a Si/SiGe HBT

The growth of the complete transistor structure by MBE meets the demand of precise control of thickness, doping and Ge content and avoids high temperature processes. For the fabrication of the device [32] a PtAu emitter metal is defined by lift off and acts as a mask for the following selective wet chemical etch which stops at the SiGe layer. The base metallisation is self-aligned with respect to the emitter because of the controlled over-etch. A dry etch gives access to the buried layer. After the collector metallisation, a second dry etch forms deep trenches that separate the contact pads from the active area leaving air bridges.

10.6.2 SiGe oscillator circuit on silicon substrate

A suitable topology for a 24 GHz DRO design for hybrid or monolithic application on a high-resistivity Si substrate is a state of the art series feedback configuration using the HBT device in common emitter configuration [34, 35]. A dielectric resonator is placed at the base of the transistor and the output port is at the collector. The microstrip circuit is designed using small signal S-parameters of the HBT, measured up to 40 GHz by means of an HP8510 network analyser, and using linear CAD software to characterise critical microstrip discontinuities such as T-junctions and radial stubs. The dielectric resonator coupled to the microstrip is modelled as a lossy parallel resonant circuit. By optimising the distance between resonator coupling locus and transistor base, and the length of the emitter feedback line, the magnitude of the reflection coefficient at the output port (collector) is maximised. The output port is designed to match the collector impedance to a 50 Ω load with respect to the small signal oscillation condition. The hybrid oscillator, fabricated on a 150 μm thick substrate (chip size 6 mm × 6 mm), is shown in Fig. 10.18. HBT chips (0.3 mm × 0.5 mm) are inserted into a hole made in the substrate and connected by gold leads to the microstrip lines. The 50 Ω termination at the transistor base applied to quench spurious oscillations is realised by a chip resistor attached to the substrate. Bias networks consist of quarter wavelength 75 Ω lines and 60 degree radial stubs. A barium-titanate dielectric resonator puck with a dielectric constant of 36 is used (diameter 2.7 mm, height 1.2 mm). At 23.2 GHz an RF power of +7 dBm with a phase noise of -92 dBc/Hz (100 kHz off carrier) is measured with an unpassivated HBT device.

10.7 Remarks and conclusions

A simplified representation of the Si/SiGe SIMMWIC technology is shown in Fig. 10.19. For the fabrication of detectors, transmitters, oscillators and switches nearly identical technology steps are used. In all devices buried layers are used. The further fabrication steps depend on the device type. For PIN diodes only n^+ and p^+ buried layers are necessary, the intrinsic i-zone is provided by the high resistivity substrate. No epitaxial process is used. Therefore, only an additional metallisation step is needed for the fabrication of PIN diodes.

394 *Silicon millimetre-wave ICs*

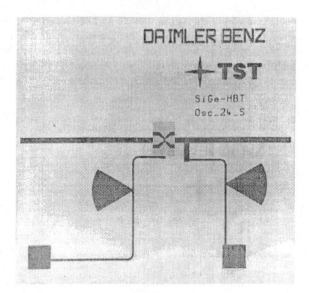

Fig. 10.18 Layout of 24 GHz HBT DRO

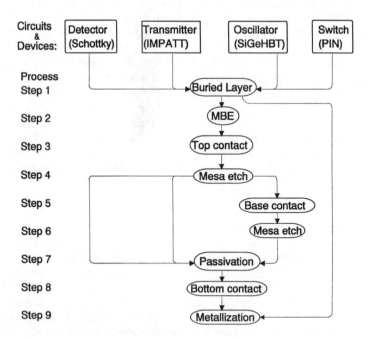

Fig. 10.19 Unified SIMMWIC technology

Different active device layers are grown by silicon molecular beam epitaxy (Si-MBE) to realise monolithic integrated Schottky barrier diode detectors, IMPATT diode oscillators and SiGe HBT oscillators. Usually the top contacts are then formed and mesa etches are performed. In the case of HBT devices, a base contact and a second mesa etch is needed. Next, the devices are passivated, the bottom contact is formed, and the final metallisation layer is patterned. Most of these steps are already compatible with standard silicon technology, and work on an integrated process is going on [36].

Using SIMMWICs with IMPATT diodes operated in a self-oscillating mixer mode [37], compact and cost effective solutions for short range sensor applications can be realised [38]. If the distance to be covered exceeds the coherence length of the free-running two-terminal oscillators some kind of frequency stabilisation has to be provided. Injection locking, subharmonic or fundamental, using a stable HBT oscillator and a multiplier is one possibility [39]. For more demanding requirements concerning the frequency stability in communication applications, a concept of analogue frequency division with a PLL is proposed [40]. If an auxiliary oscillator is necessary it can again be realised with a SiGe HBT. This device causes a kind of revolution in high frequency technology. Using two-terminal based SIMMWICs, i.e. with IMPATT diodes, Schottky and PIN diodes, the scope of applications for these chips is limited. With the three-terminal HBT the severe problem of frequency stability and regulation can now be solved. As the technology advances, maximum oscillation frequencies of more than 150 GHz can be expected, and a lot of interesting civil applications can be covered using Si/SiGe MMICs.

10.8 References

1. BUECHLER, J., KASPER, E., RUSSER, P., and STROHM, K. M.: 'Silicon high-resistivity-substrate millimeter-wave technology', *IEEE Trans. Electron Devices*, ED-33, no. 12, 1986, pp. 2047-2052
2. WIESBECK, J.: 'Berechnung der dämpfung ungeschirmter streifenleitungen', Wiss. Ber. AEG-TELEFUNKEN 45(4), 1972, pp. 162-166
3. HAMMERSTAD, E., and JENSEN, O.: 'Accurate models for microstrip computer-aided design', IEEE Int. Microwave Symp. Dig., 1980, pp. 1611-1612
4. HYLTIN, T.M.: 'Microstrip transmission on semiconductor dielectrics', *IEEE Trans. Microwave Theory Tech.*, MTT-13, 1965, pp. 777-781
5. ROSEN, A., CAULTON, M., STABILE, P., GOMBAR, A. M., JANTON, W. M., WU, C. P., CORBOY, J. F., and MAGEE, C. W.: 'Silicon as a millimeter-wave monolithically integrated substrate - a new look', RCA Rev., Vol. 42, 1981, pp. 633-660
6. STABILE, P. J., and ROSEN, A.: 'A silicon technology for millimeter-wave monolithic circuits', RCA Rev., Vol. 45, 1984, pp. 587-605
7. BÜCHLER, J.: 'Integrierte millimeterwellenschaltungen auf silizium', Phd thesis, Technische Universität München, 1989

8. STROHM, K. M., LUY, J.-F., KASPER, E., BUECHLER, J., and RUSSER, P.: 'Silicon technology for monolithic millimeter-wave integrated circuits', *Mikrowellen & HF Magazin*, Vol. 14, no. 8, 1988, pp. 750-760
9. BATTERSHALL, B.W., and EMMONS, S. D.: 'Optimization of diode structures for monolithic integrated microwave circuits', *IEEE Journal of Solid State Circuits*, SC-3, 1968, pp. 107-112
10. SCHÄFFLER, F.: 'Strained Si/SiGe heterostructures for device applications', *Solid State Electronics*, Vol. 37, 1994, pp. 765-771
11. Van der MERWE, J.H.: 'Structure of epitaxial crystal interfaces', *Surface science*, Vol. 31, 1972, p. 198
12. PEOPLE, R.: 'Physics and applications of Ge_xSi_{1-x}/Si strained-layer heterostructures', *IEEE Journal of Quantum Electronics*, Vol. 22, no. 9, 1986, pp. 1696-1710
13. READ, W. T.: 'A proposed high-frequency negative resistance diode', *Bell Syst. Tech. Journal*, Vol. 37, 1958, pp. 401 - 446
14. LUY, J.-F.: 'Transit-time devices' in *Silicon Based Millimeter Wave Devices* ed. by J.-F.Luy, P.Russer, Springer-Verlag, 1994
15. LEISTNER, D.: 'Herstellung und untersuchung von lawinenlaufzeitdioden und oszillatoren für V- und W-band-frequenzen', Phd thesis, Techn. Universität München, 1988
16. DALLE, C., and ROLLAND, P. A.: 'Read versus flat doping profile structures for the realization of reliable high-power, high-efficiency 94 GHz IMPATT sources', *IEEE Trans. Microwave Theory Tech.*, MTT-38, 1990, pp. 366-372
17. OGAWA, T.: 'Avalanche breakdown and multiplication in silicon pin junctions', *Japanese Journal of Appl. Phys.*, Vol. 4, 1965, pp. 473-484
18. STROHM, K. M., BUECHLER, J., LUY, J.-F., and SCHÄFFLER, F.: 'A silicon technology for active high frequency circuits', *Microelectronic Engineering* Vol. 19, 1992, pp. 717-720
19. PRESTING, H., BUECHLER, J., KUISL, M., STROHM, K. M., and LUY, J.-F.: 'Silicon monolithic mm-wave integrated circuit (SIMMWIC) devices mounted up-side-down on a copper heat sink integral with cavity resonator', *IEEE Trans. Microwave Theory Tech.*, Special Issue on Interconnections and Packaging, Sept. 1994
20. KUVAS, R. L., IMMORLICA, A. A., LUDINGSTON, B. W., and SZALKOWSKI, F. S.: 'Heterojunction IMPATT diodes: theoretical performance and material development studies', in Proc. 6th Biennial Cornell Elect. Eng. Conf., 1977, pp. 256-267
21. LUY, J.-F., JORKE, H., KIBBEL, H., and CASEL, A.: 'Si/SiGe MITATT diode', *Electron. Lett.*, Vol. 24, 1988, pp. 1386-1387
22. GUNN, J. B.: 'Microwave oscillation of current in III-V semiconductors', *Solid State Communic.*, Vol. 1, 1963, p. 88
23. BOSCH, B. G., and ENGELMANN, R. W.:*Gunn-effect electronics*, Electronic Engineering Series, Pitman Publishing, London, 1975
24. PRESTING, H., and LUY, J.-F.: 'SiGe Gunn device', German patent application no. P 40 19 247.4, 1990

25. HOWELL, C. M.: 'Comparison of gallium arsenide and silicon PIN diodes for high speed microwave switch circuits', MIOP Conference Proceedings, 1989
26. KLAASSEN, A., and SASSE, E.: 'E-band monolithic Si PIN diode switches', International Workshop of the German IEEE Joint MTT/AP Chapter on Silicon Based High Frequency Devices and Circuits, Günzburg, 1994
27. McDADE, J. C., and SCHIAVONE, F.: 'Switching time performance of microwave PIN diodes', *Microwave Journal*, 1974, pp. 65-68
28. STROHM, K. M., LUY, J.-F., BÜCHLER, J., SCHÄFFLER, F., and SCHAUB, A.: 'Planar 100 GHz silicon detector circuits', *Microelectronic Engineering*, Vol. 15, 1991, pp. 285-288
29. DIEUDONNÉ, J. M., ADELSECK, B., SCHMEGNER, K.-E., RITTMEYER, R., and COLQUHOUN, A.: 'Technology related design of monolithic millimeter wave Schottky diode mixers', *IEEE Trans. Microwave Theory Tech.*, MTT-40, 1992, pp. 1466-1474
30. STROHM, K. M., BÜCHLER, J., and LUY, J.-F.: 'A monolithic millimeter wave integrated silicon slot line detector', Asia Pasific Microwave Conf., 1993
31. SZE, S. M.: *Physics of Semiconductor Devices*, J.Wiley, 1981
32. GRUHLE, A.: 'SiGe-HBTs' in *Silicon Based Millimeter Wave Devices* ed. by J.-F. Luy, P. Russer, Springer Verlag, 1994
33. SCHÜPPEN, A., GRUHLE, A., ERBEN, U., KIBBEL, H., and KÖNIG, U.: '90 GHz f_{max} SiGe-HBTs', Device Research Conference, Boulder, IIIA-2, 1994 *and* LUY, J.-F., STROHM, K. M., SASSE, E., SCHÜPPEN, A., BUECHLER, J., WOLLITZER, M., GRUHLE, A., SCHÄFFLER, F., GUETTICH, U., and KLAAßEN, A.: 'Si/SiGe MMICs', *IEEE Trans. Microwave Theory Tech.*, Dec. 1994
34. GÜTTICH, U., GRUHLE, A., and LUY, J.-F.: 'A Si-SiGe HBT dielectric resonator stabilized microstrip oscillator at X-band frequencies', *IEEE Microwave and Guided Wave Lett.*, 1992, pp. 281-283
35. GÜTTICH, U., GRUHLE, A., and LUY, J.-F.: 'Dielectrically stabilized oscillators for X- and K-band frequencies with Si/SiGe HBTs'; MIOP Conference Proceedings, 1993, Network, Hagenburg, pp. 146-150
36. DIETRICH, H., HEFNER, A., ARNDT, J., SCHÜPPEN, A., KIBBEL, H., and KÖNIG, U.: 'SiGe-HBTs based on a differential epitaxy for low power applications', International Workshop of the German IEEE Joint MTT/AP Chapter on Silicon Based High Frequency Devices and Circuits, Günzburg, 1994
37. CLAASSEN, M.: 'Self-oscillating mixers', in *Silicon Based Millimeter Wave Devices*, ed. by J.-F. Luy, P. Russer, Springer-Verlag, 1994
38. BUECHLER, J., LUY, J.-F., and STROHM, K. M.: 'V- and W-band MMIC sensors for mobile applications', MTT Workshop on Microwave Sensing, Ilmenau, 1993, pp. 34-41
39. WOLLITZER, M., BUECHLER, J., and BIEBL, E.: 'Subharmonic injection locking (of) slot oscillators', *Electron. Lett.*, Vol. 29, no. 22, 1993, pp. 1958-1959
40. NÜCHTER, P., and MENZEL, W.: 'A mm-wave frequency divider', IEEE Int. Microwave Symp. Dig., 1992, pp. 695-697

Chapter 11

Measurement techniques

S. Lucyszyn

11.1 Introduction

All electronic sub-systems are made up of devices and networks. In order to simulate the overall performance of a sub-system under development, all the components that make up the sub-system must be accurately modelled. To this end, precision measurement techniques must be employed at the component level. Not only do precision measurements enable a manufacturer to check whether devices are within their target specifications and to monitor variations in parameter tolerances, due to process variations, they also allow more accurate empirical models to be extracted from the measurements and help new numerical modelling techniques to be validated. Also, the operation and performance of some experimental devices can often only be understood from accurate measurements and subsequent modelling.

Devices and networks are traditionally characterised using z, y or h-parameters. In order to measure these parameters directly, ideal open and short circuit terminations are required. These impedances can be easily realised at low frequencies. However, at microwave frequencies such impedances can only be achieved over narrow bandwidths and can also result in conditionally stable circuits becoming unstable. Fortunately, scattering- (or S-) parameters can be determined at any frequency. To perform such measurements, the device under test (DUT) should be terminated with reference impedance 'matched loads'. This enables extremely wideband measurements to be made and also greatly reduces the risk of instability. S-parameter measurements also offer the following advantages:

(1) Any movement in a measurement reference plane along a transmission line will vary the phase angle only.
(2) For a linear device or network, voltage or current and measured power are related through the measurement reference impedance.
(3) For certain passive and reciprocal structures, ideal S-parameters can be deduced from spatial considerations, enabling the measurements of the structure to be checked.

By applying a known incident voltage wave to the DUT and then measuring the reflected and transmitted voltage waves, S-parameters can be calculated from the resulting voltage ratios. The equipment used to perform this operation is called a vector network analyser (VNA) [1]. The element values associated with a small-signal equivalent circuit model of the DUT can be determined, in principle, from the S-parameter measurements (and with DC measurements) using either direct calculations, iterative optimisation, intuitive tuning or a combination of these methods. This process is generally referred to as parameter extraction.

With a monolithic microwave integrated circuit (MMIC), either a test fixture or probe station is employed to secure the MMIC and to provide a stable means of electrically connecting the MMIC to the measurement system [2]. The use of test fixtures and probe stations at ambient room temperature is reviewed and their role at thermal and cryogenic temperatures is discussed in this chapter. Finally, with the increasing need for performing non-invasive (or non-contacting) measurements, experimental field probing technologies are introduced.

11.2 Test fixture measurements

Although probe stations result in much more accurate and reproducible measurements, test fixtures are still widely used. The principal reasons are that they are very much cheaper than probe stations and they offer a greater degree of flexibility, such as facilitating larger numbers of RF ports and enabling DC bias circuitry to be located next to the chip. Also, the heat dissipation required when testing monolithic high power amplifiers can be easily provided with test fixtures. In addition, test fixtures are ideally suited when RF measurements are required during temperature cycling and when cryogenic device characterisation is required [3-5].

An illustration of a basic two-port test fixture is shown in Fig. 11.1. Most test fixtures are, in principle, based on this generic design, typically consisting of four different components: a detachable chip carrier, a rigid metal housing, connector/launchers and bond wires. The MMIC is permanently attached to the chip carrier with either conductive epoxy glue or solder. The metal housing is employed to hold the chip carrier and the connector/launchers in place. The launcher is an extension of the coaxial connector's centre conductor which passes through the housing wall to make contact with the associated chip carrier's microstrip transmission line. Bond wires are used to connect the other end of the microstrip line to the MMIC under test. The parasitic element values associated with a test fixture are typically an order of magnitude greater than those of the MMIC under test.

Before any accurate measurements can be performed, all measurement systems must first be calibrated, in order to correct for the systematic errors resulting from the various reflection and transmission losses. A calibration kit is required to perform this calibration procedure. This 'cal kit' has a number of reference electrical length/impedance standards and a computer program.

For a two-port measurement system, the calibration standards must:

(1) Define the primary reference planes
(2) Remove any phase ambiguity, using an open circuit and/or short circuit reflection standard
(3) Define the reference impedance, using either delay lines, a matched load or an attenuator standard

Some of the various combinations of different reference standards that can be employed in a calibration procedure are listed in Table 11.1. The computer program, which is initially downloaded into the VNA's non-volatile memory, contains accurate empirical models for the standards and the algorithms required to implement the chosen calibration method. Therefore, the accuracy of subsequent measurements ultimately depends on how well all the standards remain characterised.

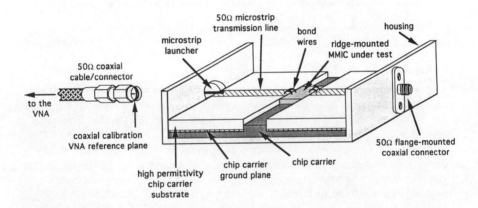

Fig. 11.1 Generic design of a two-port test fixture

The non-idealities of a measurement system are quantified using mathematical error correction models and represented by flow diagrams (also known as error adapters or boxes). The function of the calibration procedure is to solve for the error coefficients in these models by applying the uncalibrated measurements of the reference standards to a set of linear independent equations. The basic two-port calibration procedures have an 8-term error model (4-terms associated with each port) and require only 3 standards. A more accurate 12-term error model, as used in two-port coaxial calibration, takes cross-talk and the effects of impedance discrepancies at the RF switches in the VNA's test set into account. Once the calibration procedure has been performed, it can be verified by measuring separate verification standards.

Table 11.1 Common calibration methods

Method	Calibration Standard					
	Through		Reflect	Reference Impedance		
	L=0	L≠0	ρ1=ρ2	Line	Match	Atten.
TRL	O		O	O		
LRL		O	O	O		
TRM	O		O		O	
LRM		O	O		O	
TRA	O		O			O
LRA		O	O			O
TSD	O		ρ=-1	O		

11.2.1 Two tier calibration

One method of calibrating the measurement system is to split the process into two tiers [6]. Initially, a coaxial calibration is performed, where the VNA reference planes are located at the end of its cable connectors. Since test fixtures are far from ideal, a second process is required to shift the initial VNA reference planes to the MMIC under test. This second process is known as de-embedding or de-convolution [7]. To perform de-embedding it is necessary to accurately characterise the test fixture [8]. Another reason why you may need to characterise a test fixture is when multiple RF port MMICs are to be measured using a two-port VNA [9]. Here, power reflected from mismatched loads on the auxiliary ports of the MMIC can result in significant measurement errors. These errors will increase as the mismatch losses increase and/or the number of RF ports increases. As a result, MMICs that have more RF ports than the VNA require all the loads to be individually characterised and a further process of matrix renormalisation [9-13] in order to remove the effects of the mismatched loads on the auxiliary ports. Three methods which can, theoretically, be employed to characterise a test fixture are time-domain gating, in-fixture calibration, and equivalent circuit modelling.

11.2.1.1 Time-domain gating

A conventional VNA, such as the HP 8510 or Wiltron 360, can apply an inverse Fourier transform (IFT) to the frequency domain measurements of the MMIC embedded in its test fixture [14-20]. A unit impulse (or a unit step) excitation is generated mathematically from ideal frequency domain responses. The resulting reflection and transmission responses can then be analysed to provide information about the test fixture and MMIC under test.

A mathematical time-domain window, called a gate, can be applied to isolate a particular target feature, such as a connector/launcher or the bonded MMIC under test. The predicted frequency response of the discontinuity can then be displayed. The gate-start and gate-stop times, which define the -6dB time span of this filter, must be set at the baseline if low frequency distortion is to be minimised [14]. When the gate is switched on, all reflections outside the gate are ideally set to zero. In practice, however, a time filter having a non-brick-wall response is used for gating, otherwise a $\sin x/x$ weighting would be conveyed to the frequency domain.

Time-domain measurements using conventional VNAs suffer from a number of sources of errors [15, 19], some of which are listed below:

1. *Frequency-domain window errors*
    (a) **Resolution errors:** as a result of wide impulses it may be difficult to resolve the reflection responses of the MMIC and the connector/launchers when these discontinuities are too close to each other. The duration of the 'effective' impulse excitation is inversely proportional to the frequency span over which frequency domain data are measured. In addition, with the HP 8510, there is a choice of three frequency-domain window functions that can be applied prior to the IFT: *minimum*, *normal* and *maximum*. The *minimum* window has a rectangular function that produces a $\sin x/x$ impulse having the minimum duration but also the maximum sidelobe levels. The other two window functions reduce the sidelobe levels at the expense of a wider impulse.
    (b) **Dynamic range errors:** impulse sidebands limit the dynamic range of the time-domain measurements, since the sidebands of adjacent high level responses can hide low-level target responses. Therefore, a trade-off has to be made when choosing the frequency-domain window, between the desired resolution and dynamic range.

2. *Discontinuity errors*
    (a) **Masking:** if the target discontinuity is preceded by other discontinuities, that either reflect or absorb energy, these other discontinuities will remove some of the stimulus reaching the target discontinuity.
    (b) **Multi-reflection aliasing:** multiple reflections between discontinuities can cause aliasing errors if the MMIC under test is positioned mid-way between the two connector/launchers.

3. *Time-domain window errors*
    (a) **Truncation:** with wide reflections, a limited gate span may cut out some of the target response. With the HP 8510, there is a choice of four time-domain window functions (or gate shapes) that can be applied before the Fourier transform: *maximum*, *wide*, *normal* and *minimum*. The *minimum* window has the fastest roll-off and the highest sidelobe levels, while the *maximum* window has the slowest roll-off and the

404 *Measurement techniques*

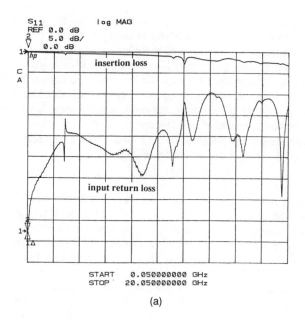

(a)

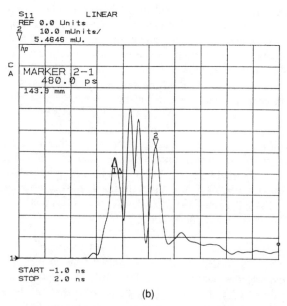

(b)

Fig. 11.2 Embedded 55 Ω MMIC through line: (a) Frequency-domain power responses; (b) Corresponding time-domain response of the input voltage reflection coefficient

minimum sidelobes levels. To minimise this source of error, the wider gate shapes are preferred.

(b) **In-gate attenuation:** for a fixed gate span, the duration of in-gate attenuation may be excessive with the wider gate shapes.

(c) **Out-of-gate attenuation:** the sidelobes of the time filter may permit earlier or subsequent reflections to have some influence on the frequency response of the target reflection. To minimise this source of error, the wider gate shapes are preferred.

(d) **Gate offset:** distortion can occur if the gate-centre is offset from the centre of the target response. As with truncation and out-of-gate attenuation, the wider gate shapes are preferred. However, with small test fixtures a trade-off has to be made when choosing the most appropriate gate shape.

(e) **Reflection/transmission switching:** if gating is performed on a voltage reflection coefficient response, the corresponding gating in a voltage transmission coefficient response will not be appropriate and the insertion loss frequency response will be meaningless.

As an example, a 2.9 mm long 55 Ω MMIC through line was assembled into a one inch square test fixture. The frequency-domain power responses are shown in Fig. 11.2 (a). The corresponding time-domain response of the input port's voltage reflection coefficient is shown in Fig. 11.2 (b). Here, the first and last peaks correspond to the reflections associated with the coaxial-to-microstrip transitions of the input and output ports, respectively. The two centre peaks correspond to the reflections associated with the microstrip-to-MMIC transitions. It will be apparent from Fig. 11.2 (b) that accurate de-embedding would not be possible using time gating. If de-embedding was attempted in the above example, the ripples in the frequency-domain responses would be smoothed out; however, this does not constitute accurate de-embedded measurements.

11.2.1.2 In-fixture calibration

In general, a quality test fixture is approximately one tenth of the price of a probe station. Suitably designed quality test fixtures can be accurately characterised using in-fixture calibration techniques. As with coaxial calibration, the most appropriate algorithms use a combination of through, reflection and delay line standards, with common methods being TRL, TSD and LRL. The main reason for employing these types of calibration is that only one discrete impedance standard is required, such as an open or short, which is relatively easy to implement. The matched load is avoided, which is advantageous since it is more difficult to fabricate non-planar 50 Ω loads to the same level of accuracy that can be achieved with low dispersion transmission lines. However, there are still significant disadvantages with in-fixture calibration:

(1) Multiple delay lines may be required for wideband calibration (any one line must introduce between about 20° and 160° of electrical delay to avoid phase ambiguity, limiting the bandwidth contribution of each line

to an 8:1 frequency range). This adds to the repeatability problem, since the launchers are continually being disturbed during calibration.

(2) A frequency invariant reference impedance must be taken from the characteristic impedance of the delay lines, and frequency dispersion in microstrip lines may not always be corrected for.

(3) The high level of accuracy is immediately lost with test fixtures that employ low cost components.

(4) The calibration substrates dictate and, therefore, restrict the location of the RF ports.

(5) For devices with more than two ports the calibration procedure must be significantly extended and all the results must be able to be stored and retrieved.

(6) The microstrip-to-MMIC transition is not taken into account.

11.2.1.3 Equivalent circuit modelling

Test fixtures built in-house tend to be simple in design, such as the type shown in Fig. 11.1, and extremely cheap to produce, costing only a small fraction of the price of a good quality commercial test fixture. Unfortunately, these non-ideal test fixtures suffer from unwanted resonances [21], poor grounding [22] and poor measurement repeatability. The problem of unwanted resonances can be clearly seen in the frequency domain responses of Fig. 11.2. Here, the resonances at 3 GHz and 12 GHz are attributed to the production grade coaxial connectors used in the test fixture. Because of poor repeatability, employing elaborate and expensive calibration techniques to characterise such fixtures would appear unjustified, because a significant measurement degradation is inherent. As an alternative, equivalent circuit models (ECMs) can provide a crude, but effective, means of de-embedding, also known as stripping. This process results in about the same level of degradation as would be found if in-fixture calibration was used with a non-ideal test fixture, but with minimal expense and greater flexibility. The ECMs can be easily incorporated into conventional frequency-domain simulation software packages. Also, ECMs based on the physical structure of the fixture have demonstrated a wide bandwidth performance and a large degree of flexibility. They can also be employed to simulate packaged MMICs. An example of an ECM for a test fixture similar to the one in Fig. 11.1 is shown in Fig. 11.3 [9]. This model has demonstrated a sufficient degree of accuracy from DC to 19 GHz for the popular Omni Spectra SMA connector/wedge-shaped launcher.

The exact nature of the ECM, the element values and the microstrip parameter data are extracted from through-line measurements of the test fixture. Both a direct microstrip through-line and an MMIC through-line should be used in order to provide more information for the parameter extraction process and to make it possible to model the microstrip-to-MMIC transition accurately. De-embedding can be carried out with most CAD packages by converting the ECM into a series of negative elements connected onto the ports of the measured data. Some CAD packages provide a 'negation' function which allows the ECM subcircuit to be directly stripped from the measured data. In either method the order of the node numbers is critical, and the de-embedding routine should be verified.

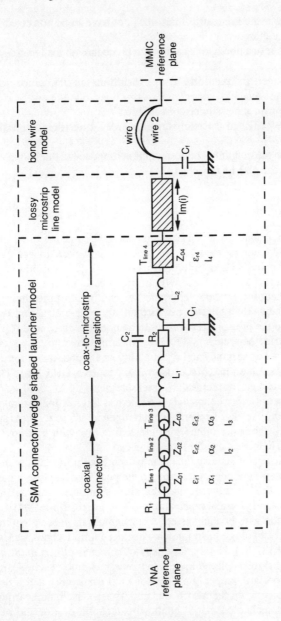

Fig. 11.3 Equivalent circuit model of a microstrip test fixture

In addition to the those already mentioned, de-embedding using equivalent circuit modelling has the following advantages:

(1) Dispersion in the microstrip lines does not have to be corrected for in the VNA's calibration.
(2) There is no restriction by the calibration procedure on the location of the RF ports.
(3) Systematic errors resulting from variations in the characteristic impedance of the chip carrier's microstrip lines, due to poor fabrication process control, can easily be corrected for.
(4) Bond wires [23] and the microstrip-to-MMIC transition can easily be modelled [24].
(5) Resonant mode coupling between circuit components, due to a package resonance, can also be modelled [21].

11.2.2 One tier calibration

Improved contact repeatability and prolonged contact lifetime are two considerations that favour the two tier process [6]. In practice, however, to achieve the best performance, in-fixture TRL or line-network-network [25] calibration is applied directly to a quality test fixture, without the need for the two tier coaxial calibration/de-embedding process. This one tier calibration procedure gives more accurate measurements than the two tier method, since de-embedding is inherently prone to errors, and the propagation of measurement errors is reduced [6]. Using this approach, the Cascade Microtech MTF26 test fixture, shown in Fig. 11.4, can give almost the same level of accuracy and repeatability as a probe station for unpackaged microwave integrated circuits (MICs) from 0.01 to 26.5 GHz [26]. The Wiltron 3680V test fixture can perform repeatable measurements up to 60 GHz. At the time of writing, a number of other companies produce test fixtures for accurate in-fixture calibration, including Hewlett Packard, Intercontinental Microwave, Argumens and Design Techniques. They are either split-block fixtures, with a removable centre section, or they use launchers attached to sliding carriages.

With the high levels of accuracy that can be achieved using quality test fixtures, the poor characterisation of bond wires, due to the poor repeatability of conventional manually operated wire bonding machines, becomes significant. Improvements in the modelling accuracy and physical repeatability of the microstrip-to-MMIC transition when using automatic wire-bonding assembly techniques have been reported [24]. In addition, flip-chip technology (also known as solder-bump technology) is currently under development [27-31]. Here, a tiny bead of solder is placed on all the MMIC bond pads and the MMIC is placed upside down directly onto the chip carrier. When heated to the appropriate temperature, the solder flows evenly and a near perfect connection is made between the MMIC pad and its associated chip carrier pad. The advantages of this technology over bond wire technology, for the purposes of measurements, are its superior contact repeatability and high characterisation accuracy of the microstrip-to-MMIC transition.

Fig. 11.4 Photograph of the MTF26 microstrip test fixture (Courtesy of Cascade Microtech Europe)

11.2.3 Test fixture design considerations

The following guidelines should be kept in mind when selecting, designing or using a test fixture:

(1) Split-block test fixtures [5, 32] are ideal for two-port in-fixture TRL calibration, since they can provide good repeatability. Here, a short standard is preferred, since significant energy may be radiated with an open standard.

(2) Side walls can form a resonant cavity. The size of the cavity should be made small enough so that the dominant resonant frequency is well above the maximum measurement frequency. Carefully placed tuning screws and/or multiple RF absorbing pads can be very effective at eliminating or suppressing unwanted modes [33].

(3) Poor grounding, due to excessively long ground paths and ground path discontinuities, must be avoided.

(4) Avoid thick chip carriers, wide transmission lines (sometimes used for off-chip RF decoupling) and discontinuities, in order to minimise the effects of surface wave propagation and transverse resonances at

410 *Measurement techniques*

millimetric frequencies. Transverse currents can be suppressed by introducing narrow longitudinal slits into the low impedance lines.
(5) Use substrates with high dielectric constant to avoid excessive radiation losses and to minimise unwanted RF coupling effects.
(6) New precision connector/launchers should be used whenever possible, and measurements should be performed below the rated cut-off frequency of the dominant TEM mode of propagation.
(7) Launchers should be separated from the DUT by at least 3 or 4 times the substrate thickness, so that any higher order evanescent modes, generated by the non-ideal coax-to-microstrip transition, are sufficiently attenuated at the DUT.

11.3 Probe station measurements

Until recently, the electrical performance of an MMIC was almost always measured using test fixtures. Nowadays, extremely accurate MMIC measurements can be achieved using probe stations. Such techniques were first suggested for use at microwave frequencies in 1980 [34], demonstrated experimentally in 1982 [35] and introduced commercially by Cascade Microtech in 1983. During the past decade there have been rapid developments in probe station measurement techniques. A photograph of the Cascade Microtech Summit 9000 analytical probe station with four positioners is shown in Fig. 11.5.

Fig. 11.5 Photograph of the Summit 9000 analytical probe station at Kings College

Today, the Cascade Microtech *W*-band wafer probing system can perform repeatable frequency-domain measurements at frequencies as high as 110 GHz [36], and GE Aerospace/GGB Industries have successfully tested discrete devices and circuits at 120 GHz [37]. In addition, a full two-port VNA has been implemented with active probes, enabling *S*-parameter measurements to be made from DC up to 120 GHz [38]. Here, high speed non-linear transmission-line (NLTL)-gated directional time-domain reflectometers (or directional samplers) were realised using GaAs MMIC technology [39]. When compared with test fixtures, probe station measurements have the following advantages:

(1) They are currently possible up to 120 GHz.
(2) They are more accurate and much more repeatable, since they introduce much smaller systematic errors.
(3) They have a simpler calibration procedure, which can be automated with on-wafer calibration and verification standards [12, 40].
(4) They enable the VNA measurement reference planes to be located directly at the probe tips.
(5) They provide a fast, non-destructive means of testing the MMIC, thus allowing chip selection prior to dicing and packaging.

Overall, the microwave probe station provides the most cost effective way of measuring MMICs when all costs are taken into account.

11.3.1 Passive microwave probe design

At frequencies greater than a few hundred megahertz, probe needles have severe parasitic impedances, due to the excessive series inductance of the long thin needles and shunt fringing capacitances. If the needles are replaced by ordinary coaxial probes which are sufficiently grounded, measurements up to a few gigahertz can be achieved. The upper frequency is ultimately limited by the poor coaxial-to-MMIC transition. A tapered coplanar waveguide (CPW) probe provides a smooth transition with low cross-talk. Cascade Microtech have developed tapered CPW probes that enable measurement to be made from DC to 65 GHz with a coaxial input, or between 75 GHz and 110 GHz with a waveguide input [36]. The maximum frequency limit for coaxial-input probes is imposed by the onset of higher order modes propagating in the conventional coaxial cables and connectors. For *W*-band operation, Hewlett Packard developed a coaxial cable and connector that has an outer screening conductor diameter of only 1 mm. An illustration of the use of this 1 mm coaxial technology to give state-of-the-art performance up to 110 GHz, with a Cascade Microtech Summit 10000 probe station, is shown in Fig. 11.6.

The tapered coplanar waveguide probe is usually made from an alumina substrate or an ultra-low-loss quartz substrate. The probe tips that make the electrical contacts consist of hard metal bumps which have been electroplated over small cushions of metal, allowing individual compliance for each contact. As the probes are over-travelled (in the vertical plane) the probe contacts wipe or 'skate' the probe pads (in the horizontal plane). One of the major limitations of tapered CPW probes is their short lifetime, since the substrate has limited compliance and the probe contacts can wear down quite quickly. As a result, the more the probe is used the more over-travel has to be applied to them. Eventually, either the probe substrate begins to crack or the probe tips fall apart. For this reason, GGB Industries developed the *Picoprobe*™. This coaxial probe is more compliant and can achieve operation between DC and 67 GHz with a coaxial input, and between 75 GHz and 120 GHz with a waveguide input [37]. From DC to 40 GHz, this probe has demonstrated an insertion loss of less than 1.0 dB and a return loss better than 18dB. However, one potential disadvantage of coaxial probes is that the isolation between probes may be limited when operating above V-band. For even better compliancy, durability, ruggedness and flexibility, Cascade Microtech have developed the *Air Coplanar*™ tipped coaxial probe [41]. From DC to 40 GHz, this probe has demonstrated an insertion loss of less than 1.0 dB, a return loss better than 10dB, and cross-talk better than -40 dB with probes separated by 100 μm in air.

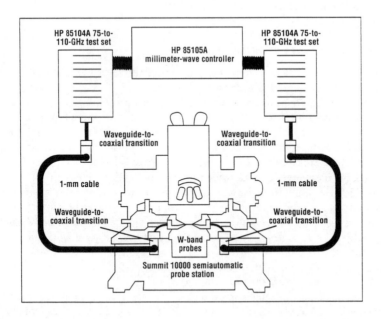

Fig. 11.6 Illustration of a 110 GHz on-wafer probing system (Courtesy of Cascade Microtech Europe)

When selecting the type of microwave probe required, it is necessary to supply the vendor with the following specifications:

(1) Footprint: ground-signal-ground (GSG) is the most common for MMICs.
(2) Pitch (distance between contacts): 200 µm is very common, although probes are commercially available with pitches ranging from 50 µm to 1200 µm.
(3) Contact width: 50 µm is typical.
(4) Coaxial connector type: APC3.5 for operation to 26.5 GHz, the Wiltron K-connector (2.92 mm) for operation to 40 GHz, APC2.4 for measurements up to 50 GHz, the Wiltron V-connector (1.85 mm) for measurements up to 67 GHz, or a waveguide flange for even higher frequencies.
(5) Launch angle, ϕ.

If the launch angle is too small, unwanted coupling between the probe and adjacent circuits on a wafer may occur. Alternatively, if the angle is too large there will not be enough skate on the probe pads. It has been found analytically and empirically that the best angle occurs when the horizontal components of the phase velocities in the probe and in the MMIC match each other [40]. Therefore:

$$\phi = \cos^{-1}\left\{\sqrt{\frac{\varepsilon_{\text{eff. probe}}}{\varepsilon_{\text{eff. MMIC}}}}\right\} \approx 30° \text{ for alumina probes and GaAs MMICs}$$

where $\varepsilon_{\text{eff. probe}}$ = effective permittivity of probe substrate

and $\varepsilon_{\text{eff. MMIC}}$ = effective permittivity of MMIC substrate.

11.3.2 Prober calibration

Before the calibration procedure can take place a planarity check must be made between the probes and the ultra-flat surface of the wafer chuck. A contact substrate, consisting of a polished alumina wafer with a gold-plated topside, is used to test that all of the three probe tips (GSG) make even contact marks in the gold.

Probe stations use a one tier calibration procedure, with the reference standards located either on-wafer or on a precision impedance standard substrate (ISS). On-wafer standards can characterise the probe-to-MMIC transition more accurately, but precision ISS standards can be fabricated to much tighter tolerances. There are a number of calibration techniques that can be used for on-wafer measurements [40, 42-47]. The LRL and TRL techniques require a minimum of two CPW delay lines and a non-critical reflection standard (usually an open or short circuit). The reference impedance is taken from the characteristic impedance of the delay lines. In practice, in order to cover a useful frequency range, it is necessary to employ a number of different delay line lengths to overcome phase ambiguity at all the measurement frequencies. This means that the probe separation has to be adjusted during the calibration procedure. For many applications, such as automated test systems, this is a major limitation and for these applications the line-reflect-match (LRM) calibration [42], developed by Cascade Microtech, is preferred to LRL/TRL. The multiple CPW delay lines required with the LRL/TRL calibration are effectively replaced by CPW matched loads, to theoretically represent an infinitely long delay line. This results in the following advantages:

(1) A very wideband calibration can be achieved.
(2) The probes can be set in a fixed position.
(3) Automatic calibration routines can be applied.
(4) Reflections and unwanted modes in long CPW delay lines are avoided.
(5) A considerable saving of ISS/wafer area can be made.

One concern with the LRM calibration technique is the absolute accuracy of the 50 Ω matched loads. These loads inevitably have some parasitic inductance and capacitance and, furthermore, may have frequency-dependent resistance due to the skin effect. Cascade's LRRM calibration is a more accurate version of the standard LRM calibration, in which load-inductance correction is incorporated by including an extra reflection standard. For the ultimate accuracy, however, there is strong support for the LRL and TRL methods: the UK National Physical Laboratory and the US National Institute of Standards and Technology are currently developing GaAs ISS wafers with calibration standards and verification components of certified quality [46, 47].

A two-port probe station traditionally uses a 12-term error model, although a 16-term error model has recently been introduced which requires five two-port calibration standards [48-49]. This more accurate model can correct for poor grounding and the additional leakage paths and coupling effects encountered with open air probing. Now, with the extremely high levels of accuracy that are possible with modern probe stations, the effects of calibration errors become more noticeable. Calibration errors can result from:

(1) ISS manufacturing variations
(2) Probe and standards degradation with use and/or time
(3) Probe placement errors due to position, pressure and planarity variations
(4) Surface wave effects on calibrations [50].

So far, it has been assumed that no de-embedding has been performed within the MMIC. However, with the extremely high level of measurement accuracy, the effects of probe pad discontinuities and associated transmission lines can become noticeable above a few gigahertz. Very accurate models for MMIC probe pads and microstrip transmission lines should be provided by the foundry that fabricates the MMIC. The metrologist must use these foundry-specific models to determine the actual measurements of the DUT. When de-embedding is performed using equivalent circuit modelling, these foundry-specific models can be easily incorporated into conventional simulation software packages.

11.3.3 Measurement errors

Even when the system has been successfully calibrated, measurement errors can still occur. Some of the more common sources of errors are as follows:

(1) Probe placement errors
(2) Temperature variation between calibration and measurement
(3) Cable-shift induced phase errors between calibration and measurement
(4) Radiation impedance changes due to the probes or wafer chuck being moved
(5) Matrix renormalisation not being performed with multiple RF port MMICs
(6) Resonant coupling of the probes into adjacent structures [51]
(7) Low frequency changes in the characteristic impedance and effective permittivity of microstrip and CPW transmission lines [51].

11.3.4 DC biasing

Depending on the nature and complexity of the device or circuit under test, DC bias can be applied to an MMIC in a number of ways:

(1) Through the RF probes, via bias-tees in the test set of the VNA
(2) Through single DC needles mounted on probe station positioners
(3) With multiple DC needles attached to a DC probe card, which may in turn be mounted on a positioner.

There is a limit to the maximum number of needles per card, but 10 is typical. One needle is normally required to provide a ground reference. Off-chip de-coupling capacitors and resistors can usually be added to the card to minimise the risk of unwanted oscillations. However, it should be stressed that the probe needle has a significant inductance, and as a result this additional bias de-coupling may not cure an oscillation problem when making RFOW measurements. With bias-tees and DC needles, the maximum DC bias voltage and current are approximately 40V and 500mA, respectively. With multiple DC needles, standard in-house DC footprints should be used wherever possible. This will reduce measurement costs considerably.

11.3.5 MMIC layout considerations

The foundry's design guidelines will define a minimum distance between the centres of the probe pad vias and the minimum distance from the vias to the edge of the active area. Generally, a particular company or institute will have standardised on a certain pad size and pitch for a particular probe-tip specification. In order to save valuable chip area, probing directly onto via-hole grounds may be considered. However, the probe tips may puncture the gold tops of the via holes, which could damage the probe tips and destroy the MMIC. It is certainly possible to use this probing method on a chip, but it is likely that the chip would fail a QA inspection afterwards. As a result, when designing MMICs for RFOW measurement, it is important to consult the foundry design guidelines on the probe pad specifications.

The location and orientation of the probe pads must also be considered. If the pads associated with one port are too close to those of another port, the very fragile probe tips are likely to get damaged if they accidentally touch one another during the probe alignment procedure. The minimum separation distance between probe tips is determined by the design rule on probe pad spacing (typically 250 µm with vias or 200 µm without vias). Also, if 3 or 4 probe positioners are attached to the probe station they will be oriented orthogonal to one another. As a result, the probe pads for a 3 or 4 port MMIC must also be orthogonal to one another, unless multi-signal probes are to be used. It should be noted that small deviations (<10°) from orthogonality can be tolerated. This may be unavoidable when there are layout restrictions.

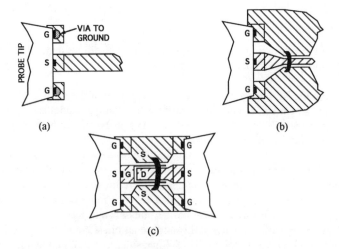

Fig. 11.7 Common launcher techniques: (a) Microstrip; (b) Coplanar waveguide; (c) Direct probing onto a device

On the MMIC, transitions (or launchers) are required to connect the probe pads to the DUT. In most cases the DUT is in the microstrip medium, and transitions from CPW-to-microstrip must be employed before and after the DUT. With reference to Fig. 11.7 (a), microstrip launchers require through-GaAs vias to provide a low inductance earth path from the probe to the MMIC's backside metallisation layer. A microstrip launcher should be long enough for the higher-order evanescent modes, resulting from the CPW-to-microstrip transition, to be sufficiently attenuated and have minimum interaction with the DUT. As a rule of thumb, the microstrip launchers should ideally have a length of 3 to 4 times the substrate thickness. With reference to Fig. 11.7 (b), when the DUT is in the CPW medium, through-GaAs vias are not required and a matched taper from the probe pads to the DUT is used. Even though this taper is very short, if the correct 50 Ω characteristic impedance is not maintained throughout the transition, significant capacitance or inductance can be introduced. In special cases, launchers are not required at all for some devices. One example of this is with a simple FET structure, as shown in Fig. 11.7 (c), where two GSG probes are placed directly onto the source-gate-source and source-drain-source pads. This approach eliminates the need for de-embedding the effects of launchers from the measurements, but the effects of the bond pads should still be considered. At this point it is important to note that, for frequencies above a few gigahertz, the equivalent circuit model of a device that has been realised in one medium, e.g. microstrip or CPW, should only be used in the design of new circuits of the same medium.

Single devices such as transistors and diodes can be biased through the bias-tees of the network analyser. However, in order to test a complete circuit using a probe station, special consideration has to be given to the layout of the DC bias pads and the design of the bias networks. When using DC needles to bias a circuit, the following points should be considered:

(1) The foundry may impose a minimum pad size and centre-to-centre pitch.
(2) For ease of DC probe card fabrication and probe alignment, the DC probe pads should be arranged in a linear array along the edge of the chip, and should be kept away from the RF pads. A common method is to have the RF probe pads on the east and west edges of the chip, and the DC bias pads on the north and south edges. If layout constraints suggest that orthogonal RF inputs and outputs would be more convenient, first check that suitable positioners are available.
(3) The bias networks of the circuit should be modelled separately for the RFOW measurement scenario to ensure that oscillations will not occur. Off-chip de-coupling capacitors cannot be placed as near to the chip as they can be in test fixture.
(4) High value resistors can be added on-chip to prevent RF leakage and catastrophic failure resulting from accidental forward biasing of varactor diodes or transistors. With varactor diodes, cold-FETs and switching-FETs, a trade-off may have to be made in the value of these bias resistors. If the resistance is too small there may not be enough isolation. If the resistance is too high the maximum switching speed may not be reached, due to an excessive R-C time constant. In practice, a minimum resistance value of ~1 KΩ should suffice for most applications.

11.3.6 Low-cost multiple DC biasing technique

Microwave probes have a useful working lifetime of between 3 and 9 months, depending on how often they are used. However, conventional DC probe cards may need to be replaced for every new MMIC design, unless standard DC probe footprints can be used. This throw-away approach is very costly, especially when the DC probe cards are supplied by a commercial vendor, since automated and precision manufacturing techniques have to be used for aligning the needles. Moreover, the cost of the cards increases with the number of needles, because the individual needles are themselves precision-made components.

At King's College, a flexible, low-cost technique has been developed for providing an MMIC under test with multiple DC bias connections. The MMIC is attached to a small gold-plated chip carrier, with either conductive epoxy glue or solder. An array of single-layer microwave capacitors is then attached to the chip carrier, in close proximity to the MMIC. BAR-CAPS™, made by Dielectric Labs. Inc., are ideal for this purpose since they are available as single-chip strips of three, four or six 100 pF shunt capacitors, each having a probe-able area of ~650×325 μm^2 and separated by ~170 µm. A gold bond wire is then used to connect the MMIC's DC probe pad to its assigned off-chip capacitor.

As an example of this technique, a microphotograph of an MMIC requiring 15 DC bias lines is shown in Fig. 11.8.

Fig. 11.8 Microphotograph of an MMIC with multiple DC biasing using the low-cost technique

It has been found that this low-cost solution has a number of important advantages for use in the R&D laboratory:

(1) Since these extra de-coupling capacitors are very close to the circuit there is less chance of unwanted oscillations.
(2) The linear array of off-chip capacitors automatically provides a standard in-house DC footprint, reducing long-term measurement costs considerably.
(3) When designing the MMIC layout, the DC probe pads do not need to be arranged in a linear array along the edges of the chip. This adds great flexibility to the layout.
(4) The probe-able area of the off-chip capacitors is approximately 15 times larger than that of the MMIC probe pads and the capacitors can withstand greater mechanical forces. As a result, in-house DC probe cards can be made by hand, because of the relaxation in the precision requirements of manufacture, reducing short-term costs considerably.

11.4 Thermal and cryogenic measurements

11.4.1 Thermal measurements

In real-life applications, microwave circuits can be exposed to temperatures other than ambient room temperature (i.e. 24°C or ~300 K). For example, some components in geostationary orbiting satellites may be exposed to temperatures in the range of -150°C to +150°C, depending on the amount of visible sunlight, the levels of localised heat generated within the satellite and the effectiveness of the thermal control sub-system. During the development of a sub-system, the levels of performance degradation while operating over a predefined temperature range must be known in order to be able to provide continuity of service. Therefore, at the MMIC design stage, the temperature dependant characteristics of all the components that make up the sub-system must first be determined. Once a prototype sub-system has been assembled, temperature-cycling is performed so that the measured levels of performance degradation can be compared with those predicted during simulation. The Cascade Microtech Summit 10600 semi-automatic probing system, in conjunction with the *Microchamber*™ enclosure, enables very fast set-up and measurements can be performed up to 65 GHz in a dark, temperature controlled and electromagnetic interference-isolated environment. This thermal probing system [52] can be seen in Fig. 11.9.

The MMIC under test sits on a temperature controlled wafer chuck, which can be subjected to temperatures ranging from -65°C to +200°C. If the RF probes and cables exhibit large temperature gradients, significant phase changes will be found, even at low microwave frequencies. As a result, an air flow purge is introduced into the chamber, in order to minimise the thermal coupling between the chuck and the probe/connector/cables. The system is calibrated for every new wafer chuck temperature setting. An LRRM calibration is used, with the ISS located on a separate thermally isolated stage. The at-temperature calibration procedure can be performed 15 minutes after the chuck temperature has been changed. This short wait corresponds to approximately three thermal time constants of the probe/connector/cables. Since all but the matched load impedance standards are insensitive to temperature, the ISS chuck temperature can be set at -5°C for a wafer chuck temperature of -65°C. This approach results in less than a 1% error in measurements between DC and 65 GHz.

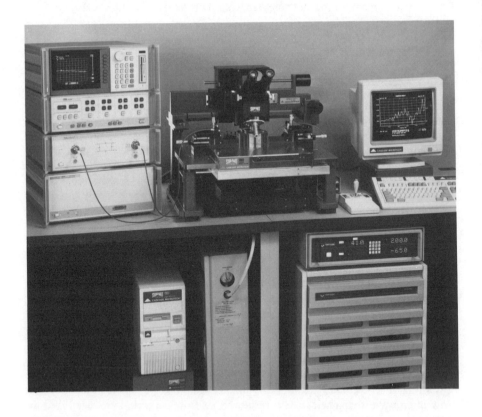

Fig. 11.9 Photograph of the Summit 10600 thermal probing system (Courtesy of Cascade Microtech Europe)

11.4.2 Cryogenic measurements

Cryogenic hybrid MICs, employing high performance active semiconductor and passive superconductor components, are being more widely used in applications ranging from radio astronomy, to space communications, to medical nuclear-magnetic-resonance scanners. Therefore, it is important to be able to determine the cryogenic temperature characteristics of these components [3-5, 53-56]. At cryogenic temperatures, the noise figures of conventional GaAs transistors are reduced dramatically from their ambient room temperature values. For example, at 10 GHz the measured noise figure of a typical 0.6×100 μm MESFET is 0.8 dB at 300 K and only 0.4 dB at 35 K [55]. With HEMT technology, electron mobility can increase by a factor of five when the lattice temperature is reduced from 300 K to 77 K [55], resulting in a considerable improvement in gain and noise performance. Furthermore, measurements made at temperatures as low as 10 K may provide information that can give a unique insight into the physics of experimental devices. Also, in addition to the advances being made in new semiconductor devices, there is considerable interest in the developments of ultra-low-loss high-temperature superconducting components, which currently have to be refrigerated below around 100 K.

The first microwave test fixture to be used in cryogenic measurements was reported in 1976 [3]. The fixture was designed to be immersed in liquid nitrogen (LN_2), which has a boiling-point of 77 K. This approach suffers from the problems of poor accuracy and poor repeatability due to the changing temperature gradients exhibited by the cable/connector/launchers, and requires a complicated calibration procedure. Accurate measurements have been reported using a TRL calibrated split-block test fixture mounted on the cold head of an RMC *Cryosystems*™ LTS-22-IR helium refrigerator [5]. This approach enables small-signal S-parameter measurements to be made at 300 K and 77 K.

Cryogenic probe stations have either the MMIC under test and the probes immersed in liquid nitrogen or a liquid cryogen-cooled copper stage with a dry nitrogen vapour curtain. The former approach suffers from poor repeatability (due to varying amounts of LN_2), a short measurement duration (in order to limit the build-up of ice formation) and a limited lifetime due to the degradation of the probes in contact with the LN_2. With the latter approach, accuracy is limited by mechanical stress, caused by the large thermal gradients between the microwave hardware and the MMIC under test and reliability is limited by moisture and the build-up of ice, which increases the wear and tear on manipulators and requires extensive re-planarisation of the mechanical apparatus. Researchers at the University of Illinois have, however, demonstrated the design and operation of a cryogenic vacuum microwave probe station, for the measurement of S-parameters from DC to 65 GHz, which minimises the problems of limited accuracy and repeatability [55]. Within a vacuum chamber, the vacuum probe station has high frequency CPW probes connected to cable feeds via a custom bellows and manipulator system. A liquid helium cryogen, with a boiling-point temperature of 4.2 K, enables measurements to be performed at temperatures as low as 20 K. The copper stage is continually fed with liquid cryogen and the system is then left to stand for 15 to 20 minutes, in order to achieve temperature equilibrium. Once the at-temperature calibration has been performed, the actual device measurements can be taken for up to four hours before having to re-calibrate.

11.5 Experimental field probing techniques

So far only invasive MMIC measurement techniques have been discussed, but experimental non-contacting methods also exist. Here, all the RF ports of the MMIC under test are terminated with matched loads. An RF signal is injected into the MMIC's input port and a micron-level field probing system is used to detect the strength of the associated fields along transmission lines and at discontinuities. As a result, field probing techniques can perform internal function and failure analysis. Field probing can detect current crowding, standing waves and unwanted modes of propagation, and S-parameters can be determined from time-domain network analysis measurements.

11.5.1 Electromagnetic-field probing

The simplest method of field detection uses a semiconductor diode. At microwave frequencies, however, it becomes difficult to match the diode because its impedance varies with power level. At low power levels, bolometers are traditionally employed for use above 1 GHz. The device is similar to a thin-film resistor, where a high resistivity bismuth film is evaporated onto metallic electrodes. When exposed to microwave radiation, the bolometer absorbs the electromagnetic energy and converts it into heat energy. As the film heats up, its resistivity decreases. Since the bolometer is inherently a square law detector, the measured voltage change across the device is proportional to the change in incident RF power. In practice, however, since the signal levels are so small, the incident microwave signal is pulsed. This causes the resistance of the bolometer to change at the pulse repetition frequency, which is usually below 100 KHz. With a DC bias current applied, the low frequency voltage signal across the bolometer is applied to a lock-in amplifier, which acts as a coherent detector. This technique exhibits a high degree of sensitivity; as an example, a 4×5 µm device with a noise equivalent power of 160 pW/Hz$^{1/2}$ has been reported [57]. With the use of conventional probe microfabrication techniques, microbolometers can be employed to detect power levels as low as a few nanowatts along MMIC transmission lines. A microbolometer probe which can be used for microstrip and CPW transmission lines is illustrated in Fig. 11.10. With a perfectly symmetrical probe positioned directly above a CPW line, the wanted CPW (or even) mode will be detected and the unwanted slotline (or odd) mode will not. As well as their simple fabrication and calibration, microbolometer probes can be designed to operate in the terahertz frequency range. Unfortunately, the attainable stability and uniformity of the resistive film does not yet appear to be sufficient for the commercial production of these probes.

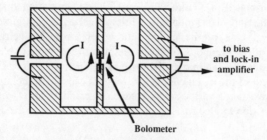

Fig. 11.10 Illustration of an electromagnetic-field probe

11.5.2 Magnetic-field probing

The simplest magnetic-field probing technique is to connect a conventional spectrum analyser to a magnetic-field probe. Using wafer probe microfabrication techniques, a miniature magnetic quadrupole antenna can be configured to match the magnetic fields associated with microstrip and CPW transmission lines, as illustrated in Fig. 11.11. Placed directly above the transmission line, the lines of magnetic flux will come up through one loop and back down through the other loop. As a result, the induced signals add. From a distance, the probe sees a near uniform magnetic field which induces signals that tend to cancel each other out. In addition to amplitude, phase measurements can also be measured. A reference signal at the same frequency, with a variable amplitude and phase, is combined with the measured signal. The measured phase is equal to the reference phase when the amplitude displayed on the spectrum analyser is at its peak. Therefore, the probe can be used to measure the amplitude and phase of currents at any node within an MMIC.

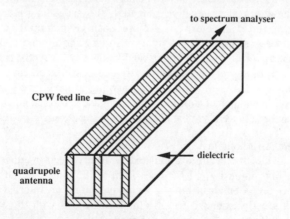

Fig. 11.11 Illustration of a magnetic-field probe

An experimental system has been reported which can operate in the 26.5 to 40 GHz frequency range [58]. Here, a 25 to 50 µm separation distance provides sufficient coupling and discrimination, while providing a negligible effect on the MMIC under test. One of the major sources of error is electrostatic pickup. Increasing the width of the loops increases the ratio of magnetic to electric coupling, but it also increases the random radiation picked up from other circuit elements. Reducing the width of the metal conductors reduces capacitive pickup, but increases the conductor's resistance and self inductance. In practice, an effective method of limiting the errors due to electrostatic pickup is to rotate the probe and average the measurements.

11.5.3 Electric-field probing

The simplest electric-field probing technique is to connect a conventional spectrum analyser to an electric-field probe. This technique was first demonstrated on MICs in 1979 [59]. The probe can be simply realised by removing a small section of the outer screening conductor and dielectric from the end of the analyser's coaxial feed line. Unfortunately, these high impedance probes have significant parasitic impedances at high microwave frequencies, which can severely perturb the operation of the circuit under test, thus causing measurement errors. Also, this technique was found to be accurate only when used with shielded transmission lines. As a result, this simple technique is unsuitable for micron-level transmission-line structures found in MMICs. Over the past decade, a number of alternative electric-field probing techniques have been investigated, with varying degrees of success:

11.5.3.1 Electron-beam probing
The voltage-contrast scanning electron microscope (SEM) was developed in the late 1960 s for detecting voltages on the conductor tracks of integrated circuits. A pulsed electron beam stimulates secondary electron emissions from the irradiated surface of metals. For conductors at a negative potential, the secondary electrons have more energy than for conductors at a more positive potential. Commercial SEMs suffer from a poor millivolt potential sensitivity and limited bandwidths of only a few gigahertz [60], although larger bandwidths have recently been reported [61]. Also, apart from its very high complexity and cost, the electron beam may effect the operation of GaAs MMICs due to charging of deep levels in the GaAs substrate. However, the major advantage of this technique is that the attainable spatial resolution that can be achieved is in the order of a few angstroms.

11.5.3.2 Photo-emissive sampling
Instead of using an electron beam to stimulated secondary electron emissions, a new approach uses a high-intensity pulsed laser beam to illuminate the surface of the metals [60]. This time-domain sampling technique offers an improved potential sensitivity and a greatly extended bandwidth. However, as with the SEM, GaAs MESFETs may be affected by the charging of deep levels.

11.5.3.3 Opto-electronic sampling

Time-domain network analysis can be performed using opto-electronic sampling techniques. Here, electrical pulses can be generated on an MMIC by illuminating DC biased photoconductive switches with a pulsed laser beam. The optical excitation of a photoconductive switch can also perform signal sampling. By comparing the Fourier transforms of the sampled incident and reflected or transmitted waveforms, the complex two-port S-parameters can be determined for the DUT [60, 62-68]. Sub-picosecond electrical pulse generation with a photoconductive switch has been reported, enabling terahertz measurement bandwidth [69]. This time-domain opto-electronic sampling technique (also known as photoconductive sampling) requires the DUT to be embedded in a single-chip GaAs test fixture. Each RF port of the DUT is connected to a test structure consisting of a 50 Ω matched load termination, photoconductive switches, DC bias lines, and a length of transmission line. These test components are not only wasteful of expensive chip space, but they must also be de-embedded from the measurements. In addition, the fabrication process of the photoconductive switches must be compatible with that of the test MMIC. However, a DC to 500 GHz measurement system has been demonstrated [68] using this technique.

11.5.3.4 Electro-optic sampling

The most promising electric-field probing technique is electro-optic sampling. A variety of non-centrosymmetric crystals, such as gallium arsenide and indium phosphide, exhibit Pockel's electro-optic effect. The presence of an electric-field will induce small anisotropic variations in the crystal's dielectric constant and, therefore, its refractive index. If a laser beam passes through this material it will experience a voltage induced perturbation in its polarisation, which is directly proportional to the change in the electric-field strength. As a result, this linear electro-optic effect can be used to provide a non-invasive means of detecting electric fields [60, 62-63, 70-78].

With internal (or direct) electro-optic probing the laser beam penetrates the GaAs MMIC in a reflection mode, as illustrated in Fig. 11.12 (a), giving good beam access and requiring only a single focusing lens [60, 62-63, 70-72, 76-78]. However, optical polishing of the MMIC substrate is required for best results. With front-side probing, the beam is reflected off the back-side ground plane metallisation, adjacent to the circuit conductor. With back-side probing, the beam is reflected off the back of the circuit conductor itself, making this scheme ideal for conventional CPW lines and slotlines. A spatial resolution of ~3 µm can be achieved with back-side probing, whereas front-side probing offers a resolution of only ~10 µm [70].

Centrosymmetric crystals, such as silicon and germanium, do not exhibit the linear electro-optic effect. Therefore, silicon MMICs must employ external (or indirect) electro-optic probing [60, 62, 73-75]. This technique uses an extremely small electric field sensor, consisting of a 40×40 µm^2 electro-optic crystal (lithium tantalate) at the end of a fused silica needle, placed in close proximity to the circuit conductor, as shown in Fig. 11.12 (b). The conductor's fringing fields can be detected by sending a laser beam down the needle and measuring the induced change in the refractive index of the crystal from the returning beam. Since the beam can be focused down to a spot size of ~5 μm in diameter, excellent spatial resolution is achieved. Also, there is no need for MMIC substrate polishing.

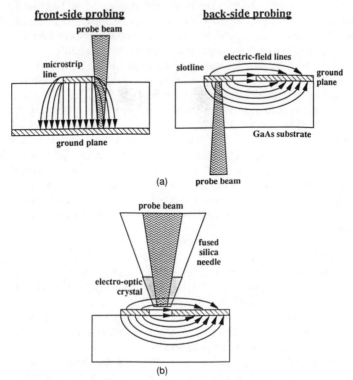

Fig. 11.12 Illustration of electric-field probe: (a) Internal; (b) External

With electro-optic probing, picosecond pulses of a sub-band-gap laser beam pass through the electric-fields associated with the MMIC's circuit conductors. After being passed through a common beam splitter, the incident and return beams are combined before being passed through a polarising beam splitter. Two photodiodes detect the intensity of the orthogonally polarised components and lock-in amplifiers are then used to determine the electric-field vectors. As a result, internal node voltage measurements can be determined and impressive 2-dimensional mappings of the amplitude [74, 76-78] and phase angles [78] of microwave fields within the MMIC can be obtained. Time-domain network analysis can also be performed using electro-optic sampling. Here, picosecond electrical pulses are applied to the input port of the MMIC under test, with the generator connected to the MMIC using traditional invasive techniques. By comparing the Fourier transform of the detected incident and reflected or transmitted waveforms, the complex two-port S-parameters can be determined. To date, a 50 to 300 GHz network analyser has been demonstrated using this technique [75].

11.5.3.4 Electrical sampling scanning-force microscopy
A number of non-invasive measurement techniques have been introduced that can perform internal function and failure analysis of MMICs. The electron-beam probing technique is well established and has excellent spatial resolution, but the temporal resolution is limited because of electron transit time effects. Optical probing techniques have a superior temporal resolution, but because of the micron-beam diameters they have a limited spatial resolution. Scanning-force microscopy, in the electrical sampling mode, is a new contactless measurement technique that has high spatial, temporal and voltage resolutions [79-80]. Here, an atomically sharp needle is mounted on one end of a cantilever. When the needle is placed a fixed working distance of between 0.1 and 0.5 µm above the MMIC under test, it will be subjected to attraction or repulsion forces, causing a detectable bending of the cantilever. This very experimental technique has so far demonstrated a spatial resolution of 0.5 µm and a bandwidth of 40 GHz [79].

11.6 Summary

A wide range of techniques has been briefly introduced for the measurement of MMICs. A summary of the main features associated with the most practical invasive techniques is given in Table II. In general, the level of accuracy and repeatability obtainable is proportional to the initial investment costs of the measurement system.

Compared with traditional invasive on-wafer measurement techniques, optical systems have so far demonstrated a lower dynamic range and inferior frequency resolution. In addition, optical techniques have complicated and lengthy calibration procedures. However, with its excellent spatial resolution and extremely wide bandwidth capabilities, electro-optic probing may become common place in the not too distant future.

	In-house test fixture	Commercial test fixture	On-wafer probe station
calibration	2-tier with ECM de-embedding	1-tier	1-tier	1-tier
accuracy	moderate	high	high	very high
repeatability	moderate	moderate	high	very high
bandwidth	wideband	wideband	wideband	ultra-wideband
flexibility	excellent	poor	poor	poor
cost	very low	low	high	very high

Table 11.2 Comparison of the invasive measurement technologies

11.7 References

1. LORCH, P.: 'Applications drive the evolution of network analyzers', *Microwaves & RF*, Jan. 1994, pp. 79-84
2. LUCYSZYN, S., STEWART, C., ROBERTSON, I. D., and AGHVAMI, A. H.: 'Measurement techniques for monolithic microwave integrated circuits', *IEE Electronics & Communication Engineering Journal*, Apr. 1994, pp. 69-76
3. LIECHTI, C. A., and LARRICK, R. B.: 'Performance of GaAs MESFET's at low temperatures', *IEEE Trans. Microwave Theory Tech.*, MTT-24, 1976, pp. 376-381
4. SMUK, J. W., STUBBS, M. G., and WIGHT, J. S.: 'Vector measurements of microwave devices at cryogenic temperatures', IEEE Int. Microwave Symp. Dig., 1989, pp. 1195-1198
5. SMUK, J. W., STUBBS, M. G., and WIGHT, J. S.: 'S-parameter characterization and modeling of three-terminal semiconductive devices at cryogenic temperatures', *IEEE Microwave and Guided Wave Letters*, Vol. 2, no. 3, Mar. 1992, pp. 111-113
6. LANE, R.: 'De-embedding device scattering parameters', *Microwave Journal*, Aug. 1984, pp. 149-156
7. RIAD, S. M.: 'The deconvolution problem: an overview', *Proc. IEEE*, Vol. 74, no. 1, Jan. 1986, pp. 82-85
8. ROMANOFSKY, R. R., and SHALKHAUSER, K. A.: 'Fixture provides accurate device characterization', *Microwaves & RF*, Mar. 1991, pp. 139-148
9. LUCYSZYN, S., MAGNIER, V., READER, H. C., and ROBERTSON, I. D.: 'Ultrawideband measurement of multiple-port MMICs using non-ideal test fixtures and a 2-port ANA', *IEE Proc. A*, Vol. 139, no. 5, Sept. 1992, pp. 235-242
10. TIPPET, J. C., and SPACIALE, R. A.: 'A rigorous technique for measuring the scattering matrix of a multiport device with a 2-port network analyzer', *IEEE Trans. Microwave Theory Tech.*, MTT-30, May 1982, pp. 661-666
11. DROPKIN, H.: 'Comments on -- A rigorous technique for measuring the scattering matrix of a multiport device with a two-port network analyzer', *IEEE Trans. Microwave Theory Tech.*, MTT-31, no. 1, Jan. 1983, pp. 79-81
12. SELMI, L., and ESTREICH, D. B.: 'An accurate system for automated on-wafer characterization of three-port devices', IEEE GaAs IC Symp. Dig., 1990, pp. 343-346
13. GOLDBERG, S. B., STEER, M. B., and FRANZON, P. D.: 'Accurate experimental characterization of three-ports', IEEE Int. Microwave Symp. Dig., 1991, pp. 241-244
14. STINEHELFER, H. E.: 'Discussion of de-embedding techniques using the time domain analysis', *Proc. IEEE*, Vol. 74, Jan 1986
15. 'Basic network measurements using the HP 8510B network analyzer system', Hewlett Packard Course Number HP 8510B+24D, Edition 3.0, Jan. 1988

16. GRONAU, G., and WOLFF, I.: 'A simple broad-band device de-embedding method using an automatic network analyzer with time-domain option', *IEEE Trans. Microwave Theory Tech.*, MTT-37, no. 3, Mar. 1989, pp. 479-483
17. DUMBELL, K. D.: 'TDR for microwave circuits', IEEE Asia-Pacific Microwave Conf. Proc., 1992, pp. 361-364
18. GRONAU, G.: 'Scattering-parameter measurement of microstrip devices', *Microwave Journal*, Nov. 1992, pp. 82-92
19. LU, K., and BRAZIL, T. J.: 'A systematic error analysis of HP8510 time-domain gating techniques with experimental verification', IEEE Int. Microwave Symp. Dig., 1993, pp. 1259-1262
20. GRONAU, G., WEISS, A., and KRAMER, O.: 'CPW test fixture characterizes microwave devices', *Microwaves & RF*, Feb. 1994, pp. 101-104
21. BURKE, J. J., and JACKSON, R. W.: 'A simple circuit model for resonant mode coupling in packaged MMICs', IEEE Int. Microwave Symp. Dig., 1991, pp. 1221-1224
22. SWANSON, D., BAKER, D., and O'MAHONEY, M.: 'Connecting MMIC chips to ground in a microstrip environment', *Microwave Journal*, Dec. 1993, pp. 58-54
23. MARCH, S. L.: 'Simple equations characterize bond wires', *Microwaves & RF*, Nov. 1991, pp. 105-110
24. NELSON, S., YOUNGBLOOD, M., PAVIO, J., LARSON, B., and KOTTMAN, R.: 'Optimum microstrip interconnects', IEEE Int. Microwave Symp. Dig., 1991, pp. 1071-1074
25. HEUERMANN, H., and SCHIEK, B.: 'The in-fixture calibration procedure line-network-network-LNN', 23rd European Microwave Conf. Proc., 1993, pp. 500-503
26. JONES, K., HARWOOD, W., and DELESSERT, D.: 'TRL-calibrated test station evaluates MICs', *Microwaves & RF*, June 1989, pp. 152-153
27. MUNNS, A.: 'Flip-chip solder bonding for microelectronic applications', *Metals and Materials*, Jan. 1989, pp. 22-25
28. WARNER, D. J., PICKERING, K. L., PEDDER, D. J., BUCK, B. J., and PIKE, S. J.: 'Flip chip-bonded GaAs MMICs compatible with foundry manufacture', *IEE Proc. H*, Vol. 138, no. 1, Feb. 1991
29. FELTON, L. M.: 'High yield GaAs flip-chip MMICs lead to low cost T/R modules', IEEE Int. Microwave Symp. Dig., 1994, pp. 1707-1710
30. JIN, H., VAHLDIECK, R., MINKUS, H., and HUANG, J.: 'Rigorous field theory analysis of flip-chip interconnections in MMICs using the FDTLM method', IEEE Int. Microwave Symp. Dig., 1994, pp. 1711-1714
31. SAKAI, H., OTA, Y., INOUE, K., YOSHIDA, T., TAKAHASHI, K., FUJITA, S., and SAGAWA, M.: 'A novel millimeter-wave IC on Si Substrate using flip-chip bonding technology', IEEE Int. Microwave Symp. Dig., 1994, pp. 1763-1766
32. 'Network Analysis: Applying the HP 8510B TRL calibration for non-coaxial measurements', Hewlett Packard Production Note 8510-8, Oct. 1987

33. WILLIAMS, D. F., PAANANEN, D. W.: 'Suppression of resonant modes in microwave packages', IEEE Int. Microwave Symp. Dig., 1989, pp. 1263-1265
34. STRID, E., and GLEASON, K.: 'A microstrip probe for microwave measurements on GaAs FET and IC wafers', GaAs IC Symp., Paper 31, Nov. 1980
35. STRID, E., and GLEASON, K.: 'A DC-12GHz monolithic GaAs FET distributed amplifier', *IEEE Trans. Microwave Theory Tech.*, MTT-30, 1982, pp. 969-975
36. GODSHALK, E. M.: 'A W-band wafer probe', IEEE Int. Microwave Symp. Dig., 1993, pp. 171-174
37. LIU, J. S. M., and BOLL, G. G.: 'A new probe for W-band on-wafer measurements', IEEE Int. Microwave Symp. Dig., 1993, pp. 1335-1338
38. YU, R., REDDY, M., PUSL, J., ALLEN, S., CASE, M., and RODWELL, M.: 'Full two-port on-wafer vector network analysis to 120 GHz using active probes', IEEE Int. Microwave Symp. Dig., 1993, pp. 1339-1342
39. RODWELL, M., ALLEN, S., CASE, M., YU, R., BHATTACHARYA, U., and REDDY, M.: 'GaAs nonlinear transmission-lines for picosecond and millimeter-wave applications', 23rd European Microwave Conf. Proc., 1993, pp. 8-10
40. BAHL, I., LEWIS, G., and JORGENSON, J.: 'Automatic testing of MMIC wafers', *Int. Journal of Microwave and Millimeter-wave Computer-aided Engineering*, Vol. 1, no. 1, 1991, pp. 77-89
41. GODSHALK, E. M., BURR, J., WILLIAMS, J.: 'An air coplanar wave probe', 24th European Microwave Conf. Proc., 1994, pp. 1380-1385
42 EUL, H. J., and SCHIEK, B.: 'Thru-match-reflect: one result of a rigorous theory for de-embedding and network analyzer calibration', 18th European Microwave Conf. Proc., 1988
43. PRADELL, L., CACERES, M., and PURROY, F.: 'Development of self-calibration techniques for on-wafer and fixtured measurements: A novel approach', 22nd European Microwave Conf. Proc., 1992, pp. 919-924
44. FERRERO, A., and PISANI, U.: 'Two-port network analyzer calibration using an unknown "thru"', *IEEE Microwave and Guided Wave Letters*, Vol. 2., no. 12. , Dec. 1992, pp. 505-507
45. PURROY, F., and PRADELL, L.: 'Comparison of on-wafer calibrations using the concept of reference impedance', 23rd European Microwave Conf. Proc., 1993, pp. 857-859
46. BANNISTER, D. J., and SMITH, D. I.: 'Traceability for on-wafer CPW S-parameter measurements', IEE Colloquium Dig. on Analysis, Design and Applications of Coplanar Waveguide, Oct. 1993, pp. 7/1-5
47. PENCE, J. E.: 'Technique verifies LRRM calibrations on GaAs substrates', *Microwaves & RF*, Jan. 1994, pp. 69-76
48. BUTLER, J. V., RYTTING, D. K., ISKANDER, M. F., POLLARD, R. D., and BOSSCHE, M. V.: '16-term error model and calibration procedure for on-wafer network analysis measurements', *IEEE Trans. Microwave Theory Tech.*, MTT-39, no. 12, Dec. 1991, pp. 2211-2217

49. SILVONEN, K. J.: 'Calibration of 16-term error model', *Electron. Lett.*, Vol. 29, no. 17, Aug. 1993, pp. 1544-1545
50. GODSHALK, E. M.: 'Wafer probing issues at millimeter wave frequencies', 22nd European Microwave Conf. Proc., 1992, pp. 925-930
51. MIERS, T. H., CANGELLARIS, A., WILLIAMS, D., and MARKS, R.: 'Anomalies observed in wafer level microwave testing', IEEE Int. Microwave Symp. Dig., 1991, pp. 1121-1124
52. D'ALMEIDA, D., and ANHOLT, R.: 'Device characterization with an integrated on-wafer thermal probing system', *Microwave Journal*, Mar. 1993, pp. 94-105
53. LASKAR, J., KOLODZEY, J.: 'Cryogenic vacuum high frequency probe station', *Journal of Vacuum Science Technology*, Sept./Oct. 1990, pp. 1161-1165
54. MESCHEDE, H., REUTER, R., ALBERS, J., KRAUS, J., PETERS, D., BROCKERHOFF, W., TEGUDE, F.-J., BODE, M., SCHUBERT, J., and ZANDER, W.: 'On-wafer microwave measurement setup for investigations on HEMT's and high Tc superconductors at cryogenic temperatures down to 20 K', *IEEE Trans. Microwave Theory Tech.*, MTT-40, no. 12, Dec. 1992, pp. 2325-2331
55. LASKAR, J., and FENG, M.: 'An on-wafer cryogenic microwave probing system for advanced transistor and superconductor applications', *Microwave Journal*, Feb. 1993, pp. 104-114
56. LASKAR, J., LAI, R., BAUTISTA, J. J., HAMAI, M., NISHIMOTO, M., TAN, K. L., STREIT, D. C., LIU, P. H., LO, D. C., NG, G. I.: 'Enhanced cryogenic on-wafer techniques for accurate InxGa1-xAs HEMT device models', IEEE Int. Microwave Symp. Dig., 1994, pp. 1485-1488
57. SCHWARZ, S. E., and TURNER, C. W.: 'Measurement techniques for planar high-frequency circuits', *IEEE Trans. Microwave Theory Tech.*, MTT-34, no. 4, Apr. 1986, pp. 463-467
58. OSOFSKY, S. S., and SCHWARZ, S. E.: 'Design and performance of a non-contacting probe for measurements on high-frequency planar circuits', *IEEE Trans. Microwave Theory Tech.*, MTT-40, no. 8, Aug. 1992, pp. 1701-1708
59. DAHELE, J. S., and CULLEN, A. L.: 'Electric probe measurements on microstrip', *IEEE Trans. Microwave Theory Tech.*, MTT-28, no. 7, July 1980, pp. 752-755
60. BLOOM, D. M., WEINGARTEN, K. J., and RODWELL, M. J. W.: 'Probing the limits of traditional MMIC test equipment', *Microwaves & RF*, July 1987, pp. 101-106
61. KUBALEK, E., and FEHR, J.: 'Electron beam test system for GHz-waveform measurements on transmission-lines within MMIC', 22nd European Microwave Conf. Proc., 1992, pp. 163-168
62. BIERMAN, H.: 'Improved on-wafer techniques evolve for MMIC testing', *Microwave Journal*, Mar. 1990, pp. 44-58

63. LEE, T. T., SMITH, T., HUANG, H. C., CHAUCHARD, E., and LEE, C. H.: 'Optical techniques for on-wafer measurements of MMICs', *Microwave Journal*, May 1990, pp. 91-102
64. HUANG, S.-L. L., CHAUCHARD, E. A., LEE, C. H., HUNG, H.-L. A., LEE, T. T., and JOSEPH, T.: 'On-wafer photoconductive sampling of MMICs', *IEEE Trans. Microwave Theory Tech.*, MTT-40, no. 12, Dec. 1992, pp. 2312-2320
65. KIM, J., SON, J., WAKANA, S., NEES, J., WILLIAMSON, S., WHITAKER, J., KWON, Y., and PAVLIDIS, D.: 'Time-domain network analysis of mm-wave circuits based on a photoconductive probe sampling technique', IEEE Int. Microwave Symp. Dig., 1993, pp. 1359-1361
66. GOLOB, L. P., HUANG, S. L., LEE, C. H., CHANG, W. H., JONES, K., TAYSING-LARA, M., and DeANNI, T.: 'Picosecond photoconductive switches designed for on-wafer characterization of high frequency interconnects', IEEE Int. Microwave Symp. Dig., 1993, pp. 1395-1398
67. ARMENGAUD, L., GERBE, V., LALANDE, M., LAJZEREROWICZ, J., CUZIN, M., and JECKO, B.: 'Electromagnetic study of an electronic sampler for picosecond pulse measurements', 23rd European Microwave Conf. Proc. 1993, pp. 751-754,
68. FRANKEL, M. Y.: '500-GHz characterization of an optoelectronic S-parameter test structure', *IEEE Microwave and Guided Wave Letters*, Vol. 4, no. 4, Apr. 1994, pp. 118-120
69. VALDMANIS, J. A., and MOUROU, G.: 'Subpicosecond electrooptic sampling: principles and applications', *IEEE Journal of Quantum Electronics*, Vol. QE-22, 1986, pp. 69-78
70. BLOOM, D. M., WEINGARTEN, K. J., and RODWELL, M. J. W.: 'Electrooptic sampling measures MMICs with polarized light', *Microwaves & RF*, Aug. 1987, pp. 74-80
71. MERTIN, W., BOHM, C., BALK, L. J., and KUBALEK, E.: 'Two-dimensional field mapping in MMIC-substrates by electro-optic sampling technique', IEEE Int. Microwave Symp. Dig., 1992, pp. 1443-1446
72. LEE, C. H., LI, M. G., HUNG, H.-L. A., and HUANG, H. C.: 'On-wafer probing and control of microwave by picosecond optical beam', IEEE Asia-Pacific Microwave Conf. Proc., 1992, pp. 367-370
73. WU, X., CONN, D., SONG, J., and NICKERSON, K.: 'Calibration of external electro-optic sampling using field simulation and system transfer function analysis', IEEE Int. Microwave Symp. Dig., 1993, pp. 221-224
74. MERTIN, W., ROTHS, C., TAENZLER, F., and KUBALEK, E.: 'Probe tip invasiveness at indirect electro-optic sampling of MMIC', IEEE Int. Microwave Symp. Dig., 1993, pp. 1351-1354
75. CHENG, H., and WHITAKER, J. F.: '300-GHz-bandwidth network analysis using time-domain electro-optic sampling', IEEE Int. Microwave Symp. Dig., 1993, pp. 1355-1358

76. HJELME, D. R., YADLOWSKY, M. J., and MICKELSON, A. R.: 'Two-dimensional mapping of the microwave potential on MMIC's using electrooptic sampling', *IEEE Trans. Microwave Theory Tech.*, MTT-41, no. 6/7, June/July 1993, pp. 1149-1158
77. DAVID, G., REDLICH, S., MERTIN, W., BERTENBURG, R. M., KOBLOWSKI, S., TEGUDE, F. J., KUBALEK, E., and JAGER, D.: 'Two-dimensional direct electro-optic field mapping in a monolithic integrated GaAs amplifier', 23rd European Microwave Conf. Proc., 1993, pp. 497-499
78. MERTIN, W., LEYK, A., DAVID, G., BERTENBURG, R. M., KOBLOWSKI, S., TEGUDE, F. J., WOLFF, I., JAGER, D., and KUBALEK, E.: 'Two-dimensional mapping of amplitude and phase of microwave fields inside a MMIC using the direct electro-optic sampling technique', IEEE Int. Microwave Symp. Dig., 1994, pp. 1597-1600
79. BOHM, C., ROTHS, C., and KUBALEK, E.: 'Contactless electrical characterization of MMICs by device internal electrical sampling scanning-force microscopy', IEEE Int. Microwave Symp. Dig., 1994, pp. 1605-1608
80. MUELLER, U., BOEHM, C., SPRENGEPIEL, J., ROTHS, C., KUBALEK, E., and BEYER, A.: 'Geometrical and voltage resolution of electrical sampling scanning force microscopy', IEEE Int. Microwave Symp. Dig., 1994, pp. 1005-1008.

Chapter 12

Advanced techniques

S. Lucyszyn and I. D. Robertson

12.1 Introduction

MMICs are finding more and more applications in the commercial sector, but fierce competition between manufacturers and between different technologies has led to immense pressure on manufacturing costs: single-chip solutions with minimum chip size are demanded. In order to produce competitive MMIC products, it is necessary to be familiar with more than just the basic techniques of MMIC design. One notable example of a commercial success is GMMT's fully integrated transceiver for wireless LAN applications [1], illustrated in Fig. 12.1. This design serves to demonstrate that extremely high levels of integration are possible as long as the designers are armed with suitable CAD packages and a full range of advanced circuit techniques.

This chapter is intended to give an insight into some of the more advanced techniques that can be used in MMIC design. The chapter was originally entitled 'future developments', but as the contents took shape it became clear that many of the techniques described here are already finding widespread application in MMIC design. The aim of this chapter is to complement the preceding chapters, which have concentrated on the established mainstream design techniques, by reviewing the more advanced techniques which have been reported in the literature. The techniques described include CPW and multilayer circuits, active filters, active coupling circuits and power dividers, microwave modulators and demodulators, linearised high power amplifiers, active antennas, and optoelectronic applications. Naturally, it is not possible to do full justice to all these topics in a single chapter, but the key principles will be described and extensive references are provided for the reader who wishes to learn more.

In many cases new circuit techniques are developed as a means of replacing established components which cannot readily be used on MMICs. For example, active isolators and circulators have been investigated as potential alternatives to conventional ferrite isolators and circulators. In other cases, the advantage of MMIC technology in allowing very complex circuits to be realised has made it possible to

436 Advanced techniques

find new solutions to familiar problems and to offer improved functionality over conventional methods. Millimetre-wave active phased array antennas, which would be difficult and expensive to construct in hybrid-MIC technology, are examples of this.

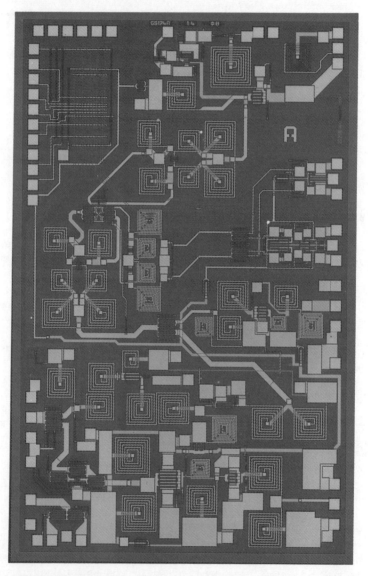

Fig. 12.1 Photograph of a single chip transceiver for 2.4 GHz wireless LAN applications
(Photograph courtesy of GEC-Marconi Materials Technology Ltd)

12.2 Uniplanar techniques

As a result of the effort to reduce chip sizes and increase operating frequency into the millimetre-wave band, there is now considerable interest in the use of 'uniplanar' techniques employing coplanar waveguide (CPW), slotline, coplanar strips (CPS) and thin-film microstrip (TFMS). The major attraction of uniplanar techniques is that through-substrate via-holes are not required for grounding purposes. The attraction of TFMS and other multi-layer techniques is that very high packing density can be achieved.

12.2.1 Transmission lines for uniplanar MMICs

12.2.1.1 Coplanar waveguide

The structure of a CPW transmission line is shown in Fig. 12.2 (a). It consists of a signal conductor placed between two ground planes, and since the dominant mode of CPW is quasi-TEM there is no low-frequency cut-off. The principal advantages of CPW are as follows:

(1) devices and components can be grounded without via-holes

(2) it suffers from much less dispersion than microstrip, making it suitable for millimetre-wave circuits

(3) a given characteristic impedance can be realised with almost any track width and gap combination

(4) a considerable increase in packing density is possible because the ground planes provide shielding between adjacent CPW lines

(5) with the back-face ground plane removed, lumped elements exhibit less parasitic capacitance

Previously, the lack of accurate design information and discontinuity models precluded CPW from being extensively used in MMICs. This situation has improved dramatically, however, and many CAD packages now have CPW elements in their libraries. These models include bends, steps, T-junctions, and many other common elements. One of the major problems with CPW is that the mode of propagation can easily degenerate from quasi-TEM into a balanced coupled-slotline mode. This happens very often at discontinuities, but can be avoided by incorporating grounding straps between the ground planes, using either air-bridges or underpasses. A second reason why CPW has not been widely used is due to the belief that it has inherently higher conduction loss than microstrip. However, it has been shown that at millimetre-wave frequencies CPW can be equal to or better than microstrip when loss and dispersion on GaAs substrates are used as a basis for comparison [2]. This comparison has shown that minimum loss for a given CPW cross section occurs at about 60 Ω impedance, whereas the minimum loss for microstrip occurs at about 25 Ω. However, the physical sizes of these minimum loss microstrip lines are much larger than the CPW type.

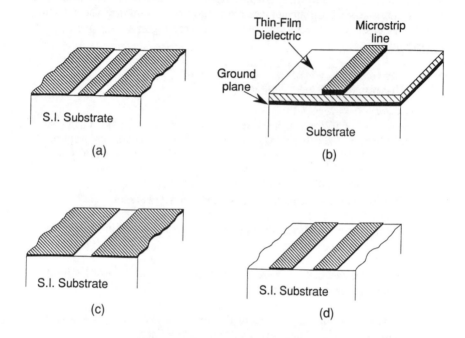

Fig. 12.2 Transmission lines for uniplanar MMICs: (a) Coplanar waveguide; (b) Thin-film microstrip; (c) Slotline; (d) Coplanar strips

12.2.1.2 Thin-film microstrip
Thin-film microstrip (TFMS) lines are essentially highly miniaturised microstrip lines, fabricated on a thin dielectric film, as shown in Fig. 12.2 (b). A conventional 50 Ω MMIC microstrip line might be 80 to 150 μm wide; a TFMS line would typically be less than 10 μm wide. Hence, TFMS circuitry can give a dramatic reduction in circuit size and, because the ground plane is on the front face of the chip, via-hole grounds are not required. The disadvantage of TFMS is its higher transmission loss, due especially to the high resistance of the very narrow signal conductors. However, by careful choice of dielectric material and thickness, the loss can be made acceptable, and NTT has demonstrated 50 Ω TFMS lines on a triple layer of polyimide. TFMS lines have already been successfully applied to the miniaturisation of branch-line couplers and a range of novel circuits [3-5].

12.2.1.3 Slotline

Slotline consists of a pair of ground planes with a narrow slot between them, as shown in Fig. 12.2 (c), and the signal propagates in a TE (transverse electric) mode. This means that slotline is not an ideal general purpose transmission-line medium, but this mode makes it very useful for circuits such as balanced mixers and amplifiers where push-pull operation is required. Also, since CPW has the quasi-TEM mode and slotline has the TE mode, a number of useful hybrid junctions and transition circuits can be produced [6]. The most popular of these is the CPW-to-slotline transition, shown in Fig. 12.3. The version shown is realised by employing a two metal layer structure, without the need for via-holes or air bridges. At the transition, the two CPW ground planes are connected with a thin metallised underpass. One of the CPW ground planes then becomes one of the slotline conductor planes. The CPW centre conductor extends over the underpass and is connected to the second slotline conductor. At high frequencies, this transition converts the quasi-TEM mode into the slotline's TE mode, yielding balun operation. This type of transition has demonstrated balun operation over a 5 to 30 GHz bandwidth [7], and has been used in miniaturised uniplanar mixers and amplifiers [8, 9].

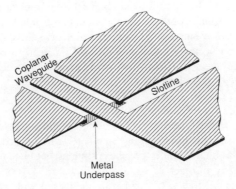

Fig. 12.3 CPW-to-slotline transition

12.2.1.4 Coplanar strips

The coplanar strip (CPS) transmission medium consists of a pair of closely coupled parallel strips as shown in Fig. 12.2 (d). It is essentially a planar equivalent to the twisted pair or Lecher line. As a balanced transmission line, it is ideally suited to balanced mixers and push-pull amplifiers. However, the lack of design information has severely restricted its use, as well as the fact that a circuit's input and output signals are usually required eventually to be in an unbalanced medium such as microstrip or co-axial line. However, as integration levels in subsystems increase this latter limitation is less severe, whereas the less critical grounding requirements make CPS an attractive medium.

12.2.2 CAD for CPW and multilayer MMICs

In order to design MMICs using CPW as the principal transmission line it is necessary to break the circuit down into building blocks which are individually available as library elements. These library elements include bends, T-junctions, cross-junctions, shunt and series lumped RLC components, tapered lines, and a number of other elements. HP-EEsof, for example, has over 20 CPW library elements in its Touchstone™ package, using models developed by Argumens in Germany. However, uniplanar MMICs often have many features which cannot be described by any library elements. For example, TFMS lines and multi-layer components have so much flexibility that standard models could not cover every situation. Hence, CAD which is based on electromagnetic modelling of arbitrary structures is an extremely important tool in the design of uniplanar MMICs. These tools have only become available relatively recently, but there is now a wide range of EM simulators available for quasi-planar circuits. The best known are *em*™ by Sonnet Software, Momentum™ by Hewlett Packard, Explorer™ by Compact Software, and LINMIC/Unisym™ by Jansen Microwave. The basic feature of these simulators is that they allow the layout of an arbitrary planar circuit to be drawn (or 'captured'), subsectioned, and then simulated electromagnetically (with one of the methods described in Chapter 4). Hence, all component interactions and higher-order modes are included in the simulation. The importance of these simulators in being able to analyse non-standard structures cannot be over emphasised, both for uniplanar MMICs and for conventional MMICs. The reason is that components on MMICs need to be packed as closely together as possible, and so it is often not a valid approach to attempt to break down a layout into standard library elements.

12.2.2.1 Xgeom™ and em™ software

In order to show how this type of planar electromagnetic simulator is used, we briefly describe here how the packages *xgeom*™ and *em*™ can be used for modelling CPW components. *em* was one of the first of these simulators to become available, and its ability to handle any number of metal and dielectric layers makes its ideal for modelling uniplanar MMICs. After capturing the circuit with *xgeom* and saving the resulting geometry file, the user may then run *em*. *em* automatically subsections the circuit and performs the electromagnetic analysis, resulting in a set of *n*-port *S*-parameters. An important feature of *em* is the push-pull (balanced, or odd-mode) port capability which allows coplanar waveguide structures to be analysed conveniently.

Fig. 12.4 shows the layout of a CPW T-junction element being displayed in *xgeom*'s window with push-pull ports (labelled -1, +1, -1, etc.) and arrows marking the reference planes, which are used for de-embedding (de-embedding removes unwanted effects occurring at the box edges). The figure also shows vias going between the upper metal layer and the lower metal underpass ground straps. The analysis of this structure is well within the capability of most workstations, with a run-time of a few minutes per frequency point, but it should be pointed out that there is a practical limit to the complexity of a layout which can be analysed. *em* is both a memory intensive and a computationally intensive programme. Small circuits are analysed quickly but the computing time and memory required to analyse a circuit increases dramatically with the number of subsections in a layout. Hence, to simulate a complete MMIC requires field-simulating only the most critical subsections of the circuit and then inputting all the S-parameter files into a circuit simulator for inclusion in the complete circuit description.

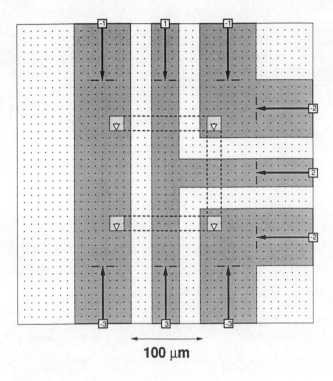

Fig. 12.4 CPW T-junction captured on *xgeom*™

12.2.3 CPW circuits

The main feature of CPW circuits is that grounding can be achieved directly on the front face without a via-hole. As well as simplifying the processing, this can also save space and improve performance. For example, in a transistor layout the area used up by the via-holes is often greater than that of the transistor itself. Furthermore, in CPW there is a considerable reduction in common-lead inductance, which is particularly important for millimetre-wave circuits. CPW distributed amplifiers operating to 100 GHz have been reported [10]. To illustrate the compact nature of CPW circuits,

442 Advanced techniques

Fig. 12.5 shows a small part of a CPW amplifier layout. It shows a bias resistor, which is decoupled to ground, and part of a meandered CPW matching line. The decoupling capacitor is directly grounded simply by placing the lower plate directly under the ground plane; the meandered line can be folded tightly together because of the shielding effect of the intervening ground plane. Note again the use of ground straps to prevent unwanted propagation modes being excited at the bend discontinuities.

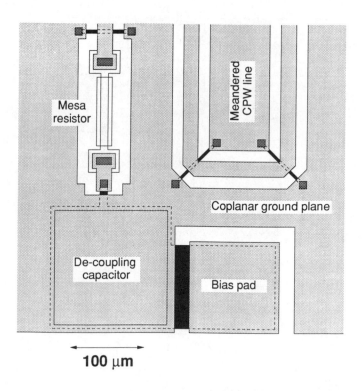

Fig. 12.5 Part of a CPW amplifier layout

12.2.3.1 Branch-line couplers

The 'reduced-size' or 'lumped-distributed' technique [11, 12] is ideal for CPW MMICs because the shunt capacitors can be implemented very easily. An impedance-transforming CPW branch-line coupler operating at 15.5 GHz has been reported [13] with an area of only 800 μm × 800 μm, which includes four sets of RFOW probe pads. The coupler's measured results showed an amplitude response of 4.7 ± 0.3 dB, an input and output return loss of better than 12 dB and 25 dB, respectively, and an isolation of better than 20 dB over the range 14-16 GHz.

12.2.3.2 Two-stage balanced Ku-band CPW MMIC amplifier

This amplifier uses the impedance-transforming coupler to reduce the complexity of the MESFET input and output matching networks. The couplers have 15 Ω output impedance, which is close to the input/output impedance of the MESFETs. Hence, a simple series line and stub matching technique can be used. This technique, along with the size advantages of CPW already discussed, leads to a very compact layout. A photograph of the amplifier is shown in Fig. 12.6. The chip size is only 2.4 mm × 1.7 mm and the measured gain is 16 dB with less than ± 0.1 dB of ripple over the range 14.75-15.25 GHz [14, 15]. As it is balanced, the amplifier has excellent input and output matches and three chips can be cascaded to give over 48 dB of gain with less than 1 dB ripple.

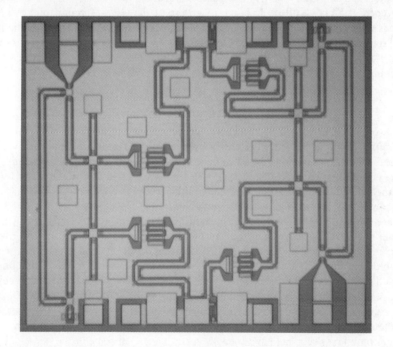

Fig. 12.6 Photograph of a Ku-band two-stage balanced amplifier using coplanar waveguide

12.2.4 Multilayer circuits

The majority of MMIC designs and layouts are still based on the traditional microstrip design techniques developed for hybrid MICs. These techniques do not necessarily take full advantage of the high integration levels achievable with integrated circuit technology. The added flexibility of MMICs over hybrid circuits is due in part to the multilayer nature of MMICs. In particular, MMICs using multiple dielectrics and metal layers enable intricate multilayer circuits to be designed. Such circuits offer the advantages of greater flexibility in layouts, very small size, more flexible means of interconnection, and lower cost. TFMS lines have already been mentioned briefly, but this technique is only one of a huge number of possible multilayer techniques. Multilayer circuits may incorporate miniature TFMS transmission lines, inverted TFMS lines, offset broadside coupling structures, multi-level spiral inductors and transformers, and low impedance CPW lines using centre conductor or ground plane underpasses. GEC-Marconi has had a stacked spiral inductor as a standard library element for many years. Now that suitable CAD tools are available, these techniques are becoming very popular, both for GaAs and silicon MMICs. Some examples of couplers and amplifiers using multilayer techniques are now described. These circuits, designed at King's College London, make use of the GEC-Marconi foundry's two metal layer process. NTT in Japan, which along with ATR was a pioneer of uniplanar and multilayer MMICs, has already developed a five metal layer MMIC process, and this makes it possible to design a wide range of new passive elements with very high packing density.

12.2.4.1 Multilayer directional couplers
The Lange coupler has an interdigitated structure in order to achieve the tight coupling required for a 3 dB directional coupler. With multilayer MMICs this tight coupling can be achieved more easily by employing one of a number of overlaid coupling structures. Among the advantages of using multilayer coupling structures are their compactness, tight coupling ability, flexibility, and also their ability to be folded or meandered. The first multilayer monolithic coupler [16] employed broad-side coupling, with one track directly above the other, and a photograph of this coupler is shown in Fig. 12.7. However, this method leads to high insertion loss because the lower metal is normally very thin due to processing limitations. The microstrip overlaid coupler was later improved [17] by using offset broad-side coupling, which enables the lower metal track to be plated-up with the upper metal. This coupler structure is shown in cross-section in Fig. 12.8. When compared to a Lange coupler this coupler achieves the same bandwidth performance while using only half of the chip area. A CPW multilayer directional coupler has also been realised [18], operating in the frequency range 12-36 GHz with an equal power imbalance of less than 1 dB. These multilayer couplers have been successfully used in some of the phase shifters described in Chapter 7.

12.2.4.2 Amplifiers using thin-film microstrip
A uniplanar amplifier can employ CPW, TFMS, and conventional lumped elements. The first amplifier demonstrated in TFMS was a distributed amplifier [19]. A cascode FET uniplanar amplifier was recently demonstrated [20] which is entirely compatible with standard foundry processing and does not need via-holes or back-face metallisation. A cascode FET topology was used in order to achieve high gain with only one amplifier stage. Matching was accomplished partly using lumped-elements and partly by using meandered thin-film microstrip (TFMS) matching stubs. Since a

TFMS line would typically be less than 10 μm wide, TFMS circuitry can give a dramatic reduction in circuit size as the stubs can be tightly meandered. The gain was over 12 dB at the centre frequency of 15.25 GHz.

In addition to enabling compact meandered lines to be used, TFMS enables extremely low impedance lines to be realised. A 1 W X-band power amplifier has been demonstrated using this advantage [21]. The circuit employed TFMS short transformers with characteristic impedances as low as 4 Ω. Such a low impedance cannot be achieved with ordinary microstrip or CPW.

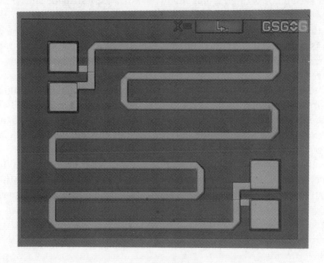

Fig. 12.7 Photograph of the first monolithic multilayer coupler

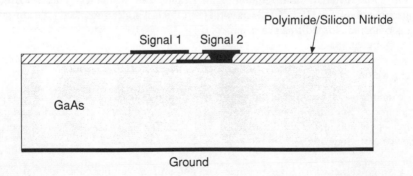

Fig. 12.8 Cross-section of a multilayer coupler using offset broadside-coupled microstrip lines [17]

12.3 Electronically tunable elements

Variable impedance devices and circuits play an important role in tunable MMIC applications. Examples include voltage controlled oscillators (VCOs) [22], analogue phase shifters [23], group delay equalisers [24], amplitude equalisers [25], and active filters [26]. In addition, with ever increasing demands for first-pass design successes, these electronically tunable elements may also be employed to provide some degree of performance control in a whole range of circuits. The elements which can be used to provide continuous electronic tuning include varactor diodes, active inductors, cold-FETs as variable resistances, and negative resistance transistor circuits.

Microwave circuits that facilitate some form of continuous frequency tuning can be divided into two categories. The first category consists of applications where the tuning range is only dictated by the frequency tuning ratio of the variable capacitor or inductor. Here, the frequency tuning ratio is simply the square root of the total capacitance ratio (TCR=C_{max}/C_{min}) or total inductance ratio (TIR=L_{max}/L_{min}). The required value of TCR can be determined from the appropriate topography and fabrication processing of the varactor diode. The required value of TIR can be achieved with the appropriate circuit topology of the active inductor. Applications that fall into this category are numerous and include VCOs and the high performance reflection-type phase shifter/delay lines described in Chapter 7.

The second category consists of applications where the tuning range is dictated by the difference in the value of the variable capacitance (ΔC) or variable inductance (ΔL).

$$\Delta C = C_{max} - C_{min} = C_{max} \cdot \left(1 - \frac{1}{\text{TCR}}\right) \qquad (12.1)$$

or

$$\Delta L = L_{max} - L_{min} = L_{max} \cdot \left(1 - \frac{1}{\text{TIR}}\right) \qquad (12.2)$$

In this latter category, the required tuning range can be achieved by selecting realisable values for C_{max} (L_{max}) and/or TCR (TIR). Applications that fall into this category include loaded-line phase shifters [27].

12.3.1 Varactor diodes

An example of a near ideal varactor diode is one that has a vertical mesa-type structure. A Schottky junction version of a mesa-type varactor diode and its equivalent circuit model are illustrated in Fig. 12.9. The major advantages of a mesa-type varactor diode are:

(1) high values of TCR are possible with non-uniform doping concentration profiles
(2) very simple equivalent circuit model

(3) bias dependency of the intrinsic elements can be described by simple expressions
(4) small parasitics and very high Q-factors, which can enable terahertz cut-off frequencies
(5) the vertical structure requires minimal area

There are two major disadvantages with mesa-type varactor diodes. The first is that not all foundry processes facilitate complex vertical structures. The second is the high cost of the fabrication processes that can facilitate such structures, especially when non-standard graded doping profiles are required for realising hyperabrupt junction characteristics. One notable example of a monolithic circuit that employs Schottky junction mesa-type varactor diodes was reported by Chen et al., in a Ku-band analogue phase shifter [28].

An example of a planar-type varactor diode is illustrated in Fig. 12.10. Here, the anode and cathode electrodes are interdigitated to realise large values of C_{max} [29, 30]. With this interdigitated planar Schottky varactor diode (IPSVD), the value of TCR can be improved with a deep active n-layer implant, since edge breakdown can be made to occur before pinch-off. Moreover, the TCR can be further increased with a non-uniform doping profile. By introducing a deeper $n+$ contact layer, the resistance to the depletion region is reduced at all bias levels, thus, increasing the Q-factor. One notable example of a monolithic circuit that employs such customised IPSVDs was reported by Dawson et al., in an X-band analogue phase shifter [30]. An interdigitated planar p-n junction varactor diode can also be implemented in monolithic technology, as demonstrated by Pao et al. in a 60 GHz analogue phase shifter [31]. There are two main disadvantages with these interdigitated varactor diodes. The first is that it is difficult to accurately characterise these devices over large bias voltage ranges and across a wide bandwidth. The second is the high cost associated with additional selective ion implantation/MBE or VPE grown layers.

A low-cost method of implementing an IPSVD is to use standard-implant MESFET processing. Here, the varactor diode can be realised by connecting together the drain and source terminations of a standard-implant MESFET, resulting in a single Schottky junction. The bias potential is then applied across the drain/source (cathode) and gate (anode) terminations. Unlike the mesa-type varactor diode, these IPSVDs do not use air-bridges or additional selective ion-implantation/MBE or VPE grown layers. As a result, these IPSVDs are less expensive to produce than the equivalent mesa-type varactor diodes, but have a limited TCR.

At King's College, a great deal of work has been undertaken to characterise experimental IPSVDs fabricated at GMMT, using their F20 foundry process [32-34]. It has been demonstrated that virtually any size of IPSVD can be very accurately characterised using an equivalent circuit model, with both forward and reverse bias potentials applied, from DC into the millimetric frequency range. These equivalent circuit models are more complicated than those of mesa-type varactor diodes, containing more extrinsic parasitic elements and having the bias dependency of the intrinsic elements described by more complicated expressions.

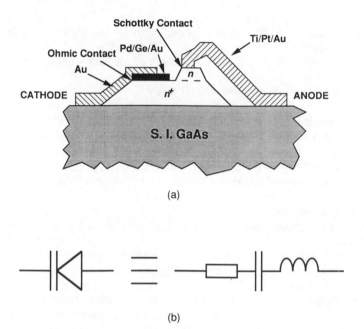

Fig. 12.9 Mesa varactor diode: (a) Cross-section; (b) Equivalent circuit model

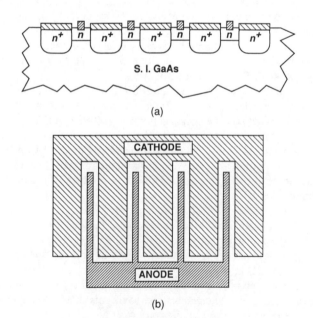

Fig. 12.10 Planar varactor diode: (a) Cross-section; (b) Layout

12.3.1.1 Heterojunction-based varactor diodes
Varactor diodes are also required for MMICs using HEMT or HBT devices. Hence, the microwave modelling of varactor diodes fabricated using heterojunction based technologies has also been investigated [35, 36]. As with the MESFET, a Schottky junction varactor diode can be realised by connecting together the drain and source terminations of a HEMT. The bias potential is then applied across the drain/source (cathode) and gate (anode) terminations. Similarly, interdigitated planar p-n junction varactor diodes can be realised by applying a bias across the collector (cathode) and base (anode) terminations of a HBT, with the emitter left open circuit.

As examples, a commercial T-gate Fujitsu HEMT and an experimental GEC-Hirst interdigitated HBT were implemented as varactor diodes and characterised up to 10 GHz. From the bias dependency associated with intrinsic elements in their one-dimensional equivalent circuit model, it was found that the performance of the HEMT varactor is comparable with that of a short-access-region MESFET varactor [35]. To increase the TCR, a double heterojunction HEMT-based varactor diode has also been investigated and applied to a 6 GHz analogue phase shifter [36]. With the standard HBT-varactor, the bias dependency of the junction capacitance can be accurately modelled using the simple expression used for the mesa-type varactor diode. However, they can suffer from low TCRs and excess series resistance due to the high sheet resistivity of the p-type layer associated with the base electrode [35].

12.3.2 Tunable active inductors

With MMIC technology, large inductances are traditionally realised using planar spiral or stacked spiral inductors. Unfortunately, these inductors can suffer from excess series resistance and undesirable high frequency resonances. Thus, spiral inductors can place a fundamental limit on the performance of a circuit. As a result, a considerable amount of interest has been shown in the use of active inductors.

The first high-Q active inductors were reported over two decades ago [37-40]. Here, the single inverted transistor topology did not facilitate independent electronic tuning of both the inductance and series resistance. The use of these inductors to realise L-C resonator type active filters was presented in two particularly notable papers [38, 39]. More recently, the active inductor demonstrated by ATR in Japan [41] has created considerable interest in active inductors for MMIC applications. Here, Hara *et al.* demonstrated the implementation of a relatively lossy broadband monolithic active inductor using a common-source cascode-FET with resistive feedback topology. Two of these active inductors were successfully employed in a broadband actively matched amplifier. Later, a significant reduction in the series resistance of the active inductor was measured with the use of a common-gate FET feedback arrangement, and a completely lossless active inductor was simulated using a common-gate cascode-FET feedback element [42].

Using the simple common-source cascode-FET with resistive feedback topology, Bastida *et al.* demonstrated the inductance tuning ability of the lossy active inductor [43]. Here, a floating cold-FET was used to implement a voltage controlled feedback

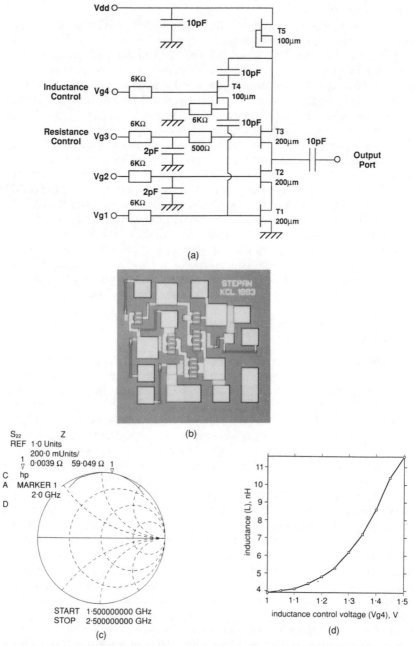

Fig. 12.11 The modified active inductor topology: (a) Circuit diagram; (b) Photograph; (c) Measured impedance on the Smith chart; (d) Tuning curve

resistor. The application of these tunable active inductors (TAIs) was demonstrated in a hybrid 180° analogue reflection-type phase shifter, having MMIC TAI reflection terminations, and in a monolithic band tunable amplifier.

A TAI has recently been presented where both the inductance and series resistance can be varied across wide ranges [26]. This tuning ability is highly desirable for realising VCOs, analogue phase shifters, adaptive active filters and high performance active filters. The topology of this TAI is illustrated in Fig. 12.11 (a). The common-source cascode-FET arrangement has been used and is implemented with T_1 and T_2. Since a high series resistance results from this basic topology, a common-gate FET feedback element is used: T_3. The gate bias of T_3 effectively controls the level of series resistance. However, this feedback transistor will only produce a negative series resistance if its gate is tied down with an additional shunt resistance, R_2. With the use of a resistive load, a cold-FET, T_4, can now be employed as a variable feedback resistor to control the value of inductance. The resistive load is implemented with an active load, T_5, which has the additional advantage of requiring minimal chip space, when compared to spiral load inductors. Finally, a large coupling capacitor was included for DC blocking purposes only.

A TAI was designed using the component values shown in Fig. 12.11 (a). The GaAs MMIC TAI was fabricated at GMMT using the F20 process. A microphotograph of the MMIC realisation is shown in Fig. 12.11 (b). The TAI has an active area of less than 1.0 mm². To illustrate its potential, the TAI was tuned to achieve its maximum Q-factor, at an arbitrary centre frequency of 2 GHz. The measured impedance is shown in Fig. 12.11 (c), demonstrating a Q-factor of more than 15 000. Here, the drain voltage, V_{dd}, is 11.0 V and the drain current is 19 mA. At 2 GHz, and with the series resistance maintained at $R<0.004\ \Omega$, the inductance was varied. The resulting tuning curve of the TAI is shown in Fig. 12.11 (d), indicating a variation in inductance of between 3.9 and 11.6 nH. Therefore, for this TAI design the TIR=3.0 and the frequency tuning ratio is 1.7. If this TAI is combined with an IPSVD, the frequency tuning ratio of the resulting active resonator will be a product of the two individual tuning ratios attributed to the TAI and IPSVD.

12.3.3 Variable resistance elements

Variable resistors are found in adaptive circuits and general microwave signal processing applications. Examples include tunable active inductors [26, 43], group delay synthesisers [24], balanced modulators [24, 44], analogue π-attenuators [45] and SPDT switches [46, 47].

In hybrid-MIC technology, a tunable resistor is traditionally realised with a current-controlled PIN diode. This implementation has the advantages of having a very large dynamic resistance range and small parasitic elements, making it ideal for realising high quality switches. Unfortunately, PIN diodes are not normally used in monolithic applications. This is because additional and, therefore, very expensive foundry processing is required to realise PIN diodes. Two examples where PIN diodes have been employed in monolithic switched-filter phase shifters were reported by

Coats *et al.* [46] and more recently Teeter *et al.* [47]. Because PIN diodes require control currents of the order of milliamperes the significant control power may prohibit its use in applications that have a limited DC power supply, such as in hand-held applications or beam forming networks employed in mobile/satellite phased array antennas.

In MMIC technology, a tunable resistor is traditionally realised with a voltage-controlled cold-FET [24, 26, 43, 44, 45]. An equivalent circuit model for a floating cold-FET is illustrated in Fig. 12.12. A standard-implant MESFET (or HEMT) is used with no DC bias applied to the drain terminal. A variable bias voltage is applied across the gate-source terminals. With a depletion-mode MESFET, when the negative bias voltage is increased from zero the depletion region within the drain-source channel expands. As a result, the drain-source channel resistance increases. When the bias voltage reaches the pinch-off voltage, the channel resistance will be very large. Ideally, the cold-FET is represented by a single channel resistance whose values at zero-bias and pinch-off are zero and infinite, respectively. In practice, the channel resistance at zero-bias is significantly larger than zero, and is inversely proportional to the size of the MESFET. The large channel resistance at pinch-off is shunted by the parasitic capacitances, and these are roughly proportional to the size of the MESFET. The equivalent circuit model shown in Fig. 12.12 assumes that the gate length is less than approximately 1µm, so that it is small enough not to have a significant shunt capacitance to ground, and that the gate terminal is perfectly isolated at all frequencies.

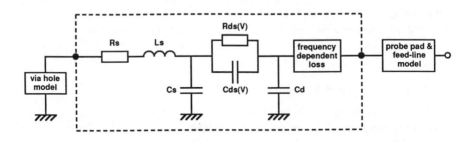

Fig. 12.12 Model for a floating cold-FET variable resistance element

An application of a floating cold-FET is in a tunable active inductor, described earlier, while an application of a shunt cold-FET is an analogue reflection-type attenuator, described in Chapter 8. With the former example, a large coupling capacitor was required on either side of the floating cold-FET, in order to ensure complete DC isolation of the drain terminal and so as not to interact with the gate bias voltage. With the latter example, large coupling capacitors do not have to be included in the reflection terminations. Instead, it may be more convenient to locate these DC blocking capacitors at the input and output ports of the modulator.

12.3.4 Tunable negative resistance elements

A lossless microwave circuit can be realised by cancelling out the dissipative losses in its resonators and transmission lines by introducing tunable negative resistance devices or circuits. Over a limited bandwidth, these devices can be represented by an equivalent negative resistance in conjunction with an inductance or capacitance. Usually, both resistive and reactive elements are tunable, but not independently. The modified active inductor [26] can provide negative resistance, and also a monolithic tunable negative resistance circuit has been demonstrated [48]. This circuit employs a FET with a series feedback capacitor in the source; negative resistance is presented at the gate terminal. Other types of negative resistance circuit have been used by Hopf *et al.* [49].

12.4 Active filters

In an integrated communications transceiver the filters and diplexers required are so large and expensive that the full benefits of MMIC technology are not readily retained when the complete RF subsystem is considered. There are many applications, however, where a low cost and/or low mass integrated circuit based module is an essential part of the enabling technology. For example, advanced satellite payloads using phased arrays may have hundreds of transceiver modules, and it is desirable to integrate the filters into the MMIC modules. Conventional passive filters are not a practical solution, however, because of the size limitations of MMICs and the low Q-factor of MMIC lumped elements. It is only at frequencies well over 20 GHz that microstrip or CPW coupled-line filters might be acceptably small. As a result, there is a great deal of activity worldwide in the area of MMIC active filters. The most important techniques which have been reported in the literature for microwave active filters are active inductor based filters, filters employing actively-coupled passive resonators, techniques using negative resistance elements, and transversal and recursive filters using FETs.

The advantages of active filters are:

(1) small size and mass
(2) low cost in mass production
(3) high selectivity
(4) easy integration with amplifiers, mixers, oscillators, etc.
(5) potential for electronic tuning

454 *Advanced techniques*

However, the drawbacks associated with active techniques are:

(1) poor noise figure
(2) non-linearity
(3) danger of oscillation
(4) complex bias circuitry and significant DC power
(5) sensitivity to fabrication tolerances
(6) environmental sensitivity

Clearly, for small size these techniques are very attractive, but at the same time they are unlikely to be used for applications such as diplexers where high power and low noise are important.

12.4.1 Active inductor based filters

The majority of the problems in passive lumped element filters arise from inductors, which are lossy and have low self-resonant frequencies. To overcome these limitations, it is possible to use active inductors in MMIC filters. The cascode-FET active inductor was described earlier as one of the types of electronically tunable component. Fig. 12.13 (a) shows a photograph of a three resonator filter using active inductors. This first design [26] achieved 0 dB insertion loss in the 100 MHz passband centred at 2 GHz. The broadband measured response of Fig. 12.13 (b) shows that the complete removal of inductors from the filter leads to an excellent spurious-free range: the 'spike' in the response is the passband, which is not fully shown because of the limited number of measurement points. The transmission response exhibits no spurious responses above the passband and the rejection is over 25 dB up to 30 GHz.

12.4.2 Actively-coupled passive resonators

The well known cascaded cells approach to active filter design involves breaking the required filter transfer function down into its constituent factors, and then realising each factor with the appropriate passive filter section. The complete filter transfer function is then achieved by cascading these sections together via amplifiers which prevent the resonators interacting. Sussman-Fort presented a second-order bandpass filter using this technique [50] and Bonetti (Comsat Labs) applied the technique to the design of a 4 to 8 GHz bandpass MMIC active filter [51].

12.4.3 Techniques using negative resistance elements and active loops

In these techniques the Q-factor of a resonator is increased by amplifying the signal within the resonator either with an active loop or by coupling negative resistance into the resonator. The active loop technique has been applied to a microstrip resonator by

Advanced techniques 455

the University of Limoges. Negative resistance elements formed from transistors have been used by Itoh [52] to achieve a 200 MHz wide bandpass filter centred on 10.5 GHz, with approximately 20 dB rejection 200 MHz from the band edges. For a single pole filter a 1 dB compression level of -13 dBm was reported, measured at the input. This shows that the power handling of active filters can be expected to be considerably worse than an amplifier using the same device, and this is due to the high peak values of voltage or current which are inevitable in a high-Q resonator.

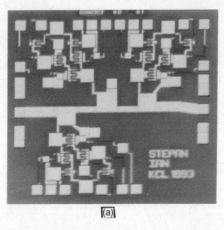

(a)

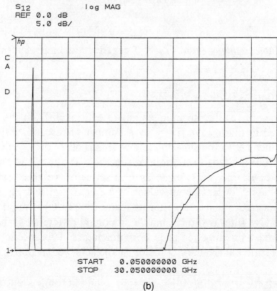

(b)

Fig. 12.13 Active filter using active inductors: (a) Photograph; (b) Broadband response

456 Advanced techniques

At King's College the use of a simple FET negative resistance circuit has been demonstrated for loss compensation. This has been applied to microstrip dual-mode ring resonator filters [48, 53] and to lumped-element monolithic filters [54]. The topology of the lumped-element monolithic filter using this negative resistance technique is shown in Fig. 12.14. This filter is far less complicated than the active inductor filter and has lower DC power consumption.

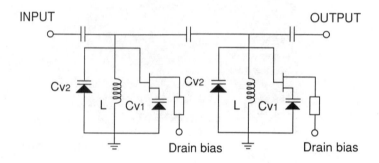

Fig. 12.14 Active filter topology employing FET negative resistance elements

12.4.4 Transversal and recursive filters

Transversal and recursive filter techniques have been taken from the digital filter domain and applied to microwave filters by a number of researchers [55, 56]. The basic transversal filter uses a transmission line with taps taken off at regular intervals, with the outputs from the taps then summed together. The result is that the output waveform at a particular instant is made up of a series of weighted samples of the past, present, and 'future' instantaneous values of the input signal. It should be noted, however, that for a true transversal filter as many as 200 taps might typically be required. The recursive filter is similar, but the important difference is that it employs a second series of taps, this time fed back from the output. This additional flexibility has the important result that arbitrary Q-factor can be achieved as positive feedback is employed.

The well known MMIC active filter work by Schindler and Tajima [57] uses a combination of conventional filtering with transversal elements to 'trim' the response. Their filter is based on the distributed amplifier, but uses bandpass structures directly integrated into the gate and drain lines. They achieved a 9.7 to 11.4 GHz passband with 3 dB insertion loss, and with 40 dB rejection approximately 1 GHz from the band edges. Work at the University of Limoges has involved using Lange couplers in transversal and recursive filters [58]. The Lange splits the signal into two paths, each of which has a certain weighting factor, and then the two paths are recombined in a second Lange coupler. The measured result of this simple filter shows a gentle bandpass response, as the number of taps is so low. It is shown that the true transversal filter has a cyclic frequency response. This feature was not apparent in Rauscher's or Schindler's filters because they used a 'hybrid' approach. Generally, it appears that transversal and recursive techniques are mainly applicable to broadband filters, such as octave bandwidths.

12.4.5 Practical applications

MMIC active filters have tremendous performance capability and can be directly integrated with other circuitry. Although they have poor noise figure and limited power handling, they are expected to be used first in applications where these performance criteria are not so important, such as in IF chains. Process variations make some form of electronic tuning essential, and varactor tuning is expected to play a vital role in narrowband active MMIC filters. Varactor diodes provide well controlled tuning characteristics, and draw no current. Hence a multi-pole filter could be tuned up by employing varactor bias potentiometers, which become analogous to the tuning screws in waveguide cavity filters. Such an approach cannot easily be avoided if brick-wall responses are required. Other tuning methods could involve control of DC bias, laser-trimming of resistors, switched spiral inductors, and the use of conveniently removable air-bridges in the circuit. With more research, it is certain that active filters will be used extensively in MMIC subsystems.

12.5 Active splitters, combiners, baluns, isolators and circulators

12.5.1 Active power dividers and combiners

The most basic form of 0° power divider consists of a pair of transistors with their inputs directly connected but with separate outputs. This is an excellent method, providing gain, near perfect amplitude and phase balance, good reverse isolation and isolation between the two (or more) outputs. The basic power combiner consists of a pair of transistors with separate inputs and a common output. However, strictly speaking this is not a true power combiner because two input signals at different frequencies would create intermodulation products as the transistor outputs are not isolated.

It has been demonstrated that MESFETs can be integrated into CPW/slotline structures to realise active power dividers and combiners (referred to as line unified or LU-FETs [59, 60]), as illustrated in Fig. 12.15. This method has been used in an active circulator [61] and an injection-locked oscillator [62].

458 Advanced techniques

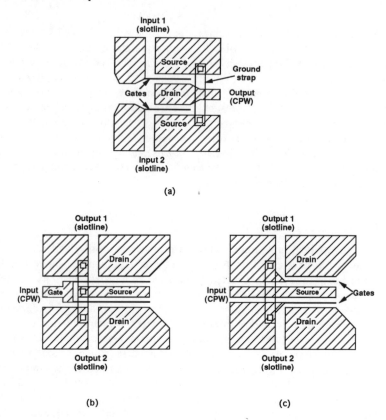

Fig. 12.15 LU-FET examples: (a) in-phase power combiner; (b) and (c) power dividers

12.5.2 Active baluns

An ideal balun is a three port device or circuit that splits a single unbalanced input signal into a balanced output signal (consisting of two mutually isolated signals that are 180° out-of-phase and of equal power) or, conversely, combines a balanced input into a single unbalanced output signal. Baluns find applications in double-balanced mixers, phase splitting-power combining analogue phase shifters and 180° bit digital phase shifters.

The most popular type of transmission-line balun is the Marchand balun. Recent monolithic implementations of this type of balun have used microstrip edge-coupled lines with lumped MIM coupling capacitors [63], Lange couplers [64, 65], and coplanar waveguide edge-coupled lines [66]. The passive balun has the advantage that it can be used for both power dividing and power combining applications and it is ideal for use at millimetric frequencies. Unfortunately, they can be prohibitively large for applications below X-band and suffer from excessive levels of insertion loss. As a

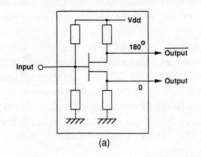

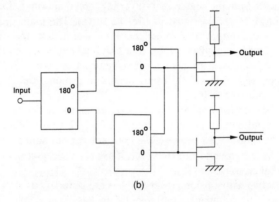

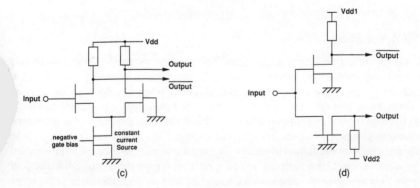

Fig. 12.16 Active baluns: (a) Source/drain; (b) Cross-coupled; (c) Push-pull; (d) Common-source/common-gate FET pair

460 Advanced techniques

result, active baluns have been developed to overcome both of these shortcomings. Active baluns are unilateral and must be designed as either power splitters or power combiners.

An example of a simple active balun power combiner was demonstrated recently in monolithic form by Jouanneau-Douard *et al.* [67]. The circuit design is based on a totem pole arrangement of two MESFETs: one in common-source configuration and the other in common-drain configuration. The two balanced input signals are applied to the two gates and the unbalanced output signal is taken from the common hot ends of the transistors. The measured results of this compact implementation demonstrated an amplitude balance of ±0.5 dB and a phase balance of ±10°, across the 1 to 18 GHz frequency range.

The simplest technique for implementing an active balun power divider is to take the balanced output from the source (non-inverting) and drain (inverting) terminations of a single MESFET, as illustrated in Fig. 12.16 (a). The main drawback of this technique is that the output port impedances and parasitics are unequal, causing significant amplitude and phase imbalances. A technique for overcoming these problems, by cross-connecting identical active baluns as shown in Fig. 12.16 (b), has been investigated and successfully implemented in a monolithic L-band bi-phase modulator by Goldfarb *et al.* [68].

Another technique for implementing an active balun power divider is to use a differential push-pull amplifier, as illustrated in Fig. 12.16 (c). With this technique imbalance can be introduced because the constant current source transistor is not connected to a virtual ground [69]. However, it has been successfully employed in an L-band double-balanced mixer [70].

The previous two active balun power dividers are generally used in low frequency applications. A high frequency implementation has been demonstrated using a common-gate/common-source pair, illustrated in Fig. 12.16 (d). This technique has been successfully employed in 3 to 18 GHz monolithic double-balanced mixer [65].

12.5.3 Active isolators

Isolators provide a unilateral signal path. This may be required in order to absorb unwanted reflected signal power or to avoid interaction within various parts of a circuit or system. Isolators are sometimes required when the ports of devices or circuits have to be protected from unwanted impedance mismatches. For instance, when the load impedance presented to an oscillator changes, the conditions necessary to maintain satisfactory operation may not be met. This may cause the amplitude and/or frequency of oscillation to change, or even a cessation of oscillation altogether. As another example, the performance of a reflection-type delay line (discussed in Chapter 7) is sensitive to impedance mismatches seen by both the input and output ports. As a consequence, undesirable ripples in the frequency responses of the relative phase shift may result when two, or more, reflection-type delay lines are cascaded. With MMIC technology, active isolators are used to buffer the output port of an oscillator or provide sufficient isolation between reflection-type delay line stages. The characteristics of an ideal active isolator include at least 20 dB of reverse isolation, constant forward insertion loss, and input and output return losses better than 15 dB, all maintained across the desired bandwidth of operation. A simple method of implementing an active isolator is to use a cascode-FET amplifier. This circuit can provide a large gain and high reverse isolation.

With classical distributed amplifiers, a constant insertion gain can be maintained across multi-octave bandwidths. However, the isolation performance degrades with increasing frequency, due to coupling through the gate-drain capacitance. A novel alternative to the distributed amplifier, for implementing a multi-octave MMIC active isolator, was proposed by Pyndiah and Bogaart [71]. The circuit consists of simple feedback amplifier with the input and output ports swapped over. A single MESFET with an active load makes this a very compact design. The measured results demonstrated a minimum isolation of 20 dB, constant insertion loss of 9 dB, and input and output return losses of 20 dB and 18 dB, respectively, all maintained within a 0.5 to 20 GHz frequency range.

12.5.4 Active circulators

As its name suggests, a circulator is a multiple-port device that can route a signal unilaterally from one RF port to another. Fig. 12.17 shows three different types of circulator. Common applications for a circulator are illustrated in Fig. 12.18. All of these can be implemented using any one of the types of circulator shown in Fig. 12.17. The circulator is used to separate two signals propagating in opposite directions. In a diplexer, these are the transmitted and received signals at port 2. In a reflection-type phase shifter, parametric amplifier or reflectometer, they are the incident and reflected waves corresponding to a terminating impedance, Z_T, at port 2. In an isolator, any reflections entering port 2 are circulated into the matched load at port 3.

A general approach to realising a classical circulator consists of a symmetrical arrangement of three directional couplers and three isolators. For an MMIC realisation, Bahl [72] proposed the use of either Wilkinson 0° couplers or Lange couplers with active isolators. Unfortunately, this approach may require a prohibitive amount of active area for applications below X-band. A highly economical solution would be to remove the directional couplers altogether and to employ the type of active isolator proposed by Pyndiah and Bogaart [71]. This approach was adopted three decades ago by Tanaka *et al.* [73], using bipolar junction transistor technology, in a very low frequency active circulator.

Until recently, there has been little interest in the active circulator. This was mainly due to its narrowband performance and the problems associated with a hybrid realisation. Nowadays, more interest is being shown in active circulators, since their bandwidths have increased considerably as a result of the advances in transistor technology. Also, active circulators are totally compatible with monolithic technology [61, 72, 74-78].

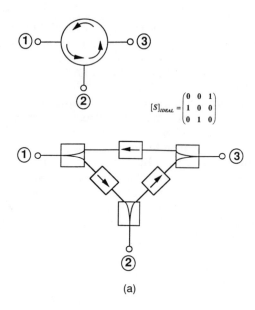
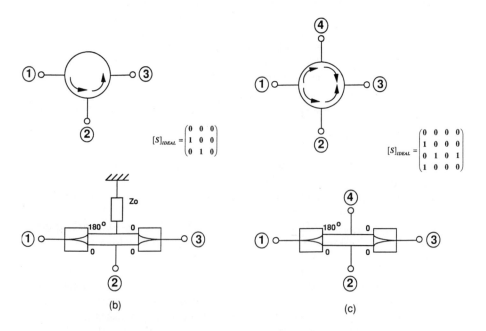

Fig. 12.17 Types of circulator: (a) Classical circulator; (b) Quasi-circulator; (c) Unilateral four-port junction

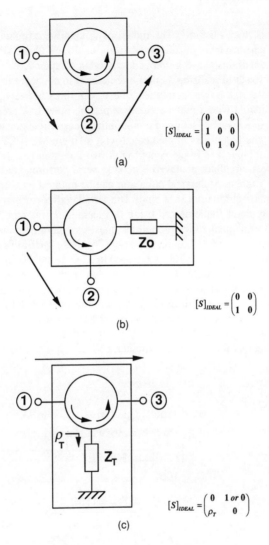

Fig. 12.18 Circulator applications: (a) Diplexer; (b) Isolator; (c) Reflection-type phase shifter/attenuator, parametric amplifier, or reflectometer

A MESFET equivalent of the classical circulator demonstrated by Tanaka *et al.* was reported by Dougherty [75]. A modified version of the topology, employing standard F20 4 × 75 μm MESFETs, is shown in Fig. 12.19. With the MESFETs operating at half I_{dss}, the simulated response for this circulator shows that 20 dB of isolation and a constant insertion loss of 3.7 dB are possible across more than a decade bandwidth. From an initial mask layout design, this circulator requires an active area of less than 1.0 mm². The model of this active circulator was used to simulate a high performance, decade bandwidth, analogue phase shifter [79]. The overall performance of this phase shifter is described in Chapter 7.

A general approach to realising a quasi-circulator is to connect the outputs from a balun to the inputs of a power combiner: one of the links between the balun and combiner is terminated with a matched load, while the other link provides the main circulating port. An active realisation of a monolithic quasi-circulator was reported by Hara *et al*. [61]. The balun was implemented with a single MESFET power divider and the power combiner was implemented with two common-gate MESFETs. This very compact quasi-circulator achieved a good general performance across a 0.1 to 15 GHz frequency range. A classical circulator can be realised by simply connecting three identical quasi-circulators in a ring. The simulated performance of a 0.1 to 10 GHz active circulator implemented using this technique was reported by Hara *et al*. [61]. A HEMT equivalent of the monolithic quasi-circulator demonstrated by Hara *et al*. has been recently reported by Kother *et al* [74]. Here, pHEMTs were used in a 40 GHz realisation, while InP-HEMTs were used in 55 GHz and 80 GHz realisations.

The topology of a unilateral four-port junction is the same as for the quasi-circulator, but with the matched load removed. Therefore, both links can provide main circulating ports. An active realisation of a monolithic-based unilateral four-port junction has been reported [77] and used in a novel reflectometer [80]. The balun was implemented with a distributed power divider and the power combiner was implemented with a distributed power combiner.

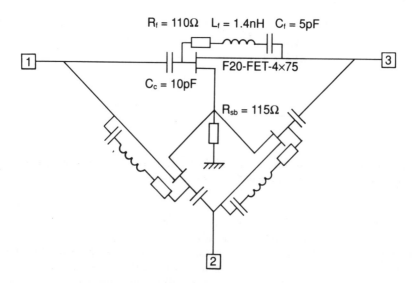

Fig. 12.19 Active circulator circuit diagram

12.6 Distributed circuits

The distributed (or travelling-wave) amplifier, described in Chapter 5, achieves ultrawideband operation by incorporating the transistor input and output capacitances into so-called 'artificial transmission lines' which are in fact low-pass ladder networks. The beauty of this technique is that matching can easily be achieved over wide bandwidths such as 1 to 20 GHz. As a result, the travelling-wave matching technique has found application in a wide range of other circuit functions such as mixers, baluns, frequency doublers, power splitters and combiners, switches, isolators and quasi-circulators, and impedance transformers. Fig. 12.20 summarises the basic circuit topologies for these functions.

The original distributed mixer [81] used single-gate FETs and employed a conventional power combiner to isolate the RF and LO signals. Monolithic distributed mixers have been reported with a wide range of different topologies. The two most important techniques are perhaps the dual-gate FET distributed mixer [82], shown in Fig. 12.20 (a), and the matrix distributed mixer [83] shown in Fig. 12.20 (b). Distributed mixers can achieve conversion gain quite readily, but they have not so far demonstrated sufficiently low noise figures to replace conventional diode mixers. Distributed active baluns have been used for wide-band mixers, the most common technique being to use common-gate and common-source FET pairs on a common input line [84] as shown in Fig. 12.20 (c). Distributed active baluns using a gate-line termination technique, as shown in Fig. 12.20 (d), have also been demonstrated [85]. The arrangement of common-gate and common-source FET pairs has also been used to realise balanced mixers [86] and the balanced frequency doubler [87] shown in Fig. 12.20 (e).

Power splitters and combiners can also be realised with the distributed amplifier structure. A distributed active splitter [88] has a single gate-line and two drain-lines, as shown in Fig. 12.20 (f). The distributed active combiner [89] has two gate-lines with the FETs connected to a common drain-line, as shown in Fig. 12.20 (g). In some respects this is not a true power combiner because input signals at different frequencies would suffer intermodulation via the drain-voltages of the FETs. Nevertheless, this effect is unimportant for many applications, and this combiner has been used successfully in distributed mixers [90], for example.

The travelling-wave switch [91] has been demonstrated with DC to 20 GHz operation, by incorporating multiple switch FETs in artificial transmission lines as shown in Fig. 12.20 (h). Other applications of the distributed amplifier structure include isolators and unilateral circuits using the reverse drain-line termination isolation properties [92], and quasi-circulators using a balanced circuit approach [77]. Finally, distributed amplifier impedance transformers using a number of topologies have been demonstrated [93].

A number of schemes have been reported for improving the performance and bandwidth of distributed amplifier circuits. These techniques, which are described in Chapter 5, are generally intended to compensate for attenuation along the artificial transmission lines, to increase the cut-off frequency for a given device capacitance, or increase the power handling in some way.

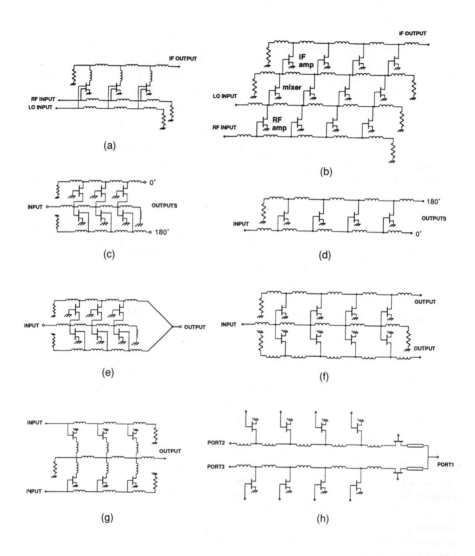

Fig. 12.20 Distributed circuits: (a) Dual-gate FET mixer; (b) Matrix mixer; (c) Active balun with common-source/common-gate FETs; (d) Gate-line termination balun; (e) Frequency doubler; (f) Power splitter; (g) Combiner; (h) Switch

12.7 Modulators, demodulators and frequency translators

This section describes the design of modulators, demodulators and frequency translators. Modulators and demodulators operating directly at the microwave (or millimetre-wave) carrier frequency are an attractive means of simplifying the hardware requirements in a communication system. Frequency translators take many forms, the most familiar of which is the RF-to-IF downconverter used in superheterodyne receivers. However, other types are often used in which the input and output signals are both at microwave frequencies. The difference in frequency may be quite large (many GHz) or very small (a few kHz). Rather different approaches are necessary in these two extremes.

Mixers are a fundamental element for the design of modulators, demodulators, transmitters and receivers, frequency translators and phase-locked loops. Double-balanced diode mixers are a ubiquitous component since they have isolation between all three ports and excellent rejection of unwanted mixing products. For applications up to a few GHz they are available very cheaply as packaged modular components from companies such as Mini Circuits. The double-balanced diode mixer is a very adaptable component which has many different uses, including:

Downconverters
Upconverters
Frequency translators
Analogue modulators and demodulators (e.g. DSBSC)
ASK modulators and demodulators
BPSK modulators and demodulators
Phase detectors

The distinction between the use of the mixer in these applications can be very subtle. Double-balanced diode mixers have this multifunctionality because of the DC-coupled IF port, the complete port-to-port isolation and the high level of rejection of unwanted mixing products. However, for individual applications only some of these parameters may be important.

On MMICs it is very hard to reproduce the traditional double-balanced diode mixer with small chip area and good performance. However, when considering the use of alternatives, such as FET mixers, it is important to realise that this multifunctionality is generally lost. The Gilbert cell multiplier is a good replacement in many applications, but the inherent unilateral nature of an active mixer means that, for example, a mixer designed for a downconverter cannot then be used in reverse for an upconverter. So, for modulators, demodulators and general frequency translator applications, the question must be first addressed as to what exactly is required from the mixer and which parameters are important. Furthermore, in some cases it may be preferable to use a completely different approach: for example, rather than using a mixer, an ASK modulator could use a switch, and a BPSK modulator could use a phase shifter.

12.7.1 Frequency translators

Frequency translators take many forms, and it is useful to separate the various types into three classifications: large-shift, small-shift, and ultra-small-shift. This classification is based on the fractional frequency shift (FFS) that is to be performed:

$$\text{FFS} = \frac{\text{translation frequency}}{\text{input centre frequency}} \qquad (12.3)$$

Large-shift frequency translators (FFS>=50%) are used as up-converters in transmitters and down-converters in superheterodyne receivers and normally use filters to remove unwanted mixer outputs. Small-shift frequency translators (FFS<50%) are employed in non-regenerative transponders (or repeaters) used in terrestrial relay stations and space communication sub-systems, where frequency separation is required to avoid the signal transmitted from the output high power amplifier from interfering with the low noise amplifier at the input. For example, in this latter application, the higher frequency uplink may be at Ku-band, while the lower frequency downlink may be at X-band. Ultra-small-shift frequency translators (FFS<1%) are employed in numerous applications, including:

(1) novel applications in communication systems, including frequency hopping, frequency-shift keying (FSK), and frequency division multiplexing [94]
(2) calibration of Doppler radars, by emulating accurate Doppler return signals
(3) smart radar jammers, by generating false Doppler return signals for providing velocity deception and for creating ghost targets [95]
(4) homodyne vector network analysers [96]
(5) frequency scanning antennas [97]

An important application of an ultra-small-shift frequency translator is direct baseband modulation of an RF carrier. Here, the input signal is the unmodulated RF carrier, the translating input is the baseband signal, and the output is a modulated RF carrier.

Fig. 12.21 shows three spectra representing these three FFS classifications: (a) large-shift, (b) small-shift, and (c) ultra-small-shift. The large-shift case is illustrated as a downconverter which might be used in a receiver. The terminology is very familiar: the local oscillator is the pump signal, the RF is the input signal, the IF is the low frequency output signal (the difference frequency, RF–LO), and the image is the mirror-image of the RF on the other side of the LO. The sum frequency, RF+LO, is easily filtered out.

With reference to Fig. 12.21 (b), in the case of small-shift frequency translators there can be considerable confusion in terminology because the local oscillator (the modulating or translating signal) is at a lower frequency than the 'IF' output signal. To further compound this confusion, the IF port of a double-balanced mixer may often be used for this low frequency pump signal. As a result is it less confusing if the RF is labelled RF_{in}, the local oscillator is still labelled LO and the 'IF' signal is labelled RF_{out}. The principle outputs of the mixer are still the sum (RF_{in}+LO) and difference frequencies (RF_{in}–LO). However, the difficulty now encountered is that because of the small LO frequency the sum and difference frequencies are now very close, being equally spaced either side of RF_{in}; they are termed the upper sideband (USB) and lower sideband (LSB), respectively. Since they are very close, they are difficult to separate with filters. Furthermore, as the translation frequency is reduced, a point is reached where the translation frequency is equal to half the bandwidth of the input RF signal and the LSB and USB meet. At this point, only a theoretical brick-wall filter could be considered. Beyond this point, there would be inherent distortion due to spectral overlap. As a result, small-shift frequency translators generally employ phasing techniques (using a pair of mixers or modulators fed with quadrature signals) to cancel out the unwanted sideband at the output. The mixer topology that achieves this cancellation of the undesired sideband is called a single sideband (SSB) modulator.

The case of the ultra-small frequency shift, illustrated in Fig. 12.21 (c), is shown as modulation of the RF carrier with a baseband signal. For ultra-small frequency translations there are alternative methods of frequency conversion, using sinusoidal and serrodyne modulators, which do not employ mixers.

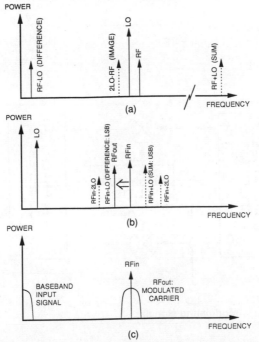

Fig. 12.21 Spectra for frequency translators: (a) Large shift; (b) Small shift; (c) Ultra-small shift

12.7.2 SSB modulators

The topology of the traditional phasing-type SSB modulator is illustrated in Fig. 12.22. It consists of two identical mixers, usually double-balanced, which are fed with phase-quadrature RF input signals. The translating signal is also fed in phase-quadrature. The outputs from the two mixers are combined with an in-phase power combiner. As shown, the SSB modulator produces only the lower sideband. If the translation frequency's 0 and 90° signals are switched then the upper sideband is produced.

In recent years, monolithic small-shift frequency translators have been demonstrated using SSB modulators of this type. One notable example was demonstrated by Parisi [98]. Here, single-balanced mixers were employed in an all lumped-element design. With an input carrier frequency of f_0=6.4 GHz and a maximum modulating frequency of 0.5 GHz a conversion loss of 7 dB, carrier suppression of 20 dB, and image suppression of 18 dB were reported. A more recent SSB modulator example was demonstrated by Baree *et al.* [99]. Here, compact common-gate/common-source balanced FET mixers were employed with a combination of distributed and lumped-element couplers. With an input frequency covering 4.25 to 13.75 GHz and a fixed LO of 2 GHz, a conversion gain of approximately 5 dB and an USB suppression of 20 dB were maintained.

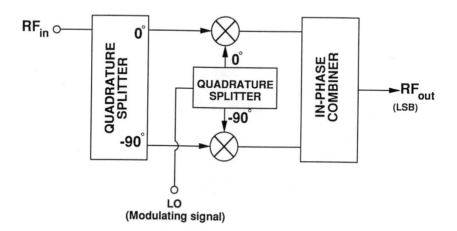

Fig. 12.22 SSB modulator topology

12.7.3 Serrodyne modulators

In practice, ultra-small-shift frequency translations can only be performed using either a non-mixing sinusoidal modulator or a serrodyne modulator. With both techniques, the generation of significant unwanted sidebands can be avoided, improving RF power efficiency (allowing a conversion loss of 0 dB) and removing the need for an unrealisable filter.

The sinusoidal modulator simply introduces a continuous linear change, with time, in its insertion phase. An increasing change causes a positive frequency translation, by an amount which is proportional to the rate of change. Conversely, a decreasing change causes a negative frequency translation. Non-mixing sinusoidal modulators have been realised using a mechanically rotating dielectric slab [100], rotating magnetic field [101], and a phase-splitting power-combining vector modulator. The first two methods are not compatible with monolithic technology.

The serrodyne (or sawtooth) modulator is a derivative of the sinusoidal modulator. Here, a klystron valve, travelling wave tube or phase shifter is modulated with a sawtooth waveform, such that one period of the sawtooth results in an induced phase shift which linearly sweeps through an integer multiple of 360°. With a 360° phase shifter, the amount of frequency translation is equal to the repetition rate of the sawtooth waveform, as illustrated in Fig. 12.23. The direction of frequency translation can be changed simply by inverting the sawtooth profile. The serrodyne modulator employing a 360° phase shifter is highly suited to monolithic implementations.

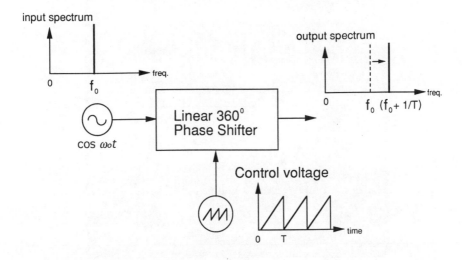

Fig. 12.23 Serrodyne modulator principle

472 Advanced techniques

If a digital phase shifter is used, unwanted sidebands are generated as a result of quantisation errors and any phase errors and amplitude-modulation in the phase shifter response. With an analogue phase shifter, the level of suppression of unwanted sidebands is very sensitive to tuning nonlinearity and amplitude-modulation. In the case of an analogue phase shifter, two different methods can be used to compensate for tuning non-linearity [102]:

(1) The non-linear tuning curve is first measured. The sawtooth is then predistorted to compensate for the tuning non-linearity. This can be accomplished digitally using coefficients stored in a look-up table. The unfiltered output from a DAC is then used directly to modulate the phase shifter.

(2) The sawtooth function deliberately does not cover the full control range, resulting in under-modulation. Garver [103] reported a 7 dB improvement in the levels of unwanted sideband suppression with 15% under-modulation.

Fig. 12.24 shows a photograph of a 24 GHz serrodyne modulator realised with a uniplanar 360° phase shifter and using under-modulation for reduced sidelobe levels [104]. More recently, a serrodyne modulator has been demonstrated in HBT technology [105].

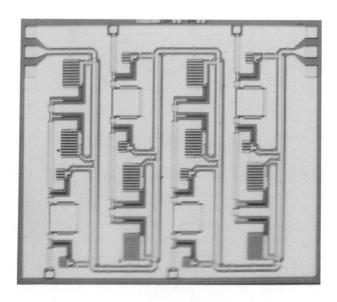

Fig. 12.24 Photograph of a uniplanar serrodyne modulator operating at 24 GHz

12.7.4 Vector modulators

There are two generic types of microwave vector modulator. The first is the phase-splitting power-combining vector modulator. The second type of vector modulator simply consists of an attenuator (α) in cascade with a phase shifter (ϕ). The phase-splitting power-combining vector modulator traditionally comes in either a four-channel or a two-channel version. With the four-channel version, a four-way orthogonal 6 dB power divider is used to create the four orthogonal channels. An individual uniphase amplitude modulator is assigned to each channel. The output signals from these amplitude modulators are then combined using a four-way in-phase 6 dB power combiner. The four-channel vector modulator has an inherent 6 dB insertion loss penalty and, therefore, usually employs active amplitude modulators that can provide enough gain to overcome this loss. A notable example of an ultra-small-shift frequency translator implemented using such active techniques was reported by Mazumder et al. [106]. With the two-channel version (the I-Q vector modulator [44, 108, 109]), a quadrature 3 dB power divider is used to create the two orthogonal channels. An individual bi-phase amplitude modulator is assigned to each channel. The output signals from these amplitude modulators are then combined using an in-phase 3 dB power combiner. The two-channel vector modulator has an inherent 3 dB insertion loss penalty. The α–ϕ vector modulator [110] can be considered as a one-channel vector modulator, with no insertion loss penalty. It can be implemented with either a uniphase amplitude modulator and a 360° variable phase shifter, as shown in Fig. 12.25, or a bi-phase amplitude modulator and a 180° variable phase shifter. The levels of AM/PM conversion and PM/AM conversion of the amplitude modulator and phase shifter, respectively, can be calibrated out by each other.

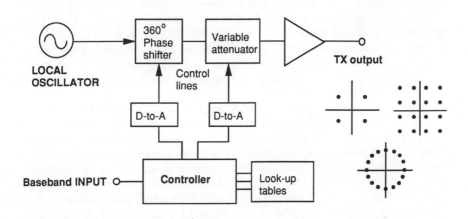

Fig. 12.25 Block diagram of an alpha-phi type vector modulator using an analogue phase shifter and attenuator

Since the vector modulator can realise any combination of signal amplitude and phase it is a very powerful control component with many applications [44, 107]. It can be used in phased array antenna beam-forming, as well as for modulation of a microwave carrier with virtually any digital modulation scheme (ASK, FSK, BSPK, QPSK, and m-QAM). By placing the vector modulator under computer or DSP control, the unit can be made switch between the different types of modulation with no change in the RF hardware.

12.7.5 Modulators and demodulators for digital communications

Direct modulation of the microwave carrier signal has been found to be a very attractive means of reducing hardware complexity in communications applications. In traditional transmitter designs, the modulator operates at an IF frequency (such as 70 MHz) and a complex chain of mixers, filters, and amplifiers is required to upconvert the modulated signal to the transmission frequency. In contrast, if the modulator is designed to operate directly at the microwave carrier frequency, the RF hardware requirement is minimal. This approach has been used for low cost VSAT terminals, and it has been shown that the output spectrum can be band-limited by pre-filtering the baseband data [111, 112]. Simple modulators for schemes such as ASK, BPSK, and FSK can be designed using a wide range of techniques: ASK modulators could use switches, BPSK modulators could use analogue or digital 180° phase shifters. FSK can be achieved with a switched VCO, or using one of the frequency translation techniques described previously.

For quadrature modulation schemes, such as QPSK and m-QAM, the I-Q vector modulator, shown in Fig. 12.26, is used widely [113, 114]. The input data is passed into a serial-to-parallel converter and split into two parallel data streams. These data streams are passed into digital-to-analogue converters (DACs) to produce quantised bi-polar PAM signals: In the case of 16-QAM, for example, each series of four bits is split into two bits for the I-channel and two bits for the Q-channel, leading to a symbol with one of four levels in each channel and 16 possible signal states in the constellation. After baseband filtering, one of the PAM signals is applied to the in-phase (I) channel input of the I-Q vector modulator, while the other is applied to the quadrature-phase (Q) channel input.

It is useful to think of a QAM signal as being two half data rate ASK signals that occupy the same bandwidth but with an orthogonal phase relationship. A special case exists with quaternary (or quadrature) phase shift keying (QPSK) as it is both four-level PSK and four-level QAM. Therefore, a QPSK signal can be generated using either a four-state variable phase shifter or implemented with two BPSK modulators. With the latter technique, the bi-phase amplitude modulators in the I-Q vector modulator can be replaced with TTL controlled binary 180° phase shifters and the DACs can be removed altogether. The traditional I/Q demodulator is illustrated in Fig. 12.27. The orthogonal output signals from the I and Q-channels are applied to baseband filters and then to analogue-to-digital converters (ADCs), the outputs from which are multiplexed to produce the originally transmitted data.

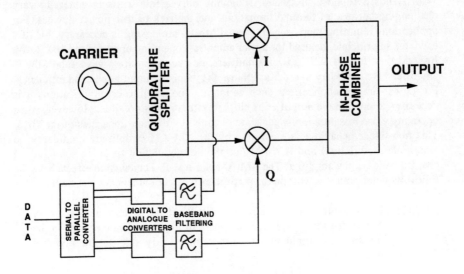

Fig. 12.26 I-Q vector modulator for digital communications

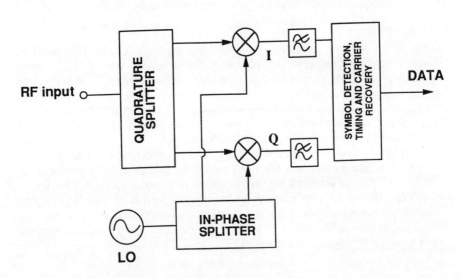

Fig. 12.27 I-Q demodulator for digital communications

12.8 Linearisers and high efficiency power amplifiers

With standard foundry processes, achieving only a few watts is generally very challenging because of thermal limitations and the lack of true power devices. For applications requiring more power, special MMIC processing is necessary. MMICs can then, in principle, be used for power amplifiers operating up to around 10 - 20 W. Beyond this, hybrid-MIC power amplifiers, or even travelling-wave tube (TWT) amplifiers, are necessary in order to achieve the required output power and efficiency. For power-limited applications such as mobile and satellite communications it is necessary to operate the amplifier as close to saturation as possible. However, when operating close to saturation, signals are distorted by the non-linearities of the HPA. This non-linear behaviour is quantified by the AM-AM (amplitude modulation to amplitude modulation) and AM-PM (amplitude modulation to phase modulation) characteristics of the amplifier. The AM-AM and AM-PM conversion effects have the following deteriorative effects on the performance of a communication system:

(1) Degradation of the bit error rate (BER) of the system. This is partly due to distorted amplitude and phase of the signalling elements in the transmitted signal constellation, and partly due to intersymbol interference.

(2) Spectral spreading of the transmitted signal, which increases undesirable interference to the adjacent channels.

(3) In frequency-multiplexed (FDM/FDMA) systems the different carrier frequencies mix together, generating intermodulation products which can also lead to interference.

These effects are pronounced in any transmission system where the HPA must be operated close to saturation for high efficiency. For modulation schemes with little or no envelope variation, such as GMSK, solid-state HPAs can be operated fairly close to 1 dB gain compression for high efficiency. However, for spectrally efficient modulation schemes like 16-QAM, which result in large carrier envelope variations, the amplifier may have to be operated several decibels below its maximum power in order to avoid problems caused by the AM-AM and AM-PM conversion. This results in a very large reduction in efficiency. There are two distinct approaches to improving this situation. One method is to adaptively control the quiescent DC bias point of the transistors [115], as shown in Fig. 12.28. This is particularly applicable to class A amplifiers and signals with envelope variations. The second method is to employ linearisation techniques. These techniques are based on compensation for the AM-AM and AM-PM conversion effects so that the overall characteristics of the linearised HPA approach those of a linear amplifier.

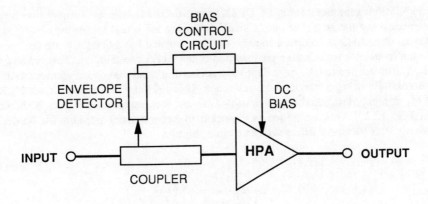

Fig. 12.28 GaAs FET power amplifier with adaptive gate bias control

12.8.1 Linearisation techniques

The three main techniques which have been used can be classed as feed-forward, feedback, and predistortion linearisers. Feedforward techniques require a considerable increase in RF hardware complexity. Predistortion techniques can be implemented at RF, IF or baseband. Baseband predistortion is carried out before modulation of the carrier and involves introducing deliberate distortion into the constellation such that the distortion introduced by the HPA restores an ideal constellation. State-of-the-art techniques use DSP processors and employ a look-up table method which incorporates spectral shaped filters and baseband predistortion. It has been shown that this technique substantially improves the performance of a digital communication system for practical CNR values [116].

RF or IF predistorters are circuits which use the non-linearity of elements such as transistors or diodes to create non-linear AM-AM and AM-PM characteristics which are the inverse of those of the HPA, so that the overall characteristics approach those of an ideal linear amplifier. For example, the lineariser may have an insertion gain which increases with increasing input power (gain expansion) in order to combat the effect of gain compression in the power amplifier. Although the maximum saturated output power cannot be increased, the power amplifier can be operated closer to saturation in order to improve its efficiency. At best the characteristics of a soft-limiter can be achieved. Systems with these types of characteristics have been found to give a considerable improvement in system performance. RF predistorters have been demonstrated both in hybrid-MIC and MMIC technology [117-120].

The final class of linearisation uses negative feedback. In microwave amplifiers, however, there is too little gain for this to be used directly. Instead, the transmitted signal is tapped off and a low-frequency component is extracted from it for feedback purposes; this could be the signal envelope, or some intermodulation products, or the signal can be demodulated to recover the baseband signal itself. For quadrature modulation schemes, such as QPSK and 16-QAM, one technique involves demodulating the amplifier output signal and using the actual transmitted baseband values of I and Q as feedback signals. The demodulated I and Q signals are fed back to the modulator for adaptive predistortion of the signal constellation. This is known as 'Cartesian feedback' or 'adaptive predistortion' and has been demonstrated successfully using analogue feedback loops [121, 122] and digital techniques [123, 124]. A highly-integrated approach using DSP and look-up table techniques is shown in Fig. 12.29. This technique is expected to become very popular for MMIC transceivers because it offers such an elegant solution.

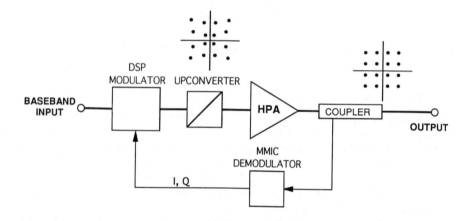

Fig. 12.29 Block diagram of a linearised amplifier using adaptive baseband predistortion with a DSP-based modulator

12.9 Active antennas and quasi-optical techniques

Advanced HEMT and HBT devices now make it possible for integrated circuits to operate at over 100 GHz. With ever increasing operating frequencies, however, it becomes more difficult to interface the circuits to the outside world, and on-chip radiating elements are an attractive solution [125]. There is a wide range of exciting new active antenna and quasi-optical techniques which are possible with integrated circuit technology. For example, on-chip radiating elements can be used for spatial power combining [126, 127] of monolithic devices. This technique overcomes the high losses of conventional power combiners and antenna feed networks, as well as the problems of interfacing to the chip. Fig. 12.30 illustrates the advantage of spatial power combining over conventional techniques. If the power amplifier and antenna are designed as separate units, as in Fig. 12.30 (a), then the power amplifier has a corporate combining network, which will entail considerable losses at millimetre-wave frequencies leading to reduced gain, output power, and efficiency. The patch antenna array has a corporate feed network which once again has significant loss, and additionally creates unwanted radiation which has undesirable effects on the radiation pattern of the array. Alternatively, if the transmitter is designed as an integrated active antenna, then in principle both of the corporate feed networks become unnecessary, as shown in Fig. 12.30 (b). For adaptive antennas each element or sub-array can have amplitude and phase control devices integrated directly at the antenna level. In principle, wafer-scale integration of millimetre-wave phased-array antennas is possible using these techniques [128-130].

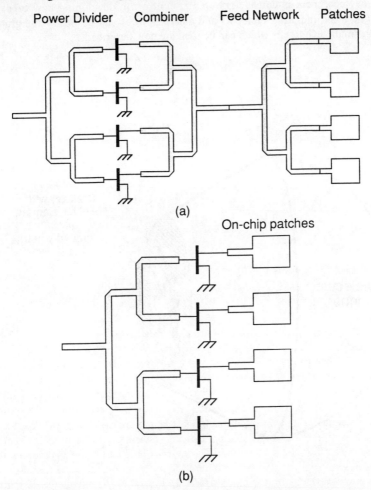

Fig. 12.30 Comparison to illustrate the advantages of active antennas: (a) Power amplifier feeding a conventional antenna array; (b) Active antenna

In addition to spatial power combining, quasi-optical techniques can be employed for imaging applications. The 'staring array' [131] employs an array of receiving elements mounted at the focal plane of a dielectric lens, as illustrated in Fig. 12.31. With this arrangement each array element has a separate beam and hence provides a pixel in the image without the need for electronic or mechanical steering or scanning. MMIC technology is ideally suited to the fabrication of the receiving element array. However, in applying *conventional* patch antenna design techniques to MMICs, one finds that the space available limits one quite severely, and that the substrate is not entirely suitable. Since the major advantage of ICs is, of course, the ready availability of active devices, there is considerable research effort in developing new types of antenna where the radiating elements and active devices can be neatly integrated. Some of the main techniques which can be used are now described.

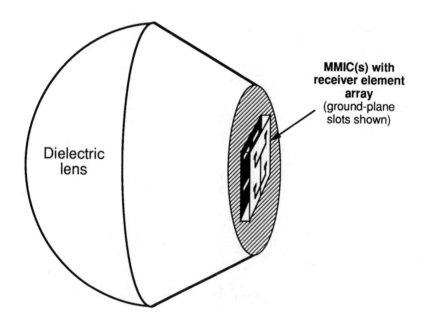

Fig. 12.31 Quasi-optical techniques: a receiver array mounted at the focal plane of a dielectric lens

12.9.1 Optimally-shaped on-chip radiators

In this technique, the aim is to derive optimum layouts whereby the devices are directly integrated with the radiating elements. This leads to very high integration and minimum losses. By minimising discontinuities in the region of the antenna, the field pattern is improved and unwanted coupling between elements reduced. Also, it is possible to incorporate impedance matching, from the radiation resistance of the antenna to the impedance of the device. This integrated approach removes unnecessary matching circuitry and the unwanted radiation that goes with it. The types of antenna which can be used are patches, slots, and Vivaldi horns (tapered slots).

12.9.2 Using the rear face of the chip

The chip area consumed by the radiating elements and feed lines represents a major drawback, even at frequencies as high as 40 GHz. To overcome this, the possibility of using the back face of the chip as an integral part of the antenna is very attractive. Configurations have been demonstrated using slots in the ground-plane with excellent results [132, 133]. The slot antenna is not, however, electrically isolated from the front-face circuitry which means that the area above the slot cannot readily be used for other circuitry and that the feed network must be carefully designed to avoid unwanted coupling effects. A number of alternative techniques have been reported involving electromagnetic coupling from one face to the other [134] or using multiple substrates bonded together [135], with either coupling through a ground-plane aperture [136] or a direct feed-through. This latter technique of using multiple substrates bonded together has been found to be particularly attractive; the patches can be placed on a low ε_r substrate for higher efficiency and bandwidth, and the feed network put on a higher ε_r substrate to reduce the size and unwanted radiation. Most commonly the two substrates share a common ground-plane, and feed-throughs are fabricated to connect the patches to the feeds. Recently [137], it has been shown that this approach can be directly applied to monolithic technology by using the thin-film microstrip (TFMS) transmission-line medium. With this technique the back-face metallisation can be patterned to form radiating elements, which are isolated from the front-face TFMS circuitry by the intervening ground plane. Interconnection between the radiating elements and TFMS feed lines can readily be achieved with through-GaAs via-holes. This monolithic antenna structure is shown in Fig. 12.32. The advantages of this technique are in its small size, the ability to optimise the semiconductor chip thickness for improved radiation efficiency and bandwidth, and the scope for employing a new range of quasi-optical techniques using the chip mounted directly onto a dielectric lens arrangement. With the increasing use of 'uniplanar' circuitry, it is expected that the use of back-face radiating elements will be of considerable importance to future millimetre-wave monolithic transmitters and receivers.

12.9.3 Multi-level techniques

MMICs employing multi-level metals allow several advanced antenna design techniques to be used, such as monolithic multi-frequency or broad-band antennas using stacked radiators. Many established techniques, such as electromagnetic and aperture coupling to patches, could also be extended to the multi-level IC. In addition, new techniques are possible which save space by 'vertically' integrating devices and radiating elements.

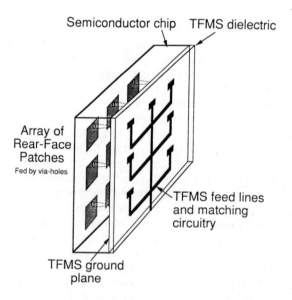

Fig. 12.32 Back-face patch antenna array

12.9.4 Antennas with novel switching and tuning functions

Because devices can be integrated so closely with the radiating element, it is possible to arrange special tuning elements (transistors or varactors) around the edge or inside the radiating element. This has been used extensively at lower frequencies [138], but is impractical with conventional patch arrays at millimetre-wave frequencies. However, it is very easy to implement switching and tuning functions on an integrated circuit. For example, in communications applications the transmitter and receiver frequencies are often too far apart for a patch antenna to cover both frequencies. Varactor diodes connected to a single patch could be switched on and off to load the patch and pull its frequency response sufficiently to cover both transmit and receive frequencies.

12.9.5 Using a patch as both a radiating and feedback element

In conventional techniques, an oscillator circuit with some type of feedback resonator would be designed to match to 50 Ω. This would then feed a patch through a matching network. From a space point of view, this is very wasteful. Instead, it is possible to use a patch, or other radiator, as both the resonator and antenna. A number of hybrid active antennas using this technique have been demonstrated [139]. Various means of injection-locking could be employed to ensure the spectral purity and coherence of the output signal; optical techniques could be used, for example.

12.10 Optoelectronics and microwaves

12.10.1 Applications

Optoelectronics and optical fibres have an increasingly important role in microwave and millimetre-wave applications. Examples of this are optically fed phased arrays [140, 141] and the use of picosecond optical pulses for ultra-broadband MMIC measurements, which was described in Chapter 11. At the same time, microwave techniques have an increasingly important role in optoelectronic systems, as data rates have gone beyond 10 Gbit/s and as ultra-short pulses are used in many applications. Examples of this are optical fibre communications [142, 143] and optical fibre sensing [144]. Other applications exist which use a happy marriage of optical and microwave techniques. Examples of this are fixed telecommunications networks with a wireless local loop and microcellular mobile communications systems employing a fibre-optic backbone (including 'radio over fibre' systems [145, 146]).

The optically-fed phased array antenna employs conventional MMIC-based transmit-receive modules, but signals are distributed to the modules using optical fibres rather than coaxial cable or hollow waveguides. This approach is feasible for active phased arrays because the optical fibres need only carry low power signals such as IF or baseband signals, control and monitoring signals and a local oscillator reference signal. The advantages of the optical feed are lower transmission loss, a reduction in mass, and the possibility of using optical techniques for signal processing and beam-forming. The disadvantage is, of course, the extra complexity because of the need for electrical/optical and optical/electrical interface components. This problem would be overcome nicely if the MMICs had optical inputs and outputs, or if direct optical control could be applied to the circuits in some way. Because GaAs is a direct band-gap semiconductor, GaAs MMICs have a major potential advantage for integrating microwave and optoelectronic devices.

Wireless local loop and microcellular mobile systems employ a fibre optic backbone, with the final communication link performed by a short microwave or millimetre-wave radio link. Since a large number of base stations are required, a number of schemes have been proposed for making the radio base stations low cost by using IF or RF subcarriers [146] and frequency division multiplexing. This approach means that the base station hardware is simpler, and does not have to cope with the 10+ Gbit/s signal that would result from a time-division approach.

12.10.2 Optically controlled circuits

The optical control of circuits allows ultra-fast switching to be performed, with a near ideal RF-to-baseband signal isolation and immunity to electromagnetic interference [147, 148]. Optical control has been applied to phase shifters, amplifiers, attenuators, switches, and oscillators. There are two distinct approaches that can be used. In direct control methods the component performing the circuit function (e.g. transistor or diode) is directly illuminated to control its electrical characteristics. With indirect methods the component which is illuminated (e.g. a photodiode) is separate from the circuit function but provides it with a control voltage or current. The advantage of the latter approach is that the photodetector component can be optimised for optical sensitivity and does not have to perform a circuit function. The photodetector-circuit interface can be optimised to improve the efficiency and linearity of the control characteristic.

The sensitivity of the circuit or device can be maximised by increasing the level of absorbed photon energy as follows:

(1) employing more powerful lasers, e.g. medium power Class IIIb laser diodes (where the optical power is >5 mW)
(2) employing collimating optical lenses
(3) increasing the proximity of the laser diode/optical lens to the MMIC
(4) selecting an optical wavelength that is less than 900 nm (for GaAs), e.g. laser diodes used in domestic CD players operate in the infra-red range of 760-780 nm.
(5) using optically transparent ohmic contacts and/or Schottky contacts [149].

12.10.2.1 Phase shifters

An optically controlled millimetre-wave bulk semiconductor phase shifter has been developed for insertion between rectangular waveguides [150]. When the absorbed energy within the semiconductor is equal to or greater than the band-gap energy level, free charge carriers (electron-hole pairs) are produced, forming a surface plasma layer in the semiconductor. As the optical illumination intensifies, the number of electron-hole pairs increases. Since the dielectric constant of the resulting electron-hole plasma region is higher than that of the rest of the guide, a sudden increase in the induced phase shift occurs, until saturation is reached.

More recently, transmission-line phase shifters implemented with a CPW line on a lossless/lossy/lossless (AlGaAs/GaAs/AlGaAs) layered semiconductor substrate have been reported [151, 152]. Here, the regions of semiconductor between the signal line and ground planes are exposed to the optical illumination, generating the electron-hole plasma in the buried GaAs layer. The conductivity of this lossy layer is increased by increasing the intensity of the optical illumination. At a fixed frequency, as the conductivity of the lossy layer is increased, the characteristic propagating mode changes from a lossy dielectric mode to a slow wave mode, due to the increase in the effective capacitance of the structure. As the conductivity is further increased, the slow wave mode eventually changes to a skin effect mode.

12.10.2.2 Switches
In the past decade, the optical control of GaAs MESFETs and HEMTs as switches has been demonstrated [147]. Plasma produced within the active layer of the semiconductor causes a photovoltaic effect across the depletion region and a photoconductive effect in the undepleted region. With a reverse gate-source bias voltage equal to the pinch-off voltage, the MESFET is in the 'off' state. By illuminating the MESFET with a sufficiently high intensity, a voltage can be generated at the Schottky junction which is enough to overcome the pinch-off voltage and switch the MESFET to the 'on' state.

12.10.2.3 Amplifiers and attenuators
FET and HEMT amplifiers can be optically controlled directly by illuminating the active region exposed in the gate-source and gate-drain gap. Both photovoltaic and photoconductive effects alter the transistor's intrinsic elements such as g_m, C_{gs}, and the channel resistance. It has been shown [147] that a change in g_m can be made the dominant mechanism, which leads to a variation in the amplifier gain. The largest gain variation is achieved with the device operated close to pinch-off and by optimising the photovoltaic effect by employing a large gate bias resistor (> 50 kΩ): Gain variation of 20 dB has been achieved in this way, but the amplifier's dynamic range is rather limited at this bias point. As an alternative, indirect methods can be used in which a photodetector converts the optical signal into an electrical signal which then controls the amplifier's DC bias. This approach has been demonstrated for the optical control of a monolithic distributed amplifier [153].

12.10.2.4 Oscillators
Oscillators can be switched, frequency-tuned, or injection-locked using optical control [147]. The key reason for controlling oscillators with optical signals is to enable large numbers of oscillators to be phase-locked to a common reference without requiring a bulky electrical distribution network. Applications include phased array antennas and active antennas employing spatial power combining [154]. Optical control can be performed directly on the oscillating device (MESFET, HEMT, HBT, Gunn or IMPATT diode) or it can be performed indirectly by illuminating an ancillary device within the resonant circuit. Subharmonic optical injection-locking of 21.5 GHz FET oscillators has been demonstrated [155].

12.10.2.5 Mixers

The separation of the RF, LO, and IF signals in a mixer is a perennial problem. In an optically-pumped mixer the LO is introduced into the mixing device by illuminating it with a laser beam which has been modulated by the LO signal. Alternatively, the LO can be directly applied as an electrical signal and the RF can be modulated onto the laser beam. The advantage is that high isolation of the signals can be achieved.

In optical mixing, two lasers with a 'small' frequency separation can be used to create microwave or millimetre-wave signals from the beat frequency [156]. Alternatively, this approach may be used with an optical phase-locked loop (OPPL) to generate microwave or millimetre-wave intensity modulated light [141].

12.10.3 Circuits for optoelectronic applications

A common application of MMICs in optoelectronic applications is the use of ultra-broadband amplifiers in high-speed optical communications systems [157]. MMICs find use in the receiver side, amplifying the signal from a photodiode, and in the transmitter side as medium power drivers for optical modulators. Links operating at 10 Gbit/s have become commonplace, and systems operating at 40 Gbit/s are already being developed. Two amplifier topologies are particularly suited to providing the required bandwidth: the distributed amplifier and the Darlington-pair feedback amplifier. Distributed amplifiers (with MESFETs or HEMTs) have the disadvantage that they cannot easily be DC-coupled and that it is hard to maintain good gain flatness down to very low frequencies. As a result, the Darlington-pair approach has become increasingly popular, since with HBT technology this can operate to frequencies beyond 20 GHz. The advantages of employing HBT technology are that DC coupling and >20 GHz bandwidth can be achieved and that transimpedance amplifiers can be designed, these being best suited for amplifying the current from p-i-n photodiode receivers. MMICs have also been used for pre-detection pulse-shaping: This has been demonstrated both with passive lumped-element equalising filters [158] and using transversal-style filtering embedded directly within the distributed amplifier structure [159].

In applications where optical control of MMICs is attractive, such as large phased-arrays, the concept of MMICs with optical inputs and outputs as shown in Fig. 12.33 has been proposed as a means of miniaturising the overall system [140]. The III-V compounds are highly suited to this because the active devices can be used as input ports, through direct optical control, and on-chip external electro-optic modulators can provide optical output ports. However, it should not be forgotten that the advanced device engineering made possible by MBE and MOCVD means that silicon-germanium technology is by no means excluded from the optical domain.

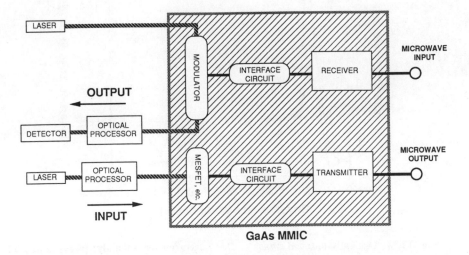

Fig. 12.33 MMIC with optical inputs and outputs (after Herczfeld [140])

12.10.4 MMIC measurement anomalies

In addition to being useful for the control of circuits, optical effects can cause errors and anomalies in measurements. Wafer probers have a powerful light source for the microscope, and this can have a significant effect on a circuit. As an example, the effect of the prober lamp on a monolithic 90° phase shifter has been investigated [160]. The principle light source used consisted of a 150 W Schott halogen lamp, fed via a short fibre optic cable and a simple focusing lens placed 35 mm from the MMIC. The resulting errors in both the relative phase shift and insertion loss are shown in Fig. 12.34. It can be seen that the level of optical control is dependant on the DC bias potential applied to the IPSVDs. For amplifiers, a significant gain variation can be observed and the prober light can be used to improve results in many cases. However, it should always be remembered that the MMIC will be in the dark when packaged or assembled into a subsystem. Hence, all device characterisation and circuit measurements should also be performed in the dark.

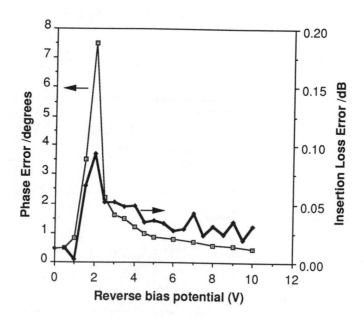

Fig. 12.34 Optically-induced phase shifter measurement anomaly: phase shift and insertion loss error against reverse bias potential

12.11 References

1. DEVLIN, L. M., BUCK, B. J., CLIFTON, J. C., DEARN, A. W., and LONG, A. P.: 'A 2.4 GHz single chip transceiver', IEEE Microwave and Millimeter-Wave Monolithic Circuits Symposium Digest, June 1993, pp. 23-26
2. JACKSON, R. W: 'Considerations in the use of coplanar waveguide for millimeter-wave integrated circuits', *IEEE Trans. Microwave Theory Tech.*, MTT-34, Dec. 1986, pp. 1450-1456
3. NAKAMOTO, H., TOKUMITSU, T., and AKAIKE, M.: 'A monolithic, port-interchanged rat-race hybrid using a thin-film microstrip line crossover', 19th European Microwave Conf. Proceedings, 1989, pp. 311-316
4. HIRAOKA, T., TOKUMITSU, T., and AKAIKE, M.: 'Very small wide-band MMIC magic-T's using microstrip lines on a thin dielectric film', *IEEE Trans. Microwave Theory Tech.*, MTT-37, Oct. 1989, pp. 1569-1575
5. TOYODA, I., HIROTA, T., HIRAOKA, T., and TOKUMITSU, T.: 'Multilayer MMIC branch-line coupler and broad-side coupler', IEEE Microwave and Millimeter Wave Monolithic Circuits Symposium Digest, 1992, pp. 79-82

6. HIROTA, T., TARUSAWA, Y., and OGAWA, H.: 'Uniplanar MMIC hybrids- A proposed new MMIC structure', *IEEE Trans. Microwave Theory Tech.*, MTT-35, June 1987, pp. 576-581
7. GILLICK, M., and ROBERTSON, I. D.: 'Accurate characterization of an ultrawideband monolithic CPW-to-slotline transition', The IEEE Asia-Pacific Microwave Conference, Adelaide, August 1992, pp. 279-282
8. CHEN, T. H., TON, T. N., DOW, G. S., NAKANO, K., LIU, L. C. T., and BERENZ, J.: 'A Q-band monolithic balanced resistive HEMT mixer using CPW/slot-line balun', IEEE GaAs IC Symp. Dig., 1990, pp. 177-180
9. MURAGUCHI, M., HIROTA, T., MINAKAWA, A., IMAI, Y., ISHITSUKA, F., and OGAWA, H.: '26 GHz-band full MMIC transmitters and receivers using a uniplanar technique', IEEE MTT-S Int. Microwave Symp. Dig., 1990, pp. 873-876
10. MAJIDI-AHY, R., *et al.*: '5-100 GHz InP coplanar waveguide MMIC distributed amplifier', *IEEE Trans. Microwave Theory Tech.*, MTT-38, Dec. 1990, pp. 1986-1993
11. VOGEL, R. W.: 'Analysis and design of lumped- and lumped-distributed-element directional couplers for MIC and MMIC applications', *IEEE Trans. Microwave Theory Tech.*, MTT-40, Feb. 1992, pp. 253-262
12. HIROTA, T., MINAKAWA, A., and MURAGUCHI, M.: 'Reduced-size branch-line and rat-race hybrids for uniplanar MMIC's', *IEEE Trans. Microwave Theory Tech.*, MTT-38, Mar. 1990, pp. 270-275
13. GILLICK, M., ROBERTSON, I. D., and JOSHI, J. S.: 'Design and realisation of reduced-size impedance transforming uniplanar MMIC branchline coupler', *Electron. Lett.*, Vol. 28, no. 16, 30th July 1992, pp. 1555-1557
14. GILLICK, M., ROBERTSON, I. D., and JOSHI, J. S.: 'Miniaturised two-stage balanced Ku-band CPW MMIC amplifier using impedance transforming couplers', *Electron. Lett.*, Vol. 29, no. 8, April 1993, pp. 670-671
15. GILLICK, M., ROBERTSON, I. D., and JOSHI, J. S.: '15 GHz balanced CPW amplifier using impedance-transforming lumped-distributed branch-line couplers', *Proc. IEE*, Part-H, Vol. 141, no. 4, August 1994, pp. 241-245
16. ROBERTSON, I. D., and AGHVAMI, A. H: 'Novel coupler for gallium arsenide monolithic microwave integrated circuit applications', *Electron. Lett.*, Vol. 24, no. 25, Dec. 1988, pp. 1577-1578
17. LUCYSZYN, S., and ROBERTSON, I. D.: 'An improved multi-layer quadrature coupler for MMICs', 21st European Microwave Conference Proceedings, Stuttgart, September 1991, pp. 1154-1158
18. GILLICK, M., ROBERTSON, I. D., and JOSHI, J. S.: 'A 12-36 GHz MMIC 3dB coplanar waveguide directional coupler', 22nd European Microwave Conference Proceedings, Espoo, Finland, Aug. 1992, pp. 724-728
19. TOKUMITSU, T., HIRAOTA, T., NAKAMOTO, H., and TAKENAKA, T.: 'Multilayer MMIC using a 3μm × 3-layer dielectric film structure', IEEE MTT-S Int. Microwave Symp. Dig., 1990, pp. 831-834

20. LUCYSZYN, S., ROBERTSON, I. D., and AGHVAMI, A. H: 'Monolithic analogue phase shifter and cascode FET amplifier using uniplanar techniques', 24th European Microwave Conference Proceedings, Cannes, France, September 1994, pp. 554-559
21. GILLICK, M., and ROBERTSON, I. D.: 'An X-band monolithic power amplifier using low characteristic impedance thin-film microstrip transformers', *IEEE Microwave and Guided Wave Lett.*, Aug. 1992, pp. 328-330
22. KARACAOGLU, U., LUCYSZYN, S., and ROBERTSON, I. D.: 'Modelling and design of X-band MMIC varactor-tuned VCOs', IEE Colloquium Digest on Modelling, Design, and Application of MMICs, London, June 1994, pp. 13/1-5
23. LUCYSZYN, S., and ROBERTSON, I. D.: 'Synthesis techniques for high performance octave bandwidth 180° analogue phase shifters', *IEEE Trans. Microwave Theory Tech.*, MTT-40, Apr. 1992, pp. 731-740
24. LUCYSZYN, S., and ROBERTSON, I. D.: 'Analog reflection topology building blocks for adaptive microwave signal processing applications', *IEEE Trans. Microwave Theory Tech.*, MTT-43, Mar. 1995, pp. 2617-2622
25. LABARRE, F., CAZAUX, J. L., GOLDZTEJN, C., PICHON N., and SOULARD, M.: 'A new concept: An electronically tunable MMIC flatness corrector', IEEE MTT-S Int. Microwave Symp. Dig., 1991, pp. 285-287
26. LUCYSZYN, S., and ROBERTSON, I. D.: 'Monolithic narrow band filter using ultra high-Q tunable active inductors', *IEEE Trans. Microwave Theory Tech.*, MTT-42, no. 12, Dec. 1994, pp. 2617-2622
27. DAWIRS, H. N., and SWARNER, W. G.: 'A very fast, voltage-controlled, microwave phase shifter', *Microwave Journal*, June 1962, pp. 99-107
28. CHEN C. L., *et al.*: 'A low-loss Ku-band monolithic analog phase shifter, *IEEE Trans. Microwave Theory Tech.*, MTT-35, Mar. 1987, pp. 315-320
29. PHILIPPE, P., EL-KAMALI, W., and PAUKER, V.: 'Physical equivalent circuit model for planar Schottky varactor diode', *IEEE Trans. Microwave Theory Tech.*, MTT-36, Feb. 1988, pp. 250-255
30. DAWSON, D. E., CONTI, A. L., LEE, S. H., SHADE, G. F., and DICKENS, L. E.: 'An analog X-band phase shifter', IEEE Microwave and Millimeter-Wave Monolithic Circuits Symp. Dig., 1984, pp. 6-10
31. PAO, C. K., *et al.*: 'V-band monolithic phase shifters', IEEE GaAs IC Symp. Dig., Nov. 1988, pp. 269-272
32. LUCYSZYN, S., GREEN, G., and ROBERTSON, I. D.: 'Accurate millimeter-wave large signal modeling of planar Schottky varactor diodes', IEEE MTT-S Int. Microwave Symp. Dig., 1992, pp. 259-262
33. LUCYSZYN, S., LUCK, J., GREEN, G., and ROBERTSON, I. D.: 'Enhanced modelling of interdigitated planar Schottky varactor diodes', IEEE Asia-Pacific Microwave Conf. Dig., Aug. 1992, pp. 273-278
34. LUCYSZYN, S., GREEN, G., and ROBERTSON, I. D.: 'Interdigitated planar Schottky varactor diodes for tunable MMIC applications', IEEE/ESA European Gallium Arsenide Applications Symp., 1992

35. LUCYSZYN, S., ROBERTSON, I. D., and AGHVAMI, A. H: 'Microwave modelling of varactor diodes fabricated using heterojunction based technologies', IEEE Workshop on High Performance Electron Devices for Microwave and Optoelectronic Applications, London, Oct. 1993
36. CAZAUX., J.-L., PAVLIDIS, D., NG, G.-I., and TUTT, M.: 'The use of double heterojunction diodes in monolithic phase shifters', GaAs IC Symp., 1992
37. CRYSTAL, E. G., HO, R. Y. C., ADAMS, D. K., and COHN, S. B.: 'Microwave synthesis techniques', Stanford Research Institute Annual Report, Section 5, Contract DAAB07-68-0088, Sept. 1968
38. ADAMS, D. K., and HO, R. Y. C.: 'Active filters for UHF and microwave frequencies'. *IEEE Trans. Microwave Theory Tech.*, MTT-17, no. 9, Sept. 1969, pp. 662-670
39. SNYDER, R. V., and BOZARTH, D. L.: 'Analysis and design of a microwave transistor active filter', *IEEE Trans. Microwave Theory Tech.*, MTT-18, no. 1, Jan. 1970, pp. 2-9
40. FLIEGLER, E.: 'Operating criteria for active microwave inductors', *IEEE Trans. Microwave Theory Tech.*, MTT-19, no. 1, Jan. 1971, pp. 89-91
41. HARA, S., TOKUMITSU, T., TANAKA, T. and AIKAWA, M.: 'Broad-band monolithic microwave active inductor and its application to miniaturized wide-band amplifiers', *IEEE Trans. Microwave Theory Tech.*, MTT-36, no. 12, Dec. 1988, pp. 1920-1924
42. HARA, S., TOKUMITSU, T., and AIKAWA, M.: 'Lossless, broadband monolithic microwave active inductors', IEEE MTT-S Int. Microwave Symp. Dig., 1989, pp. 955-958
43. BASTIDA, E. M., DONZELLI, G. P., and SCOPELLITI, L.: 'GaAs monolithic microwave integrated circuits using broadband tunable active inductors', 19th European Microwave Conf. Proc., Sept. 1989, pp. 1282-1287
44. DEVLIN, L. M., and MINNIS, B. J.: 'A versatile vector modulator design for MMIC', IEEE MTT-S Int. Microwave Symp. Dig., 1990, pp. 519-522
45. FROST, R. D., FISHER, D. A, and PECK, D. E: 'A GaAs MMIC voltage-controlled phase shifter', *Microwave Journal*, Aug. 1991, pp. 87-94
46. COATS, R., KLEIN, J., PRITCHETT, S. D., and ZIMMERMANN, D.: 'A low loss monolithic five-bit PIN diode phase shifter', IEEE MTT-S Int. Microwave Symp. Dig., 1990, pp. 915-918
47. TEETER, D., WOHLERT, R., COLE, B., JACKSON, G., TONG, E., SALEDAS, P., ADLERSTEIN, M., SCHINDLER, M., and SHANFIELD, S.: 'Ka-band GaAs HBT PIN diode switches and phase shifters', IEEE MTT-S Int. Microwave Symp. Dig., 1994, pp. 451-454
48. KARACAOGLU, U., ROBERTSON, I. D., and GUGLIELMI, M.: 'Microstrip bandpass filters with MMIC negative resistance circuits for loss compensation', IEEE MTT-S Int. Microwave Symp. Dig., June 1994
49. HOPF, B. P., WOLFF, I., and GUGLIELMI, M.: 'Coplanar MMIC active bandpass filters using negative resistance circuits', *IEEE Trans. Microwave Theory Tech.*, MTT-42, Dec. 1994, pp. 2598-2602

50. SUSSMAN-FORT, S.: 'A realisation of a GaAs microwave active filter', *IEEE Trans. Microwave Theory Tech.*, MTT-38, no. 10, Oct. 1990, pp. 1524-1526
51. BONETTI, R. R., et al.: 'An MMIC active filter with 60-dB rejection', IEEE MTT-S Int. Microwave Symp. Dig., 1992, pp. 1195-1198
52. CHANG, C., and ITOH, T.: 'Microwave active filters based on coupled negative resistance method', *IEEE Trans. Microwave Theory Tech.*, MTT-38, no. 12, Dec. 1990, pp. 1879-1884
53. KARACAOGLU, U., ROBERTSON, I. D., and GUGLIELMI, M.: 'A dual-mode microstrip ring resonator filter with active devices for loss compensation', IEEE MTT-S Int. Microwave Symp. Dig., June 1993, pp. 189-192
54. KARACAOGLU, U., and ROBERTSON, I. D.: 'High selectivity varactor-tuned MMIC bandpass filter using lossless active resonators', IEEE Microwave and Millimeter-Wave Monolithic Circuits Symposium Digest, May 1994, pp. 237-240
55. JUTZI, W.: 'Microwave bandwidth active transversal filter concept with MESFETs', *IEEE Trans. Microwave Theory Tech.*, MTT-19, no. 9, Sept. 1971, pp. 760-767
56. RAUSCHER, C.: 'Microwave active filters based on transversal and recursive principles', *IEEE Trans. Microwave Theory Tech.*, MTT-33, no. 12, Dec. 1985, pp. 1350-1360
57. SCHINDLER, M., and TAJIMA, Y.: 'A novel active filter with lumped and transversal elements', *IEEE Trans. Microwave Theory Tech.*, MTT-37, no. 12, Dec. 1989, pp. 2148-2153
58. BILLONET, L., et al.: 'Design concept for microwave recursive and transversal filters using lange couplers', IEEE MTT-S Int. Microwave Symp. Dig., 1992, pp. 925-928
59. TOKUMITSU, T., HARA, S., TAKENAKA, T., and AIKAWA, M.: 'Divider and combiner line-unified FET's as basic circuit function modules - Part 1', *IEEE Trans. Microwave Theory Tech.*, MTT-38, no. 9, Sept. 1990, pp. 1210-1217
60. TOKUMITSU, T., HARA, S., TAKENAKA, T., and AIKAWA, M.: 'Divider and combiner line-unified FET's as basic circuit function modules - Part 2', *IEEE Trans. Microwave Theory Tech.*, MTT-38, no. 9, Sept. 1990, pp. 1218-1226
61. HARA, S., TOKUMITSU, T., and AIKAWA, M.: 'Novel unilateral circuits for MMIC circulators', *IEEE Trans. Microwave Theory Tech.*, MTT-38, Oct. 1990, pp. 1399-1406
62. TOKUMITSU, T., KAMOGAWA, K., TOYODA, I., and AIKAWA, M.: 'A novel, injection locked oscillator MMIC with combined ultra-wide-band active combiner/divider and amplifier', IEEE Microwave and Millimeter-Wave Monolithic Circuit Symp. Dig., 1994, pp. 143-146
63. BRINLEE, W. R., PAVIO, A. M., and VARIAN, K. R.: 'A novel planar double-balanced 6-18 GHz MMIC Mixer', IEEE Microwave and Millimeter-Wave Monolithic Circuit Symp. Dig., 1994, pp. 139-142

64. TSAI, M. C.: 'A new compact wideband balun', IEEE Microwave and Millimeter-Wave Monolithic Circuit Symp. Dig., 1993, pp. 123-125
65. TSAI, M. C., SCHINDLER, M. J., STRUBLE, W., VENTRESCA, M., BINDER, R., WATERMAN, R., and DANZILIO, D.: 'A compact wideband balanced mixer', IEEE Microwave and Millimeter-Wave Monolithic Circuit Symp. Dig., 1994, pp. 135-138
66. MAAS, S. A., and CHANG, K. W.: 'A broadband, planar, doubly balanced monolithic Ka-band diode mixer', IEEE Microwave and Millimeter-Wave Monolithic Circuit Symp. Dig., 1993, pp. 53-55
67. JOUANNEAU-DOUARD, M., BROUZES, H., BIONAZ, S., and LEVY, D.: 'A 1 to 18 GHz out of phase combiner', IEEE Microwave and Millimeter-Wave Monolithic Circuit Symp. Dig., 1992, pp. 83-86
68. GOLDFARB, M. E., COLE, P. J. B., and PLATZKER, A.: 'A novel MMIC biphase modulator with variable gain using enhancement-mode FETs suitable for 3 V wireless applications', IEEE Microwave and Millimeter-Wave Circuits Symp. Dig., 1994, pp. 99-102
69. MAAS, S.: 'MMIC mixers for commercial applications', IEEE MTT-S Workshop on Designing RF Signal Processing MMICs for Commercial Communication Systems, 1994, pp. 99-106
70. JEAN, P., PAUKER, V., and DAUTRICHE, P.: 'Wide-band monolithic GaAs phase detector for homodyne reception', IEEE MTT-S Int. Microwave Symp. Dig., 1987, pp. 169-171
71. PYNDIAH, R., and BOGAART, F. V. D.: 'Novel multioctave MMIC active isolator (1-20 GHz)', *Electron. Lett.*, Vol. 25, no. 21, Oct. 1989, pp. 1420-1422
72. BAHL, I. J.: 'On the design of an active circulator', *Microwave and Opt. Tech. Lett.*, Vol. 1, no. 1, Mar. 1988, pp. 18-20
73. TANAKA, S., SHIMOMURA, N., and OHTAKE, K.: 'Active circulators—the realization of circulators using transistors', *Proc. IEEE*, Mar. 1965, pp. 260-267
74. KOTHER, D., HOPF, B., SPORKMANN, T., WOLFF, I., and KOBLOWSKI, S.: 'Active CPW MMIC circulator for the 40 GHz band', 24th European Microwave Conference Proceedings, Cannes, Sept. 1994, pp. 542-547
75. DOUGHERTY, R.: 'Circulate signals with active devices on monolithic chips', *Microwaves & RF*, June 1989, pp. 85-86 & 89
76. POLACEK, G. A.: 'Stable bias yields active MMIC circulator', *Microwaves & RF*, Nov. 1990, pp. 132-137
77. ROBERTSON, I. D., and AGHVAMI, A. H: 'Novel monolithic ultra-wideband unilateral 4-port junction using distributed amplification techniques', IEEE MTT-S Int. Microwave Symp. Dig., Albuquerque, 1992, pp. 1051-1054
78. KATZIN, P., AYASLI, Y., REYNOLDS, L., and BEDARD, B.: '6 to 18 GHz MMIC circulator', *Microwave Journal*, May 1992, pp. 248-256
79. LUCYSZYN, S., and ROBERTSON, I. D.: 'Decade bandwidth MMIC analogue phase shifter', IEE Coll. Dig. on Multi-octave Microwave Circuits, London, Nov. 1991, pp. 2/1-6

80. ROBERTSON, I. D., and AGHVAMI, A. H: 'A novel reflectometer using a wideband monolithic active unilateral 4-port junction', 23rd European Microwave Conference Proceedings, Madrid, 1993, pp. 296-298
81. TANG, O. S. A., and AITCHISON, C. S.: 'A practical microwave travelling wave MESFET Gate Mixer', IEEE MTT Int. Microwave Symp. Dig., 1985, pp. 605-608
82. HOWARD, T. S., and PAVIO, A. M.: 'A dual-gate 2-18 GHz monolithic FET distributed mixer', IEEE Microwave and Millimeter-wave Monolithic Circuits Symposium Digest, 1987, pp. 27-30
83. ROBERTSON, I. D., and AGHVAMI, A. H: 'A novel 1 to 15 GHz matrix distributed mixer', 21st European Microwave Conference Proceedings, Stuttgart, September 1991, pp. 489-494
84. PAVIO, A. M., et al.: 'Double balanced mixers using active and passive techniques', *IEEE Trans. Microwave Theory Tech.*, MTT-36, Dec. 1988, pp. 1948-1957
85. ROBERTSON, I. D., and AGHVAMI, A. H: 'A novel wideband MMIC active balun', 20th European Microwave Conference, Budapest, Hungary, September 1990
86. ROBERTSON, I. D., READER, H. C., and AGHVAMI, A. H: 'Operating modes in wideband monolithic distributed FET mixers', The IEEE Asia-Pacific Microwave Conference, Adelaide, August 1992, pp. 747-750
87. PAVIO, A. M., et al.: 'A distributed broadband monolithic frequency multiplier', IEEE MTT-S Int. Microwave Symp. Dig., June 1988, pp. 503-504
88. BARTA, G., et al.: 'Surface-mounted GaAs active splitter and attenuator MMICs used in a 1-10 GHz leveling loop', *IEEE Trans. Microwave Theory Tech.*, MTT-34, no. 12, December 1986, pp. 1569-1575
89. GERARD, R. E. J.: 'Multisignal amplification', U.S. Patent no. 4423386, The Marconi Company Limited, Chelmsford, England
90. ROBERTSON, I. D., and AGHVAMI, A. H.: 'A practical distributed FET mixer for MMIC applications', IEEE MTT-S Int. Microwave Symp. Dig., 1989, pp. 1031-1032
91. TAJIMA, Y., et al.: 'Broadband GaAs FET 2x1 switches', IEEE GaAs IC Symp. Dig., 1984, pp. 81-84
92. LEISTEN, O. P., COLLIER, R. J., and BATES, R. N.: 'Distributed amplifiers as duplexer/low crosstalk bidirectional elements in S-band', *Electron. Lett.*, Vol. 24, no. 5, March 1988, pp. 264-265
93. CIOFFI, K. R.: 'Active broadband impedance transformations using distributed techniques', IEEE MTT-S Int. Microwave Symp. Dig., June 1989, pp. 1043-1046.
94. LUCYSZYN, S., ROBERTSON, I. D., and AGHVAMI, A. H.: 'Novel applications in microwave communication systems for small-shift frequency translators', IEEE International Conf. on Telecommunications, Dubai, Jan. 1994
95. MADNI, A. M., and WAN, L. A.: 'Solid-state preamplifier/frequency translator finds velocity deception applications', *Microwave Systems News*, Oct. 1983, pp. 71-86

96. SCHIEK, B., and EUL, H.-J.: 'Network analysis with a single sideband generator of a high image suppression', 21st European Microwave Conf. Proc., Sept. 1991, pp. 503-508
97. SKOLNIK, M. I. (Ed.): *Radar handbook*, McGraw-Hill Book Co., N.Y., Chapter 13: 'Frequency scanned arrays', 1970
98. PARISI, S. J.: 'Monolithic, lumped element, single sideband modulator', IEEE MTT-S Int. Microwave Symp. Dig., 1992, pp. 1047-1050
99. BAREE, A. H., ROBERTSON, I. D., and BHARJ, J. S.: 'A compact wideband MMIC image-rejection mixer chip', IEEE Asia-Pacific Microwave Conference Proceedings, 1994, pp. 241-244
100. FOX, A. G.: 'An adjustable waveguide phase shifter', *Proc. IRE*, Vol. 35, Dec. 1947, pp. 1489-1498
101. BOYD, C. R.: 'Analogue rotary-field ferrite phase shifters', *Microwave Journal*, Dec. 1977, pp. 41-43
102. LUCYSZYN, S., PILCHEN, Y., ROBERTSON, I. D., and AGHVAMI, A. H.: 'Ku-band serrodyne frequency translator using wideband MMIC analogue phase shifters', 23rd European Microwave Conf. Proc., Madrid, Sept. 1993, pp. 819-822
103. GARVER, R. V.: '360° varactor linear phase modulator', *IEEE Trans. Microwave Theory Tech.*, Mar. 1969, pp. 137-147
104. LUCYSZYN, S., ROBERTSON, I. D., and AGHVAMI, A. H.: '24 GHz serrodyne frequency translator using a 360° analog CPW MMIC phase shifter', *IEEE Microwave and Guided Wave Lett.*, Vol. 4, no. 3, Mar. 1994, pp. 71-73
105. KUSHNER, L., *et al.*: 'An 800 MHz monolithic GaAs HBT serrodyne modulator', IEEE GaAs IC Symp. Dig., October 1994
106. MAZUMDER, S. R., TSAI, T. L., and TSAI, W. C.: 'A frequency translator using dual-gate GaAs FETs', IEEE MTT-S Int. Microwave Symp. Dig., 1983, pp. 346-348
107. LUCYSZYN, S., and ROBERTSON, I. D.: 'Novel serrodyne modulator for advanced communications systems', Microwaves 94 Conference Digest, October 1994, pp. 103-106
108. TRUITT, A., CERNEY, J., and MASON, J. S.: 'A 0.3 to 3 GHz monolithic vector modulator for adaptive array systems', IEEE MTT-S Int. Microwave Symp. Dig., 1989, pp. 421-422
109. PYNDIAH, R., and BOGAART, F. V. D.: 'MMIC technology sets a new state of the art of microwave vector modulators', Workshop on MMICs for Space Applications, ESTEC, Mar. 1990
110. NORRIS, G. B., BOIRE, D. C., ST. ONGE, G., WUTKE, C., BARRATT, C., COUGHLIN, W., and CHICKANOSKY, J.: 'A fully monolithic 4-18 GHz digital vector modulator', IEEE MTT-S Int. Microwave Symp. Dig., 1990, pp. 789-792
111. CANNISTRARO, V., LIOU, J.-C., and McCARTER, S.: 'Direct modulation lowers VSAT equipment costs', Microwaves & RF, August 1990, pp. 99-101
112. KUMAR, S.: 'Directly modulated VSAT transmitters', *Microwave Journal*, April 1990, pp. 255-264

113. TELLIEZ, I., *et al.*: 'A compact monolithic microwave demodulator-modulator for 64-QAM digital radio links', *IEEE Trans. Microwave Theory Tech.*, MTT-39, no. 12, Dec. 1991, pp. 1947-1954
114. PYNDIAH, R., JEAN, P., LEBLANC, R., and MEUNIER, J.-C.: 'GaAs monolithic direct (1-2.8) GHz Q.P.S.K. modulator',19th European Microwave Conference Proceedings, 1989, pp. 597-602
115. SALEH, A. A. M., and COX, D. C.: 'Improving the power-added efficieny of FET amplifiers operating with varying envelope signals', *IEEE Trans. Microwave Theory Tech.*, MTT-31, Jan. 1983, pp. 51-56
116. AGHVAMI, A. H., and GEMIKONAKLI, O.: '16-ary QAM system for low data rate satellite services', *Electron. Lett.*, Vol. 25, no. 16, August 1989, pp. 1055-1057
117. INADA, R., *et al.*: 'A compact 4 GHz linearizer for space use', IEEE MTT-S Int. Microwave Symp. Dig, 1986, pp. 323-326
118. MOOCHALLA, S., *et al.*: 'An integrated Ku-band linearizer driver amplifier for TWTAs with high gain and wide bandwidth', 14th AIAA International Communications Satellite Systems Conference, 1992, pp. 167-174
119. KUMAR, S.: 'Power amplifier linearization using MMICs', *Microwave Journal*, Apr. 1992, pp. 96-104
120. KUMAR, M., WHARTENBY, J. C., and WOLKSTEIN, H. J.: 'Predistortion linearizer using GaAs dual-gate MESFET for TWTA and SSPA used in satellite transponders', *IEEE Trans. Microwave Theory Tech.*, MTT-33, no. 12, Dec. 1985, pp. 1479-1488
121. PETROVIC, V.: 'Reduction of spurious emission from radio transmitters by means of modulation feedback', IEE Conference on Radio Spectrum Conservation Techniques, Sept. 1983, pp. 44-49
122. BATEMAN, A., HAINES, D. M., and WILKINSON, R. J.: 'Linear transceiver architectures', IEEE Vehicular Teachnology Conference Proc., 1988, pp. 478-484
123. CAVERS, J. K.: 'Amplifier linearisation using a dual predistorter with fast adaptation and low memory requirements', *IEEE Trans. on Vehicular Technology*, Vol. 39, no. 4, Nov. 1991, pp. 374-382
124. WRIGHT, A. S., and DURTHER, W. G.: 'Experimental performance of an adaptive digital linearized power amplifier', IEEE MTT-S Int. Microwave Symp. Dig, 1992, pp. 1105-1108
125. NIGHTINGALE, S. J., *et al.*: 'A 30 GHz monolithic single-balanced mixer with integrated dipole receiving element', IEEE Microwave and Millimetre-wave Monolithic Circuits Symposium Digest, 1985, pp. 74-77
126. YORK, R. A., and COMPTON, R. C.: 'Quasi-optical power combining using mutually synchronised oscillator arrays', *IEEE Trans. Microwave Theory Tech.*, MTT-39, no. 6, June 1991, pp. 1000-1009
127. CHANG, K., *et al.*: 'Experiments on injection locking of active antenna elements for active phased arrays and spatial power combiners', *IEEE Trans. Microwave Theory Tech.*, MTT-37, no. 7, July 1989, pp. 1078-1083

128. POZAR, D. M., and SCHAUBERT, D. H.: 'Comparison of architectures for monolithic phased array antennas', *Microwave Journal*, March 1986, pp. 93-104
129. ITOH, T.: 'Planar quasi-optical circuit technology', 20th European Microwave Conference Proceedings, Budapest, September 1990, pp. 83-88
130. WANG, H., *et al.*: 'Monolithic Q-band active array module and antenna arrays', *Applied Microwave*, Vol. 5., no. 1, Winter 1993, pp. 88-102
131. ALDER, C. J., *et al.*: 'Lens-fed microwave and millimetre-wave receivers with integral antennas', 20th European Microwave Conference Proceedings, Budapest, September 1990, pp. 449-453
132. ROY, L., STUBBS, M. G., and WIGHT, J. S.: 'A 30 GHz, HEMT, active antenna structure in MMIC technology', 22nd European Microwave Conference Proceedings, August 1992, pp. 236-241
133. KAWASAKI, S., and ITOH, T.: 'Optical control of active integrated antenna', 22nd European Microwave Conference Proceedings, August 1992, pp. 697-701
134. MENZEL, W., and GRABHERR, W.: 'A microstrip patch antenna with coplanar feed line', *IEEE Microwave and Guided Wave Lett.*, November 1991, pp. 340-342
135. JAMES, J. R., and HALL, P. S. (Editors): *Handbook of Microstrip Antennas*, Peter Peregrinus Ltd., London, 1989
136. POZAR, D. M.: 'Microstrip antenna aperture coupled to a microstrip line', *Electron. Lett.*, Vol. 21, no. 2, Jan. 1985, pp. 49-50
137. ROBERTSON, I. D.: 'Millimetre-wave back-face patch antenna for multilayer MMICs', *Electron. Lett.*, Vol. 29, no. 9., 29th April 1993, pp. 816-817
138. NAVARRO, J. A., HUMMER, K. A., and CHANG, K.: 'Active integrated antenna elements', *Microwave Journal*, January 1991, pp. 115-126
139. KARACAOGLU, U., AUJLA, N. S., ROBERTSON, I. D., and WATKINS, J.: 'An active patch antenna topology based on negative resistance FET oscillator design', 23rd European Microwave Conference Proceedings, Spain, September 1993
140. HERCZFELD, P. R.: 'Hybrid photonic-microwave systems and devices', *IEICE Trans. on Electronics*, Vol. E76-C, no. 2, Feb. 1993, pp. 191-197
141. SEEDS, A.: 'Optical technologies for phased array antennas', *IEICE Trans. on Electronics*, Vol. E76-C, no. 2, Feb. 1993, pp. 198-206
142. MIDWINTER, J. E.: 'The threat of optical communications', *IEE Electronics & Communication Engineering Journal*, Feb. 1994, pp. 33-38
143. COCHRANE, P., HEATLEY, D. J. T., SMYTH, P. P., and PEARSON, I. D.: 'Optical telecommunications–future prospects', *IEE Electronics & Communication Engineering Journal*, Aug. 1993, pp. 221-232
144. UTTAMCHANDANI, D.: 'Fibre-optic sensors and smart structures: developments and prospects', *IEE Electronics & Communication Engineering Journal*, Vol. 6, no. 5, Oct. 1994, pp. 237-246
145. SHIBUTANI, M., *et al.*: 'Feasibility studies on an optical fiber feeder system for microcellular mobile communications systems', IEEE International Conference on Communications, June 1991, pp. 1176-1181

146. OGAWA, H., POLIFKO, D., and BANBA, S.: 'Millimeter-wave fiber optics systems for personal radio communications', *IEEE Trans. Microwave Theory Tech.*, MTT-40, no. 12, Dec. 1992, pp. 2285-2292
147. SEEDS, A. J., and DE SALLES, A. A.: 'Optical control of microwave semiconductor devices', *IEEE Trans. Microwave Theory Tech.*, MTT-38, no. 5, May 1990, pp. 577-585
148. ROSSEK, S. J., and FREE, C. E.: 'Optical control of microwave signals using GaAs FETs', *IEE Electronics & Communication Engineering Journal*, Feb. 1994, pp. 21-30
149. BASHAR, S. A., and REZAZADEH, A. A.: 'Transparent contacts to optoelectronic devices using indium tin oxide (ITO)', Proceedings of the IEEE Int. Conf. on Telecommunications, Dubai, Jan. 1994, pp. 259-261
150. VAUCHER, A. M., STRIFFLER, C. D., and LEE, C. H.: 'Theory of optically controlled millimeter wave phase shifters', *IEEE Trans. Microwave Theory Tech.*, MTT-31, Feb. 1983, pp. 209-216
151. KAISER, M., et al.: 'Optically controlled phase shift of Schottky coplanar lines', *Electron. Lett.*, Vol. 25, no. 23, Nov. 1989, pp. 1575-1577
152. CHEUNG, P., NEIKIRK, D. P., and ITOH, T.: 'Optically controlled coplanar waveguide phase shifters', *IEEE Trans. Microwave Theory Tech.*, MTT-38, no. 5, May 1990, pp. 586-595
153. PAOLELLA, A., and HERCZFELD, P. R.: 'Optical gain control of a GaAs MMIC distributed amplifier', *Microwave and Opt. Tech. Lett.*, Vol. 1, no. 1, Mar. 1988, pp. 13-16
154. LIN, J., and ITOH, T.: 'Active integrated antennas', *IEEE Trans. Microwave Theory Tech.*, MTT-42, no. 12, Dec. 1994, pp. 2186-2194
155. DARYOUSH, A. S.: 'Optical syncronization of millimetre-wave oscillators for distributed architectures', *IEEE Trans. Microwave Theory Tech.*, MTT-38, no. 5, May 1990, pp. 467-476
156. NI, D. C., FETTERMAN, H. R., and CHEW, W.: 'Millimeter-wave generation and characterization of a GaAs FET by optical mixing', *IEEE Trans. Microwave Theory Tech.*, MTT-38, no. 5, May 1990, pp. 608-614
157. IMAI, Y., SANO, E., and KAZUYOSHI, A.: 'Design and performance of wideband GaAs MMICs for high-speed optical communication systems', *IEEE Trans. Microwave Theory Tech.*, MTT-40, no. 2, Feb. 1992, pp. 185-190
158. DARWAZEH, I., LANE, P. M., MARNANE, W. P., MOREIRA, P., WATKINS, L. R., CAPSTICK, M. H., and O'REILLY, J. J.: "GaAs MMIC optical receiver with embedded signal processing', *IEE Proc.*, Part-G, Vol. 139, no. 2, 1992, pp. 241-243
159. MOREIRA, P., DARWAZEH, I. Z., and O'REILLY, J. J.: 'Novel optical receiver design using distributed amplifier pulse shaping network', IEE Colloquium on Optical Detectors and Receivers, Savoy Place, Oct. 1993, Digest no. 1993/173, pp. 8/1-8/4
160. LUCYSZYN, S.: 'Ultra-wideband high performance reflection-type phase shifters for MMIC applications', University of London PhD Thesis, Jan. 1993, pp. 228-230

Index

1/f noise 355
absorptive SPDT switch 313
Academy™ 129
active antenna 478
active balun 297, 458, 460, 465
active filters 453
active isolator 460
active matching 177
active power divider 457
active circulator 278, 457, 461, 464
active coupling structures 244
active inductor 449, 454
active load 153
active techniques 16
adaptive predistortion 478
Air Coplanar™ 412
AM noise 358
analogue attenuators 332
analogue delay line 266
anti-phase splitter 242
applications 12
asymmetric coupled microstrip 121

back annotation 98, 100
back-face patch antenna 482
balanced amplifier 142
balanced attenuator 334
balun 74, 87, 186
baseband noise 357
bath-tub via 180
beam-forming 12
bi-CMOS 11
bias chokes 151

Boire phase shifter 295
broadband tuneable oscillators 337

CAD 91
calibration 401
capacitance ratio 343
capacitively-coupled distributed
 amplifier 187
capacitors 51, 74
Cartesian feedback 478
cascaded-match reflection-type phase
 shifter 267
cascode distributed amplifier 173
cavity resonator 343
centre-tapped transformer 244
circuit locus 356
class A amplifiers 151, 181
class B amplifiers 186
cluster matching 180, 185
cold-FET 332, 452
commercial software packages 129
common-gate configuration 177
common-gate inductive feedback 341
common-source capacitive feedback
 341
communications 14
component yield 60
conductor loss 349
constant-R networks 171
conversion loss 207
conversion loss matrix 206, 208
coplanar strips 439
coplanar waveguide (CPW) 87, 437

cost 6
couplers 82
CPW circuits 441
CPW probes 411
CPW resonator 346
CPW-to-slotline transition 439
cryogenic measurements 421
cuprate-based materials 349

Darlington pair 178, 486
DBS 14
DC biasing 150, 418
DC blocks 169
DC-coupled amplifiers 178
de-embedding 402, 408
decoupling capacitors 155
defect densities 61
delay line 259
design integration 92
device geometry 140
device line 356
dielectric resonator 343
digital phase shifters 284
digitally controlled attenuator 316
diode mixers 209
DIODEMX 218, 225
direct step on wafer 29
directional couplers 444
distributed active splitter 465
distributed amplifier 4, 163, 486
distributed matching techniques 19
distributed mixer 465
double-balanced mixer 202, 245
drain-line loss 167
DRC 107
drift region impedance 378
DRO 393
dual-gate FET 233, 282, 317

edge-coupled microstrip resonators 349
electrical sampling scanning-force microscopy 427
electro-optic sampling 425
electromagnetic field analysis 109
electron-beam lithography 29
electron-beam probing 424
EMSIM™ 129
em™ by Sonnet Software 440
epitaxial growth 28

equivalent circuit model 30, 44
equivalent circuit parameters 35
error model 401
expansion function 113
Explorer™ 122, 129
extrinsic transconductance 31

feedback 142
feedback amplifier 162
feedback oscillator 339
FET mixers 231
FET switch 307
field probing 422, 423
flicker noise 341
flicker noise corner frequency 346
flip-chip mounting 180
f_{max} 32, 43
F_{min} 32
foundries 7
foundry component library 108
foundry-defined electrical models 107
four-port FET 101
frequency drift 337
frequency instabilities 337
frequency translator 468
f_t 31, 41

GaAs MESFET 28
GaAs/AlGaAs heterojunction 34
gain circles 138
gain compression 340, 359
Galerkin method 114
GAT1 2
gate-line loss 167
GDSII 98
GMMT F20 foundry process 55
Gummel plot 40

harmonic mixer 203
harmonic balance method 351
Harmonica™ 129
HBT 10, 36, 342
helical transmission-line resonator 346
HEMT 10, 33
heterojunction transit-time diode 384
heterojunction bipolar transistor (HBT) 10, 36, 342
HFSS™ 129
hierarchical design 95, 105

high electron mobility transistor (HEMT) 10, 33
high resistivity silicon 371
history 2
HTS resonators 343

I-Q vector modulator 283, 475
image recovery mixer 203
image rejection mixer 203
IMPATT diode 337, 376, 379
Impulse™ 129
inductors 52, 68
InP-based HBT 48
intelligent-vehicle highway systems 15
interdigital capacitors 74
intermediate frequency 207
ion implantation 28

K-factor 138

L-networks 144
Lange coupler 125, 241
lattice matched 34
launchers 410
layout 98
layout-based CAD 98
LC tuned circuit 343
Libra™ 100, 129
line-reflect-match (LRM) 414
linearisation techniques 477
LINMIC+™ 117, 130
load pull measurements 182
loaded-line phase shifter 289
lossy matching 161
low noise block 14
low noise oscillators 344
LU-FETs 457
lumped-element matching 17

Marchand balun 458
matching networks 144
matrix distributed amplifier 174
maximum available gain (MAG) 139
maximum frequency of oscillation 43
maximum oscillation frequency 32
maximum stable gain (MSG) 138
MDS™ 129
measurement errors 415
measurements 399

mesa resistor 54, 77
MESFET 10, 28
microbolometer probe 422
microcellular systems 15
Microchamber™ 419
microstrip 82
microstrip resonator 346
microwave probe 413
MIM capacitor 51, 74
minimum noise figure 32
mixer 201
MODAMP™ 178
models 107
modulation noise 361
molecular beam epitaxy 33
Momentum™ 129
Mott diode 223
multi-dielectric medium 117
multilayer circuits 443
multilayer coupler 445
multi-octave tuneable YIG oscillators 343
multi-strip problem 112
mushroom gate 34
MVDS 15

negative resistance 340
negative resistance element 453, 454
negative resistance oscillators 339
noise matching 190
noise factor 189
noise figure 189, 216
noise measure 189
non-linear current method 352
non-linear impulse response 352

Octopus™ 130
OmniSys™ 129
optically controlled circuits 484
optically fed phased array antenna 483
optimum noise impedance 359
optoelectronic sampling 425
optoelectronics 483
orthogonality condition 357
overlay capacitors 74

package lid 108
parametric converter 201
passive components 50

PCM 106
PEM model 349
phase noise 337
phase error 361
phase shifter 259
phased-array antenna 11
phasing condition 339
photo-emissive sampling 424
Picoprobe™ 412
PIN diodes 387
planar interdigitated Schottky-barrier diodes 348
Plessey Research (Caswell) 2
polyimide 57, 118
power amplifier 180
power combining 180, 182
probe cards 419
probe pads 416
probe stations 410
prober calibration 413
probe yield 60
process control 60
pseudomorphic HEMT 34
pump 201
pump power 210

quarter-wave transformer 147
quasi-circulator 464
quasi-optical techniques 478

radar 14
radiation loss 344
radio over fibre 483
rat race 237, 238
reactively matched amplifier 158
reflection coefficient analysis 357
reflection-type attenuator 332
reflection-type delay line 265, 286
reflection-type phase shifter 262, 288
reflective SPST switch 313
reliability 9
reproducibility 8
resistors 54, 77
RFOW testing 158
ring resonators 349
rooftop functions 113

S-parameters 400
satellite TV 14

schematic capture 94
Schottky diodes 389
Scope™ 129
segmented dual-gate FET 317
selectivity 339
self-aligned HBT 37
self-biasing 153
self-oscillating mixer 395
Serenade™ 129
series feedback 142, 192
serrodyne modulator 471
SFPMIC+™ 130
Shark™ 130
short transformer. 147
Si/SiGe material system 373
Si/SiGe MMICs 372, 395
SiGe (silicon germanium) 11, 49, 374
SiGe Gunn device 385
SiGe heterojunction bipolar transistor 390
silicon bipolar 11
silicon nitride 57
SIMMWICs 372
simulator customisation 93
simultaneous match LNA 191
simultaneous match reflection coefficients 141
single drift diode 380
single-balanced mixer 202
single-ended mixer 202, 216
slot antenna 481
slotline 87, 439
smart antennas 305
SMART Libraries 99
Sonnet™ 122
SPDT switch 315, 319
spectral domain approach 109
spectral operator expansion technique 115
spiral inductors 68
spiral transformer 73
split gate-line distributed amplifier 188
SPST switch 314
spurious oscillations 343
SSB modulator 470
stability 134, 135, 159
stability circles 136
stacked spiral 70
stacked bias 154

staring array 480
statistical process control 62
Stingray™ 130
stub matching 147
Success™ 129
Super Compact™ 129
Super Spice™ 129
superconductive resonator 343
superlattice 385
surface resistance 349
surface-state 342
switch FET 309
switch FET model 311
switched attenuator 318
switched bridged-T attenuator 321
switched scaled FET 321
switched T- or π-attenuators 323
switched-coupler 90° bit phase shifter 297
switched-filter phase shifter 291
switched-line phase shifter 285
switches 305, 388
system simulation 131

tapered drain-line 189
technologies 9
temperature instability 344
temperature coefficients 344
temperature compensation 344
test fixture 400, 408, 409
TFMS 444
thermal measurements 419
thermal noise 355
thin-film resistors 78
thin-film microstrip 87, 438, 444, 481
throughput yield 60
time-domain analysis 130
time-domain gating 402
time-domain methods 350
total capacitance ratio 446
Touchstone™ 129
transconductance 30, 40
transducer power gain 138
transition frequency 31, 355

transitions 87
transposed flicker noise 346
transposition coefficient 358
transposition gain 355
transversal and recursive filters 456
travelling wave amplifier 127
TRL 405, 408
TVRO 14
two-dimensional electron gas 33

unconditional stability 137
underpass 70
unilateral transducer gain, 138
uniplanar techniques 87, 437
UNISIM™ 123, 130
unit gate-width 140
upconversion coefficient 359
user model capability 130

van der Pol oscillator 352
varactor diode 343, 446
variable gain amplifier 317
variable resistor 451
vector modulator 473
via-holes 79
visual inspection 60
voltage controlled oscillator 342
Volterra non-linear transfer functions 351
Volterra series 351

wafer 57
wiggly lines 276
Wilkinson power splitter 242
wireless local loop 484
wireless LANs 15, 436

X and Y numbers 106
xgeom™ 441

YBCO 349
yield 60
YIG resonator 343